Student Study Guide/ Solutions Manual

to accompany

General, Organic, and Biochemistry

Fifth Edition

Katherine J. Denniston
Towson University

Joseph J. Topping
Towson University

Timothy M. Dwyer
Towson University

Boston   Burr Ridge, IL   Dubuque, IA   Madison, WI   New York   San Francisco   St. Louis
Bangkok   Bogotá   Caracas   Kuala Lumpur   Lisbon   London   Madrid   Mexico City
Milan   Montreal   New Delhi   Santiago   Seoul   Singapore   Sydney   Taipei   Toronto

The McGraw·Hill Companies

Student Study Guide/ Solutions Manual to accompany
GENERAL, ORGANIC, AND BIOCHEMISTRY, FIFTH EDITION
KATHERINE J. DENNISTON, JOSEPH J. TOPPING, AND TIMOTHY M. DWYER

Published by McGraw-Hill Higher Education, an imprint of The McGraw-Hill Companies, Inc., 1221 Avenue of the Americas,
New York, NY 10020. Copyright © 2007 by The McGraw-Hill Companies, Inc. All rights reserved.

 Recycled/acid free paper
This book is printed on recycled, acid-free paper containing 10% postconsumer waste.

1 2 3 4 5 6 7 8 9 0 QPD/QPD 0 9 8 7 6 5

ISBN-13: 978-0-07-282849-8
ISBN-10: 0-07-282849-8

www.mhhe.com

Contents

1 *Chemistry: Methods and Measurement*

Learning Goals

1 Describe the interrelationship of chemistry with other fields of science and medicine.
2 Discuss the approach to science, the scientific method, and distinguish among the terms hypothesis, theory, and scientific law.
3 Describe the properties of the solid, liquid, and gaseous states.
4 Provide specific examples of physical and chemical properties and physical and chemical change.
5 Distinguish between intensive and extensive properties.
6 Classify matter as element, compound, or mixture.
7 Distinguish between data and results.
8 Learn the major units of the measure in the English and metric systems, and be able to convert from one system to the other.
9 Report data and results using the proper number of significant figures and scientific notation.
10 Use appropriate experimental quantities in problem solving.
11 Calculate the density of an object from mass and volume data and calculate the specific gravity of an object from its density.

Introduction

The subject matter of chemistry deals with all material substances as well as all the changes that these materials undergo. The basic tools of chemistry will enable you to further your understanding of how chemistry works.

1.1 The Discovery Process

Learning Goal 1:	Describe the interrelationship of chemistry with other fields of science and medicine.

Chemistry is the study of matter and the changes that matter undergoes. **Matter** is anything that has mass and occupies space. The changes that matter undergoes always involve either gain or loss of energy. **Energy** is nonmaterial, and is the ability to do work (to accomplish some change). Thus, a study of chemistry involves matter, energy, and their interrelationship.

1

Major Areas of Chemistry

The **major areas of chemistry** include **biochemistry**, the study of matter associated with living things; **organic chemistry**, the study of matter principally composed of two elements, carbon and hydrogen; **inorganic chemistry**, the study of matter that consists of all of the other elements and their combinations; **analytical chemistry**, the analysis of matter to determine its composition; and **physical chemistry**, a discipline that attempts to explain the way matter behaves.

The Scientific Method

Learning Goal 2:	**Discuss the approach to science, the scientific method, and distinguish among the terms hypothesis, theory, and scientific law.**

The **scientific method** consists of six interrelated processes:

1. **Observation**. The description of the properties of a substance is a result of observation. The measurement of the temperature of a liquid or the size or mass of a solid results from observation.
2. **Formulation of a question**. The observation is the basis of the question.
3. **Pattern recognition**. Discerning a cause-and-effect relationship may give rise to a scientific **law**. A scientific law is a description of the orderly behavior observed in nature.
4. **Developing theories.** Observation of a phenomenon calls for some explanation. The process of explaining observed behavior begins with a **hypothesis**—an educated guess. If this hypothesis survives extensive testing, it may attain the status of a **theory**. A theory is a hypothesis supported by testing (experimentation) that explains scientific facts and is capable of predicting new facts.
5. **Experimentation**. The heart of the scientific method is the verification of theories. This verification process results from conducting carefully designed experiments intended to reinforce or refute the model system, the theory, or the hypothesis.
6. **Summarizing information**. Many phenomena have common causes and explanations. A scientific law summarizes and clarifies large amounts of information.

Models in Chemistry

A model of a chemical unit or system is often used to make ideas more clear. A good model is based on everyday experience and provides a great deal of information in a simple fashion. One example is the model of the molecule methane, found on page six of the textbook.

1.2 Matter and Properties

Matter and Physical Properties

Properties (characteristics) of matter may be classified as either physical or chemical. **Physical properties** enable us to identify different kinds of matter without changing the identity (chemical composition) of the sample. Examples of physical properties include color, odor, taste, melting and boiling temperatures, and compressibility.

Three states of matter exist: the gaseous state, the liquid state, and the solid state. These states are distinguishable by differences in physical properties. They include gases, liquids, and solids.

1. **The gaseous state.** Gases have a very low mass per unit volume, density, because the individual particles that comprise the gas are separated by large distances. Consequently, gases can be compressed (pushed into a smaller volume) or expanded to a larger volume.
2. **The liquid state.** Particles that make up a liquid are much closer together than in the vapor state. Thus, liquids expand and contract only slightly, and the density of liquids is much greater than that of gases.
3. **The solid state.** Solids are characterized by particles that are very close together. Attractive forces between the particles are strong enough to provide a rigid shape or structure to the bulk material. The proximity of particles prevents significant expansion or compression. All common materials, except water, follow this order of decreasing particle separation.

Water is the most common example of a substance that may exist in all three states over a reasonable temperature range. The conversion of ice to liquid water or liquid water to the gaseous state is an example of a **physical change**. A physical change does not alter the composition or identity of the substance undergoing change.

Matter and Chemical Properties

Chemical properties result in a change in composition and can only be observed through chemical reactions. The process of photosynthesis is a common example of chemical change.

Learning Goal 4: Provide specific examples of physical and chemical properties and physical and chemical change.

Example 1

Which of the following are physical changes?

1. An iron nail rusts.
2. A block of ice melts.
3. Water on the floor evaporates.
4. Food is digested in the small intestine.
5. Gasoline undergoes combustion.

Answer

Items 2 and 3

The terms *rust*, *digest*, and *combustion* all indicate chemical change. The terms *melt*, *boil*, *freeze*, and *evaporate* all indicate a physical change.

Intensive and Extensive Properties

Learning Goal 5:	Distinguish between intensive and extensive properties.

An **intensive property** is independent of the quantity of the substance. For example, the density of a drop of water is exactly the same as the density of a liter of water. Mass and volume are extensive properties. An **extensive property** depends on the quantity of a substance.

Classification of Matter

Learning Goal 6:	Classify matter as element, compound, or mixture.

All matter can be classified as either a pure substance or a mixture (see Figure 2.3 in the textbook). A **pure substance** is a form of matter that has identical composition and physical and chemical properties throughout. A **mixture** is a combination of two or more pure substances in which the combined substances retain their identity.

A mixture may be either homogeneous or heterogeneous matter. A **homogeneous mixture** has uniform composition. Its particles are well mixed, or thoroughly intermingled. A **heterogeneous mixture** has a nonuniform composition.

Pure substances may also be subcategorized as elements or compounds. An **element** is a pure substance that cannot be converted into a simpler form of matter by any chemical reaction. A **compound** is a substance resulting from the combination of two or more elements in a definite, reproducible fashion.

Example 2

Which of the following are mixtures?

1. soup
2. tap water
3. wine
4. blood

Example 2 continued

Answer
All, since they contain more than one pure substance.

1.3 Measurement in Chemistry

Data, Results, and Units

> **Learning Goal 7:** Distinguish between data and results.

A scientific experiment produces **data**. Each piece of data is the result of a single measurement. Examples include the mass of a sample and the time required for a chemical reaction to occur. Mass, length, volume, time, temperature, and energy are the most common types of data obtained from chemical experiments.

Results are the outcome of an experiment. Data and results may be identical, but more often several pieces of data are combined to produce a result.

A **unit** defines the basic quantity of mass, volume, time, and so on. A number not followed by the correct unit conveys no useful information.

English and Metric Units

> **Learning Goal 8:** Learn the major units of measure in the English and metric systems, and be able to convert from one system to the other.

The English system of measurement has as its most commonly used unit of **weight** the standard pound (lb), its fundamental unit of length, the standard yard (yd), and its basic unit of volume, the standard gallon (gal). The English system is not used in scientific work primarily because of the difficulty involved in converting from one unit to another.

The metric system is a decimal-based system; it is inherently simpler and less ambiguous. In the metric system there are three basic units. **Mass** is represented as the gram, length as the meter, and volume as the liter. Any subunit or multiple unit contains one of these units preceded by a prefix indicating the power of ten by which the base unit is to be multiplied to form the subunit or multiple unit. The most common metric prefixes are shown in Table 1.1 of the textbook and should be memorized.

Unit Conversion: English and Metric Systems

A **conversion factor** or series of conversion factors relates two units when converting from one unit to another. The use of these conversion factors is referred to as the **factor-label method**.

This method is used either to convert from one unit to another within the same system, or convert units from one system to another.

Conversion of Units within the Same System

The factor-label method is a self-indicating system; only if the factor is set up properly will the correct units result. Several commonly used English system conversion factors are included in Table 1.2 of the textbook. Most of these should already be familiar to you.

Conversion of units within the metric system may be accomplished using the factor-label method as well. Unit prefixes, which dictate the conversion factor, make conversion of units easy (refer to the textbook, Table 1.1).

Example 3

Convert 8.62 gallons into pints.

Work

$$8.62 \text{ gallons} \times \frac{4 \text{ quarts}}{1 \text{ gallon}} \times \frac{2 \text{ pints}}{1 \text{ quart}} = 68.96 \text{ pints}$$

Notice how the unit of gallon(s) cancels each other out, and the unit of quart(s) also cancel each other out, leaving "pints" as the final unit.

Study hint: You may be familiar with other ways to solve this problem, most notably by using ratios. Please spend some time becoming comfortable using this facto[r] method, however. As problems become more complex, you may be more likely to make errors using methods other than the factor-label method. When a problem is set up using the factor-label method, all of the units cancel properly, leaving the unit that you want, and if the conversion factors are correct, you can be assured that your answer is correct.

Answer

69.0 pints (Three significant figures are needed.)

Example 4

Convert 16.4 kilometers into centimeters.

Work

$$16.4 \text{ kilometers} \times \frac{1000 \text{ meters}}{1 \text{ kilometer}} \times \frac{100 \text{ centimeters}}{1 \text{ meter}} = 1.64 \times 10^6 \text{ cm}$$

Answer

1.64×10^6 centimeters (Three significant figures are needed.)

Example 5

How many kilometers are in 1.568×10^6 centimeters?

Work

$$1.568 \times 10^6 \text{ cm} \times \frac{1 \text{ m}}{100 \text{ cm}} \times \frac{1 \text{ km}}{1000 \text{ m}} = 15.68 \text{ km}$$

$$1.568 \times 10^6 \text{ cm} \times \frac{1 \text{ m}}{1 \times 10^2 \text{ cm}} \times \frac{1 \text{ km}}{1 \times 10^3 \text{ m}} = 1.568 \times 10^1 \text{ km}$$

Answer

1.568×10^1 km, or 15.68 km (Four significant figures are needed.)

Conversion of Units from One System to Another

The conversion of a quantity expressed in units of one system to an equivalent quantity in the other system (English to metric or metric to English) requires a bridging conversion unit. For example:

Quantity	English		Metric
Mass	1 pound	=	454 grams
	2.2 pounds	=	1 kilogram
Length	1 inch	=	2.54 centimeters
	1 yard	=	0.91 meters
Volume	1 quart	=	0.946 liters
	1 gallon	=	3.78 liters

The conversion may be represented as a three-step process:

1. Conversion from units stated in the problem to a bridging unit.
2. Conversion to the other system using the bridge.
3. Conversion within the desired system to units required by the problem.

Example 6

A patient's weight is 129 pounds. What is the patient's weight in kilograms? (Use 454 grams = 1 pound and 1 kilogram = 1000 grams in your set-up.)

Example 6 continued

Work

$$129 \text{ pounds} \times \frac{454 \text{ grams}}{1 \text{ pound}} \times \frac{1 \text{ kilogram}}{1000 \text{ grams}} = 58.666 \text{ kg}$$

Answer

58.6 kilograms (Three significant figures are needed. See Section 1.4.)

1.4 Significant Figures and Scientific Notation

Learning Goal 9:   **Report data and results using the proper number of significant figures and scientific notation.**

Modern pocket calculators often generate more digits than are merited, given the data used in the calculation. It is important that the result of these calculations be reflective of the level of certainty of the measurement. Proper use of significant figures, scientific notation, and rounding-off are essential.

Significant Figures

Significant figures are all digits in a number representing data or results that are known with certainty, *plus the first uncertain digit*.

The number of significant figures associated with a measurement is determined by the measuring device. Conversely, the number of significant figures reported is an indication of the sophistication of the measurement itself.

Recognition of Significant Figures

Only significant digits should be reported as data or results. The six rules enumerated below describe the assignment of significant figures.

> **Rule 1:**   All nonzero digits are significant.
>
> **Rule 2:**   The number of significant digits is independent of the position of the decimal point.
>
> **Rule 3:**   Zeros located between nonzero digits are significant.
>
> **Rule 4:**   Zeros at the end of a number (often referred to as trailing zeros) are significant if the number contains a decimal point.
>
> **Rule 5:**   Trailing zeros are insignificant if the number does not contain a decimal point.
>
> **Rule 6:**   Zeros to the left of the first nonzero integer are not significant; they serve only to locate the position of the decimal point.

Scientific Notation

Very large numbers may be represented with the proper number of significant figures using **scientific notation**. Scientific notation, also referred to as exponential notation, involves the representation of a number as a power of ten; the usual convention shows the decimal point in standard position—to the right of the leading digit.

To convert a number greater than 1 to scientific notation, the original decimal point is moved x places to the left, and the resulting number is multiplied by 10^x. The exponent (x) is a positive number equal to the number of places the original decimal point was moved.

To convert a number less than 1 to scientific notation, the original decimal point is moved x places to the right, and the resulting number is multiplied by 10^{-x}. The exponent ($-x$) is a negative number equal to the number of places the original decimal point was moved.

Example 7
Convert $30,000,000 \times 10^{-4}$ centimeters into correct scientific notation, and use the proper number of significant digits.

Work

$30,000,000 \times 10^{-4} \text{cm} - 3 \times 10^7 \times 10^{-4} \text{cm} = 3 \times 10^3 \text{cm}$
(Notice that only one significant digit, the "3", is present.)

Answer
3×10^3 cm

Error, Accuracy, Precision, and Uncertainty

Error is the difference between the true and measured value. **Accuracy** is the degree of agreement between the true and measured value. High levels of accuracy correlate with small values of error. **Precision** is the measure of agreement of replicate measurements. **Uncertainty** is the degree of doubt in a single measurement.

Significant Figures in Calculation of Results

In the process of addition or subtraction, the position of the decimal point in the quantities being combined determines the number of significant figures in the answer.

In multiplication and division, this is not the case: the decimal point position is irrelevant. What is important is the number of significant figures. Remember: *The answer can be no more precise than the least precise number from which the answer is derived.*

Example 8

Do the following multiplication and division problems using scientific notation. Also, round off the answer to the proper number of significant digits.

$$\frac{68.0 \times 10^{-5}}{4.61 \times 10^{-6}} \times \frac{16 \times 10^2}{3.6} = ?$$ *1600*

Work

Start by writing each part in scientific notation form:

$$\frac{6.80 \times 10^1 \times 10^{-5}}{4.61 \times 10^{-6}} \times \frac{1.6 \times 10^1 \times 10^2}{3.6} = ?$$

$$\frac{6.80 \times 10^{-4}}{4.61 \times 10^{-6}} \times \frac{1.6 \times 10^3}{3.6} = ?$$

Then multiply as below. First, combine the exponent parts.

$$\frac{(6.80)(1.6) \times 10^{-4} \times 10^3}{(4.61)(3.6) \times 10^{-6}} = \frac{(6.80)(1.6) \times 10^{-1}}{(4.61)(3.6) \times 10^{-6}} = \frac{10.88 \times 10^{-1}}{16.596 \times 10^{-6}} = ?$$

Then, divide and round off to two significant figures, since 3.6 contains only two significant figures.

$$0.6555795 \times \frac{10^{-1}}{10^{-6}} = 6.6 \times 10^{-1} \times \frac{10^{-1}}{10^{-6}} = 6.6 \times 10^{-1} \times 10^5 = 6.6 \times 10^4$$

Answer

6.6×10^4

When a number is raised to a power, $n^x = y$, the number of significant figures in the answer (y) is identical to the number contained in the original term (n).

Defined or counted numbers do not determine the number of significant figures. The quantity being converted, not the conversion factor, determines the number of significant figures.

Exact and Inexact Numbers

Inexact numbers have uncertainty, illustrated by the doubt in the final significant figure. In contrast, **exact numbers**, a result of counting, have no uncertainty and do not limit the number of significant figures when used in calculations.

Rounding Off Numbers

A generally accepted rule for rounding off states that if the first digit dropped is 5 or greater, we raise the last significant digit to the next higher number. If the first digit dropped is 4 or less, the last significant digit remains unchanged.

Example 9
Round off 16.00468 grams to two significant figures.

Answer
The first zero past the 6 will not increase the value. Thus, the answer is 16 grams.

1.5 Experimental Quantities

Learning Goal 10:   Use appropriate experimental quantities in problem solving.

Mass

Mass describes the quantity of matter in an object. The terms *weight* and *mass*, in common usage, are often considered synonymous. In fact, they are not. Weight is the manifestation of the force of gravity on an object.

$$\text{weight} = \text{mass} \times \text{acceleration due to gravity}$$

The common conversion units for mass are as follows:

$$1 \text{ gram (g)} = 1 \times 10^{-3} \text{ kilogram (kg)} - 1/454 \text{ pound (lb)}$$

The **atomic mass unit (amu)** is a convenient way to represent the mass of very tiny quantities of matter, such as individual atoms and molecules.

$$1 \text{ atomic mass unit (amu)} = 1.661 \times 10^{-24} \text{ g}$$

Length

The standard metric unit of length is the meter. Large distances are measured conveniently in kilometers, and smaller distances are measured in millimeters or centimeters. Common conversions for length are as follows:

$$1 \text{ meter (m)} = 1 \times 10^{2} \text{ centimeters (cm)} = 3.94 \times 10^{1} \text{ inches (in)}$$

In the same way that atomic mass units represent mass on an atomic scale, either the nanometer (nm) or the angstrom (Å) is a convenient measure of distance.

$$1 \text{ nm} = 10^{-7} \text{ cm} = 10^{-9} \text{ m}$$

$$1 \text{ Å} = 10^{-8} \text{ cm} = 10^{-10} \text{ m}$$

Volume

The standard metric unit of volume is the liter. A liter is the volume occupied by 1000 grams of water at 4° Celsius. A volume of 1 liter also corresponds to:

1 liter (1) = 1000 milliliters (ml) = 1.06 quarts (qt)

Time

The standard metric unit of time is the second. The need for accurate measurement of time by chemists is necessary in many applications.

Temperature

Temperature is the degree of "hotness" of an object. Many substances, such as mercury, expand as their temperature increases, and this expansion provides us with a way to measure temperature and temperature changes. The height of the mercury in a thermometer is proportional to the temperature. A mercury thermometer may be calibrated, or scaled, in different units, just like a ruler. Three common temperature scales are Fahrenheit (°F), Celsius (°C) and Kelvin (K). Two convenient reference temperatures used to calibrate a thermometer are the freezing and boiling temperatures of water. Conversion from one temperature scale to another may be accomplished as shown below.

Fahrenheit to Celsius: $\qquad °C = \dfrac{(°F - 32)}{1.8}$

Celsius to Fahrenheit: $\qquad °F = 1.8°C + 32$

Celsius to Kelvin: $\qquad K = °C + 273$

Example 10
What value, in °F, corresponds to 310.0 Kelvin?

Work
$\qquad K = °C + 273$
$\qquad\qquad$ or
$\qquad °C = K - 273$
$\qquad °C = 310.0 - 273$

Example 10 continued

$$°C = 37.0$$

(The rules of significant figures require that the result be expressed to the nearest tenth of a degree, since 310.0 is to the nearest tenth of a degree.)

$$°F = 1.8°C + 32$$
$$°F = 1.8(37.0) + 32$$
$$°F = 98.6$$

Answer
310.0 Kelvin = 98.6°F

(Note that "273" and "32" are exact numbers used in the conversion equations, and are not used to determine the number of significant figures.)

Energy

Energy, the ability to do work, may be categorized as either **kinetic energy**, the energy of motion, or **potential energy**, the energy of position. Kinetic energy is energy in process, while potential energy is stored energy. All energy is either kinetic or potential. Energy may also be classified according to form. The principal forms of energy include light, heat, electrical, mechanical, and chemical energy. Remember the following characteristics of energy:

1. In conventional chemical reactions, energy cannot be created or destroyed.
2. Energy may be converted from one form to another.
3. Energy conversion always occurs with less than 100% efficiency.
4. All chemical reactions involve either a gain or loss of energy.

Energy absorbed or liberated in chemical reactions is often in the form of heat energy. Heat energy may be represented in units of calories or Joules, their relationship being

$$1 \text{ calorie (cal)} = 4.18 \text{ Joules (J)}$$

One calorie is defined as the amount of heat energy required to change the temperature of 1 gram of water by 1°C.

Concentration

Concentration is a measure of the number of particles of a substance, or the mass of those particles, that are contained in a specific volume. Concentration is used to describe mixtures of different substances, such as the number of red blood cells in a specified volume of blood.

Density and Specific Gravity

> **Learning Goal 11:** Calculate the density of an object from mass and volume data and calculate the specific gravity of an object from its density.

Density, the ratio of mass to volume, $\quad d = \dfrac{\text{mass}}{\text{Volume}} = \dfrac{m}{V} = \dfrac{\text{gram}}{\text{milliliter}}$

is independent of the amount of material and is characteristic of a substance; each substance has a unique density.

Values of density are often related to a standard, well-known reference, the density of pure water at 4°C. This "referenced" density is the **specific gravity**. Note that specific gravity has no units.

$$\text{specific gravity} = \frac{\text{density of object (g/ml)}}{\text{density of water (g/ml)}}$$

Example 11

A solid block that has a mass of 1267.4 grams was found to have the following measurements: length = 9.86 cm, width = 46.6 mm, and height = 0.224 m. What is its density?

Work

1. Convert each measurement into centimeters:

   length = 9.86 cm
   width 46.6 mm = 4.66 cm
   height 0.224 m = 22.4 cm

2. Then use the equation for the volume of a solid:

   V_{solid} = length x width x height

   V_{solid} = (9.86 cm) (4.66 cm) (22.4 cm) = 1.0292262 x 10^3 cm^3
   (Use only three significant figures.)

   V_{solid} = 1.03 x 10^3 cm^3

Example 11 continued

3. Then use the density equation:

$$d = \frac{m}{V} = \frac{1267.4 \text{ g}}{1.03 \times 10^3 \text{ mL}}$$

$$d = \frac{1.2674 \times 10^3 \text{ g}}{1.03 \times 10^3 \text{ mL}}$$

$$d = 1.23 \text{ g/mL}$$

Answer

The density equals 1.23 grams/milliliter.

Example 12

If the density of carbon tetrachloride is 1.59 grams/mL, what is the mass (in grams) of 2.00 liters of carbon tetrachloride?

Work

$$2.00 \text{ liters} \times \frac{1000 \text{ mL}}{1 \text{ liter}} \times \frac{1.59 \text{ grams}}{1 \text{ mL}} = 3.18 \times 10^3 \text{ grams}$$

Answer

3.18×10^3 grams (Three significant figures are needed.)

Example 13

Calculate the mass in grams of 25 mL of mercury. The density of mercury is 13.5 g/mL.

Work

Use mass = $(d)(V)$, or use the factor-label method as below:

$$25 \text{ mL of mercury} \times \frac{13.5 \text{ g of mercury}}{1 \text{ mL of mercury}} = 337.5 \text{ g}$$

(Convert to scientific notation and round off to two significant figures.)

Answer

3.4×10^2 g

Example 14
Calculate the volume in mL of a liquid (ethyl alcohol) that has a density of
0.789 g/mL and a mass of 5.555 g.

Work

Use $V = \dfrac{m}{d}$, or use the factor label method as below:

$$5.555 \text{ g of liquid} \times \frac{1 \text{ mL of liquid}}{0.789 \text{ g of liquid}} = 7.0405576 \text{ mL}$$

(Round off to three significant figures.)

Answer
7.04 mL

Glossary of Key Terms in Chapter One

accuracy (1.4) the nearness of an experimental value to the true value.

chemical property (1.2) property of a substance that relates to the substance's participation in a chemical reaction.

chemical reaction (1.2) a process in which atoms are rearranged to produce new combinations.

chemistry (1.1) the study of matter and the changes that matter undergoes.

compound (1.2) a substance that is characterized by constant composition and can be chemically broken down into elements.

concentration (1.5) a measure of the quantity of a substance contained in a specified volume of solution.

data (1.3) a group of facts resulting from an experiment.

density (1.5) mass per unit volume of a substance.

element (1.2) a substance that cannot be decomposed into simpler substances by chemical or physical means.

energy (1.1) the capacity to do work.

error (1.4) the difference between the true value and the experimental value for data and results.

extensive property (1.2) a physical property that depends on the quantity of the substance.

gaseous state (1.2) a physical state of matter characterized by the lack of fixed shape or volume and ease of compressibility.

heterogeneous mixture (1.2) a mixture of two or more substances and is characterized by nonuniform composition.

homogeneous mixture (1.2) a mixture of two or more substances and is characterized by uniform composition.

hypothesis (1.1) attempt to explain observations in a common-sense way.

intensive property (1.2) a property of a substance that is independent on the quantity of the substance.

kinetic energy (1.5) the energy resulting from motion of an object [K.E. = $\frac{1}{2}$(mass)(velocity^2)].

law (1.1) a summary of a large quantity of information.

liquid state (1.2) a physical state of matter characterized by the fixed volume and absence of a fixed shape.

mass (1.5) a quantity of matter.

matter (1.1) the material component of the universe.

mixture (1.2) a material composed of two or more substances.

physical change (1.2.) a change in the form of a substance but not in its chemical composition. No chemical bonds are broken in a physical change.

physical property (1.2) a characteristic of a substance that can be observed without the substance undergoing change (examples include color, density, and melting and boiling points).

potential energy (1.5) stored energy or energy due to position or composition.

precision (1.4) the degree of agreement between replicate measurements of the same quantity.

properties (1.2) physical or chemical characteristics of matter.

pure substance (1.2) a substance with constant composition.

results (1.3) the outcome of a designed experiment, often calculated from individual bits of data.

scientific method (1.1) the process of studying our surrounding that is based on experimentation.

scientific notation (1.4) a system used to represent numbers as powers of ten.

significant figures (1.4) all digits in a number known with certainty and the first uncertain digit.

solid state (1.2) a physical state of matter characterized by its rigidity and fixed volume and shape.

specific gravity (1.5) the ratio of the density of a substance to the density of water at a specified temperature

temperature (1.5) a measure of the relative "hotness" or "coldness" of an object.

theory (1.1) a hypothesis supported by extensive testing that explains and predicts facts.

uncertainty (1.4) the degree of doubt in a single measurement.

unit (1.3) a determinate quantity (of length, time, etc.) adopted as a standard of measurement.

weight (1.5) the force exerted on an object by gravity.

Self Test for Chapter One

1. Calculate the density (in normal unit form) of a liquid with a volume of 348 mL and a mass of 0.3546 kg.

2. $\dfrac{68.0 \times 10^{-3}}{4.61 \times 10^{-1}} = ?$

3. $\dfrac{10.6 \times 10^{-4}}{0.0641 \times 10^{-2}} \times 16.4 \times 10^{-7} \times \dfrac{0.111 \times 10^{4}}{10.0 \times 10^{2}} = ?$

4. 50.0 miles/hour = ? cm/second

5. If one atom of uranium weighs 238.0 atomic mass units, and one atomic mass unit is equal to 1.66×10^{-24} gram, what is the mass, in grams, of one atom of uranium?

6. The speed of light is 186,000 miles per second. What is its speed in cm/second?
   USE: 5280 feet = 1 mile; 12 inches = 1 foot; 2.54 cm = 1 inch

7. $-4.6\,°F = ?\ °C$

8. If the density of carbon tetrachloride is 1.59 g/mL, what is the mass of 2.65 liters of carbon tetrachloride in grams?

9. $-40.0\,°C = ?\ °F$

10. What is the specific gravity of an object that weighs 13.35 g and has a volume of 25.00 mL? The density of water under the same conditions is 0.980 g/mL.

11. The ability to do work describes what term?

12. A scientific experiment produces what information?

13. Each piece of data is the individual result produced by what process?

14. Which system of measurement is a decimal-based system?

15. What factors are used in the factor-label method?

16. Which temperature system does not use a degree sign? *Kelvin*

17. 98.6°F = ? K   *353.42*

18. Round off 0.00369865 to two significant figures.   *3.6*

19. Which of the following are physical properties of an object?

    a. combustibility
    b. color
    c. melting point
    d. gas phase
    e. chlorine combines with hydrogen to form HCl
    f. density
    g. volume
    h. $S + O_2 \rightarrow SO_2$

20. Which of the following are mixtures?

    a. NaCl
    b. salt in water
    c. a soft drink
    d. tap water
    e. blood
    f. cake mix

21. What are the three states of matter?   *solid, liquid, gas*

22. Which state of matter will expand to fill any container?   *gas*

23. Which state of matter has no definite shape or volume?   *liquid*

24. List six examples of physical properties of matter.   *V, mass, d, shape, color*

25. Salt and pepper is an example of what type of mixture?   *Homogeneous*

Vocabulary Quiz for Chapter One

1. The study of matter and the changes that matter undergoes is referred to as *Chem.*_____.

2. _____ is a group of facts resulting from an experiment.

3. A suggested explanation of observed behavior of our surroundings is a(n)___*scientific method*.

4. The __*Gas*___ state is a physical state of matter characterized by the lack of fixed shape or volume and ease of compressibility.

5. Stored energy is called _*potential*_.

6. The outcome of a scientific experiment, often calculated from data, is the _____ of that experiment.

7. An _____ property depends on the quantity of matter.

8. The ratio of the density of a substance to the density of water at the same temperature is called the _*s.g.*___.

9. A(n) __*Theory*__ is a hypothesis developed to explain observed behavior of matter that has been verified by using the scientific method.

10. _____ is a force exerted on an object by gravity.

2 *The Structure of the Atom and the Periodic Table*

Learning Goals

1. Describe the important properties of protons, neutrons, and electrons.
2. Calculate the number of protons, neutrons, and electrons in any atom.
3. Distinguish among atoms, ions, and isotopes.
4. Trace the history of the development of atomic theory, beginning with Dalton.
5. Explain the critical role of spectroscopy in the development of atomic theory and in our everyday lives.
6. State the basic postulates of Bohr's theory, its utility, and its limitations.
7. Recognize the important subdivisions of the periodic table: periods, groups (families), metals, and nonmetals.
8. Use the periodic table to obtain information about an element.
9. Describe the relationship between the electronic structure of an element and its position in the periodic table.
10. Write electron configurations for atoms of the most commonly occurring elements.
11. Use the octet rule to predict the charge of common cations and anions.
12. Use the periodic table and its predictive power to estimate the relative sizes of atoms and ions, as well as relative magnitudes of ionization energy and electron affinity.

Introduction

In this chapter we will learn some of the properties of atomic particles and the experiments that form the basis of our current understanding of atomic structure. The atomic structure of each element is unique, but there exist structural similarities among these elements.

The **periodic law** states that properties of elements are periodic functions of their atomic numbers. The periodic table results from this relationship. The periodic table is an organized "map" of the elements that relates their structure to their chemical and physical properties. The chemical and physical properties of elements follow directly from the electronic structure of the atoms that make up these elements. Familiarity with the periodic table allows prediction of the structure and properties of the various elements, and serves as the basis for understanding chemical bonding.

2.1   Composition of the Atom

Learning Goal 1:	Describe the important properties of protons, neutrons, and electrons.

Electrons, Protons, and Neutrons

The basic structural unit of an element is the **atom**, which is the smallest unit of an element that retains the chemical properties of that element. An atom is composed of three primary particles: the **electron**, the **proton,** and the **neutron**. These particles are located in one of two distinct regions:

1. The **nucleus** is a small, dense, positively charged region in the center of the atom. The nucleus is composed of positively charged protons and uncharged neutrons.
2. Surrounding the nucleus is a diffuse region of negative charge populated by electrons, the source of the negative charge. Electrons are tiny in comparison to the protons and neutrons. The properties of these particles are summarized in Table 2.1 of the textbook.

The number of protons determines the identity of the atom. When the number of protons is equal to the number of electrons, the atom is neutral, because the charges are balanced and effectively cancel.

The **atomic number (Z)** is equal to the number of protons in the atom, and the **mass number (A)** is equal to the sum of the protons and neutrons (the mass of the electrons is so small as to be insignificant).

Isotopes

Isotopes are atoms of the same element which have different masses due to different numbers of neutrons (different atomic mass).

Inspection of the periodic table reveals that the atomic mass of many of the elements is not an integral number. For example the atomic mass of chlorine is actually 35.45 amu, not 35.00 amu. The existence of isotopes accounts for this difference. A natural sample of chlorine is principally composed of two isotopes, $^{35}_{17}Cl$ and $^{37}_{17}Cl$, in approximately a 3:1 ratio, and the tabulated atomic mass is the weighted average of the two isotopes.

Certain isotopes of elements emit particles and energy (**radioactivity**, Chapter 9) that may be useful in tracing the behavior of biochemical systems. These isotopes otherwise behave identically to any other isotope of the same element.

Learning Goal 2:	Calculate the number of protons, neutrons, and electrons in any atom.

Example 1
Calculate the number of protons, neutrons, and electrons found in carbon–14 ($^{14}_{6}C$).

Example 1 continued

Work

The atomic number of carbon (from the periodic table) is 6. Thus, there are 6 protons and 6 electrons in one atom of carbon. The 14 in carbon–14 means that this specific isotope of carbon has a mass number of 14. The mass number of an atom is equal to the sum of the protons and neutrons:

mass number = number of protons + number of neutrons

number of neutrons = mass number – number of protons

Thus, for carbon–14:

number of neutrons = 14 – 6 = 8

Answer

Carbon–14 contains 6 protons, 6 electrons, and 8 neutrons.

Learning Goal 3:   Distinguish among atoms, ions, and isotopes.

Example 2

Calculate the number of protons, neutrons, and electrons in the three isotopes of hydrogen:

$$^1_1H \qquad\qquad ^2_1H \qquad\qquad ^3_1H$$

normal hydrogen   deuterium   tritium

Work

Since all are hydrogen atoms, they each contain only 1 proton and 1 electron, because the atomic number of hydrogen is 1.

number of neutrons = mass number – number of protons

The mass number of normal hydrogen is 1. Thus, it contains no neutrons.
The mass number of deuterium is 2. Thus, it contains 1 neutron in its nucleus:
number of neutrons in deuterium = 2 – 1 = 1
The mass number of tritium is 3. Thus, it contains 2 neutrons in its nucleus:
number of neutrons in tritium = 3 – 1 = 2

Answer

The three isotopes each have 1 proton and 1 electron. Normal hydrogen has no neutrons. Deuterium has 1 neutron; tritium has 2 neutrons.

Ions

Ions are charged particles that result from a gain of one or more electrons by the parent atom (forming negative ions, or **anions**) or a loss of one or more electrons from the parent atom (forming positive ions, or **cations**).

For simplification, the atomic and mass numbers are often omitted. For example, the hydrogen cation would be written as H^+ and the anion as H^-.

Example 3

How many protons and electrons are present in each of the following atoms and ions?

$$H \quad + \quad 1e^- \quad \rightarrow \quad H^-$$

Work

The H with no charge means that it is an atom; atoms have no charge, since they have the same number of protons and electrons. The H^- is an ion that has a negative one charge. This means it has one more electron than it has protons. You can see in the reaction that the hydrogen atom (given as H) has gained one electron to form the ion H^-.

ATOM	**ION**
$H \quad + \quad 1e^- \quad \rightarrow \quad H^-$	

$+ 1 \ = $ proton	1 proton = +1
$- 1 \ = $ electron	2 electrons = –2
0 \ charge	charge = –1

Answer

The hydrogen atom contains 1 proton and 1 electron. The –1 hydrogen ion contains 1 proton and 2 electrons.

Remember that if an ion is positive, it was formed by the loss of the required number of electrons from the given atom.

$$Mg \rightarrow 2e^- + Mg^{2+}$$ • Two electrons were lost by the magnesium atom.
$$K \rightarrow 1e^- + K^+$$ • One electron was lost by the potassium atom.

A negative ion, on the other hand, is formed by a gain of one or more electrons by the given atom.

$$S + 2e^- \rightarrow S^{2-}$$
$$Cl + 1e^- \rightarrow Cl^-$$

2.2   Development of the Atomic Theory

Learning Goal 4:   Trace the history of the development of atomic theory, beginning with Dalton.

Dalton's Theory

The first experimentally based theory of atomic structure was proposed by John Dalton in the early 1800s. Dalton postulated that:

1. All matter consists of tiny particles called atoms.
2. Atoms cannot be created, divided, destroyed, or converted to any other type of atom.
3. Atoms of a particular element have identical properties.
4. Atoms of different elements have different properties.
5. Atoms combine in simple, whole-number ratios.
6. Chemical change involves joining, separating, or rearranging atoms.

Postulates 1, 4, 5, and 6 are presently regarded as true. The discovery of the processes of nuclear fusion, fission, and radioactivity (Chapter 9) have disproved the postulate that atoms cannot be created or destroyed. Postulate 3, that the atoms of a particular element are identical, was disproved by the later discovery of isotopes.

Evidence for Subatomic Particles: Electrons, Protons, and Neutrons

Although Dalton pictured atoms as indivisible, various experiments, particularly those of William Crookes and Eugene Goldstein, indicated that the atom is composed of charged (+ and −) particles.

J. J. Thomson demonstrated the electrical and magnetic properties of cathode rays (Figures 2.4 and 2.5 in the textbook). Crookes observed rays, which he called cathode rays, emanating from the cathode (− charge) of an evacuated (vacuum) tube. Further experiments showed that the ability to produce cathode rays is a characteristic of all materials. In 1897, Thomson announced that cathode rays were streams of negative particles of energy. These particles are electrons.

Similar experiments, conducted by Goldstein, led to the discovery of protons, particles equal in charge to the electron, but opposite in sign.

The neutron has a mass virtually equal to that of the proton, and zero charge. The neutron was first postulated in the early 1920s, but it was not until 1932 that James Chadwick demonstrated its existence.

Evidence for the Nucleus

In the early 1900s, it was believed that protons and electrons were uniformly distributed throughout the atom. However, an experiment by Hans Geiger led Ernest Rutherford (in 1911) to propose that the majority of the mass and positive charge of the atom was actually located in a small, dense region, the nucleus, and that the small, negatively charged electrons spread across a much larger, diffuse area outside of the nucleus.

2.3 Light, Atomic Structure, and the Bohr Atom

Learning Goal 5:	Explain the critical role of spectroscopy in the development of atomic theory and in our everyday lives.

Light and Atomic Structure

The study of the interaction of light and matter is termed spectroscopy. Light, electromagnetic radiation, travels at a speed of 3.0×10^8 m/s, the speed of light. Light is made up of many wavelengths. Collectively, this range of wavelengths is the electromagnetic spectrum. Samples of elements emit certain wavelengths of light when an electrical current is passed through the sample. Different elements emit a different pattern (different wavelengths) of light.

Learning Goal 6:	State the basic postulates of Bohr's theory, its utility, and its limitations.

The Bohr Atom

Bohr and his contemporaries were puzzled by the emission spectrum of hydrogen. He developed a hypothesis to explain his observations. This evolved into Bohr's atomic theory, summarized below.

1. Atoms can absorb and emit energy via **promotion** of electrons to higher **energy levels** and **relaxation** to lower levels. These levels are referred to as Bohr **orbits**.
2. Energy that is emitted upon relaxation is observed as a single wavelength of light.
3. These spectral lines are a result of electron transitions between "allowed levels" in the atom.
4. The allowed levels are **quantized** energy levels, or orbits.
5. Electrons are found only in these energy levels.
6. The highest-energy orbits are located farthest from the nucleus.
7. Atoms absorb energy by excitation of electrons to higher energy levels producing the **excited state**.
8. Atoms release energy by **relaxation** of electrons to lower energy levels, the **ground state**.
9. Energy differences may be calculated from the wavelengths of light emitted during these electronic transitions.

Modern Atomic Theory

Although Bohr's concept of principal energy is still valid, restriction of electrons to fixed orbits is too rigorous. All current evidence shows that electrons do *not*, in fact, orbit the nucleus. We now speak of the *probability* of finding an electron in a *region* of space within the principal energy level, referred to as an **atomic orbital**. The rapid movement of the electron spreads the charge into a *cloud* of charge. This cloud is more dense in certain regions, the **electron density** being proportional to the probability of finding the electron at any point in time. Insofar as these atomic orbitals are part of the principal energy levels, they are referred to as sublevels. In Chapter 3, we will see that the orbital model of the atom can be used to predict how atoms can bond together to form compounds. Furthermore, electron arrangement in orbitals enables us to predict various chemical and physical properties of these compounds.

2.4 The Periodic Law and the Periodic Table

Numbering Groups in the Periodic Table

Two systems currently exist to number the groups in the periodic table. The older system uses Roman numerals followed by the letter A for the main-group elements or the letter B for the transition elements. The newer system, recommended by I.U.P.A.C., simply numbers the groups from 1 to 18. Both numbering systems are shown on the periodic table in your text.

Periods and Groups

Learning Goal 7:	Recognize the important subdivisions of the periodic table: periods, groups (families), metals, and nonmetals.

A horizontal row of elements in the periodic table is referred to as a **period**. The periodic table consists of seven periods, six of which contain 2, 8, 8, 18, 18, and 32 elements. The seventh period is still incomplete but potentially holds 32 elements. Note that the lanthanide series is a part of period six and the actinide series is a part of period seven.

The columns of elements in the periodic table are called **groups** or families. The elements of a particular family share many similarities in physical and chemical properties that are related to similarities in electronic structure. The various groups are labeled with Roman numerals, and each is subtitled with the letter A or B.

Group A elements are called **representative elements**, and Group B elements are **transition elements.** Certain families have common names as well as a Roman numeral-letter designation. Group IA elements are also known as the **alkali metals**; Group IIA, as the **alkaline earth metals**; Group VIIA, as the **halogens**; and Group VIIIA, as the **noble gases**.

Metals and Nonmetals

A bold zig-zag line runs from top to bottom of the table beginning to the left of boron (B) and ending between polonium (Po) and astatine (At). This line acts as the boundary between **metals**, to the left, and **nonmetals,** to the right. Elements straddling the boundary, such as Ge and As, have properties intermediate between metals and nonmetals and are often termed **metalloids**.

Atomic Number and Atomic Mass

The atomic number (Z, number of protons in the nucleus; the nuclear charge) and the atomic mass of each element are available from the periodic table. More detailed periodic tables may provide information such as electron arrangement, relative sizes of atoms, and most probable ion charges.

Learning Goal 8:   **Use the periodic table to obtain information about an element.**

Example 4

For each of the following symbols, provide the name of the element, its atomic number (Z), and the family to which it belongs.

1. Li   4. Si   7. I
2. Ra   5. N   8. Cu
3. Al   6. S

Answer

1. Li: lithium; Z = 3; family = IA   5. N: nitrogen; Z = 7; family = VA
2. Ra: radium; Z = 88; family = IIA   6. S: sulfur; Z = 16; family = VIA
3. Al: aluminum; Z = 13, family = IIIA   7. I iodine; Z = 53; family = VIIA
4. Si: silicon; Z = 14; family = IVA   8. Cu: copper; Z = 29; family = IB

2.5 Electron Arrangement and the Periodic Table

Learning Goal 9:   **Describe the relationship between the electronic structure of an element and its position in the periodic table.**

The most important factor in chemical bonding is the arrangement of the electrons in the atoms that are combining. The periodic table provides us with a great deal of information about the electron arrangement, or electronic structure, of atoms.

Valence Electrons

Outermost electrons in an atom are **valence electrons**. For representative elements, the number of valence electrons in an atom corresponds to the number of the family in which the atom is found. Metals tend to have fewer valence electrons, and nonmetals tend to have more valence electrons.

The energy levels are symbolized by n, with the lowest energy level assigned a value of $n = 1$.

1. Each energy level may contain up to a fixed maximum number of electrons.
   Two general rules of electron configuration are based upon the periodic law.
2. The number of valence electrons in a neutral atom equals the group number for all representative (A group) elements.

3. The energy level (n = 1, 2, etc.) in which the valence electrons are located corresponds to the period in which the element may be found.

Helium is an exception to rule 1, above. It cannot have eight valence electrons, since all of its electrons are in the n = 1 level, which has a maximum capacity of only two electrons.

The Quantum Mechanical Atom

Erwin Schröedinger described electrons in atoms in probability terms, and developed equations that emphasize the wavelike character of electrons. His theory, often described as quantum mechanics, incorporates Bohr's principle energy levels, which are made up of one or more sublevels. Each sublevel contains one or more atomic orbitals.

Energy Levels and Sublevels

The principal energy levels are designated n = 1, 2, 3, and so forth. The number of possible sublevels in a principal energy level is also equal to n. When n = 1, there can be only one sublevel; n = 2 allows two sublevels, and so forth. The total electron capacity of a principal level is $2(n)^2$.

The sublevels, or subshells, increase in energy: $s < p < d < f$

Both the principal energy level and type of sublevel are specified when describing the location of an electron. For example: $1s$, $2s$, $2p$. The first principal energy level (n = 1) has one possible subshell, $1s$. The second principal energy level (n = 2) has two possible subshells: $2s$ and $2p$. The third principal energy level (n = 3) has three possible subshells: $3s$, $3p$, and $3d$. The fourth principal energy level (n = 4) has four possible subshells: $4s$, $4p$, $4d$, and $4f$.

An **orbital** is a specific region of a subshell containing a maximum of two electrons. The s sublevel contains only one orbital, the p sublevel contains three orbitals, the d sublevel contains five orbitals, and the f sublevel contains seven orbitals. Each orbital may be empty, contain one electron, or be filled, containing two electrons. The s, p, d, and f sublevels have maximum capacities of 2, 6, 10, and 14 electrons.

Each type of orbital has its own characteristic shape:

*   The s orbital is spherically symmetrical; a model appears as a Ping-Pong ball with its center corresponding to the intersection of imaginary x, y, and z coordinates.
*   There exist three kinds of p orbitals, each identical in shape (often modeled as a dumbbell). They differ only in their orientation in space, along the hypothetical x-axis, (p_x), y-axis, (p_y), and z-axis, (p_z).
*   The d and f orbitals are more complex; the textbook focuses exclusively on s and p orbitals in subsequent discussion.

Each atomic orbital has a maximum capacity of two electrons. The electrons are perceived to *spin* on an imaginary axis, and the two electrons in the same orbital must have opposite spins, clockwise and counterclockwise. Two electrons in one orbital that possess opposite spins are referred to as *paired* electrons.

Electron Configuration and the Aufbau Principle

Learning Goal 10: Write electron configurations for atoms of the most commonly occurring elements.

The arrangement of electrons in atomic orbitals is referred to as the atom's **electron configuration**. We may represent the electron configuration of atoms of various elements using the aufbau, or building-up, principle. According to this principle, electrons fill the lowest-energy orbital that is available first. By knowing the order of filling of atomic orbitals, lowest to highest energy, you may write the electron configuration for any element.

Example 5
Name the two elements that have electrons only in the first energy level.

Answer
hydrogen and helium

Example 6
List the sublevels in order of lowest energy to highest energy.

Answer
$s < p < d < f$

Example 7
Write the symbols of all the possible sublevels found in the fourth principal energy level ($n = 4$).

Answer
$4s$, $4p$, $4d$, and $4f$

Example 8
Complete the following statement about an orbital:
Each orbital may be empty, _____, or be _____.

Answer
Each orbital may be empty, contain one electron, or be filled, containing two electrons.

Shorthand Electron Configurations

Electron configurations may be abbreviated by substituting the symbol for a noble gas (in brackets, [ ]) for the part of the electron configuration represented by that noble gas. For example, the electron configuration of strontium, Sr, would be quite long: $1s^2 2s^2 2p^6 3s^2 3p^6 4s^2 3d^{10} 4p^6 5s^2$. However, the noble gas that is located before Sr on the periodic table is Kr, which has the following electron configuration: $1s^2 2s^2 2p^6 3s^2 3p^6 4s^2 3d^{10} 4p^6$. The electron configuration of Sr, then, can be abbreviated as follows:

Sr $[Kr] 5s^2$

2.6 The Octet Rule

Elements in the last family, the noble gases, have either two valence electrons (helium) or eight valence electrons (neon, argon, krypton, xenon, and radon). Their most important property is their extreme stability. A full $n = 1$ energy level (as in helium) or an outer octet of electrons is responsible for this unique stability.

Atoms of elements in other groups are more reactive than the inert gases because they are, in the process of chemical reaction, trying to achieve a more stable "noble gas" configuration by gaining or losing electrons. This is the basis of the **octet rule**. In chemical reactions they will gain, lose, or share the minimum number of electrons necessary to achieve a more stable energy state.

Ion Formation and the Octet Rule

Learning Goal 11:	Use the octet rule to predict the charge of common cations and anions.

    Metallic elements tend to form positively charged ions called **cations**. Positive ions are formed when an atom loses one or more electrons. These ions are more stable than their corresponding neutral atoms. The ion is isoelectric (that is, it has a similar number of electrons) with its nearest noble gas neighbor and has an octet of electrons in its outermost energy level. Nonmetallic elements tend to gain electrons to become isoelectronic with the nearest noble gas element, forming negative ions referred to as **anions**.

Example 11

Show how a sodium atom will form a sodium ion.

Answer

1. Na $\rightarrow$ $Na^+ +$ $1e^-$

     Atom   Ion

2. Na (atom); it has $11e^-$
3. Na^+ (ion); it has $10e^-$

Example 12

Show how an oxygen atom will form an oxide ion.

Answer

1. O $+$ $2\,e^- \rightarrow$ O^{2-}

     oxygen atom   oxide ion

2. O (atom); it has $8e^-$
3. O^{2-} (ion); it has $10e^-$

    The transition metals tend to form positive ions by losing electrons, just like the representative metals. However, the transition elements are characterized as "variable valence" elements; depending upon the type of substance they react with, they may form more than one stable ion. For example, iron has two stable ionic forms: Fe^{2+} and Fe^{3+}.

2.7 Trends in the Periodic Table

Learning Goal 12:	Use the periodic table and its predictive power to estimate the relative sizes of atoms and ions, as well as the relative magnitudes of ionization energy and electron affinity.

Atomic Size

The size of the atom will be determined principally by two factors:

1. The energy level (n-level) in which the outermost electron(s) is located increases as we go down a group (recall that the outermost n-level correlates with period number).

2. As the magnitude of the positive charge of the nucleus increases, its "pull" on all of the electrons increases, and the electrons are drawn closer to the nucleus.

Consequently, atomic size increases down a group and decreases across a period

Ion Size

Three generalizations can be made about the size of ions:

1. Positive ions (cations) are smaller than the parent atom.

2. Negative ions (anions) are larger than the parent atom.

3. Ions with multiple positive charge (such as Cu^{2+}) are even smaller than their corresponding monopositive ion (Cu^+); ions with multiple negative charge (such as O^{2-}) are larger than their corresponding less-negative ion.

Ionization Energy

The energy required to remove an electron from an isolated atom in the gas phase is the **ionization energy**. The magnitude of the ionization energy correlates with the strength of the attractive force between the nucleus and the outermost electron.

1. As we go down a group, the ionization energy decreases, since the atom's size is increasing. The outermost electron is progressively farther from the nuclear charge and hence easier to remove.

2. As we go across a period, atomic size decreases, as the outermost electrons are closer to the nucleus, more tightly held, and more difficult to remove. Therefore, the ionization energy must increase.

A correlation does indeed exist between trends in atomic size and ionization energy. Atomic size decreases from bottom to top of a group and left to right in a period. Ionization energies increase in the same periodic way. Note also that ionization energies are highest for the noble gases; this accounts for the extreme stability and nonreactivity of the noble gases.

Electron Affinity

The energy released when a single electron is added to a neutral atom in the gaseous state is known as the **electron affinity**. Electron affinity is a measure of the ease of forming negative ions. A large value of electron affinity (energy released) indicates that the atom becomes more stable as it becomes a negative ion (through the process of gaining an electron).

Periodic trends for electron affinity are as follows:
1. Electron affinities generally decrease as we go down a group.
2. Electron affinities generally increase as we go across a period.

Be aware that exceptions to these general trends do exist.

Glossary of Key Terms in Chapter Two

alkali metal (2.4) occupy group IA (1) of the periodic table.
alkaline earth metal (2.4) occupy group IIA (2) of the periodic table.
anion (2.1) a negatively charged atom or group of atoms.
atom (2.1) the smallest unit of an element that retains properties of that element.
atomic mass (2.1) the mass of the atom expressed in atomic mass units.
atomic number (2.1) the number of protons in the nucleus of an atom. It is a characteristic identifier of an element.
atomic orbital (2.3 and 2.5) a specific region of space where an electron may be found.
cathode rays (2.1) a stream of electrons that are given off by the cathode (negative electrode) in a cathode ray tube.
cation (2.1) a positively charged atom or group of atoms.
electromagnetic radiation (2.3) energy that is propagated as waves at the speed of light.
electromagnetic spectrum (2.3) the complete range of electromagnetic waves.
electron (2.1) a negatively charged particle outside of the nucleus of an atom.
electron affinity (2.7) the energy released when an electron is added to an isolated atom.
electron configuration (2.5) the arrangement of electrons in atomic orbitals.
electron density (2.3) the probability of finding the electron at a particular location.
energy level (2.3) one of the numerous atomic regions where electrons may be found.
group (2.4) any one of 18 vertical columns of elements. (Note that a group is often referred to as a *family.*)
halogen (2.4) occupy group VIIA (17) of the periodic table.
ion (2.1) an electrically charged particle that is formed by the gain or loss of an electron.
ionization energy (2.7) the energy needed to remove an electron from an atom in the gas phase.
isoelectronic (2.6) atoms, ions, and molecules containing the same number of electrons.
isotope (2.1) atom of the same element that differs in mass because it contains different numbers of neutrons.
mass number (2.1) the sum of the number of protons and neutrons in an atom.
metal (2.4) element located on the left side of the periodic table (left of the "staircase" boundary).
metalloid (2.4) an element along the "staircase" boundary between metals and nonmetals; they exhibit both metallic and nonmetallic properties.
neutron (2.1) an uncharged particle, with the same mass as the proton, found in the nucleus of an atom.
noble gas (2.4) elements in group VIIIA (18) of the periodic table.
nonmetal (2.4) element located on the right side of the periodic table (right of the "staircase" boundary).
nucleus (2.1) the small, dense center of positive charge in the atom.
octet rule (2.6) a rule predicting that atoms form the most stable molecules or ions when they are surrounded by eight electrons in their highest occupied energy level.
period (2.4) any one of seven horizontal rows of elements in the periodic table.
periodic law (2.4) a law stating that properties of elements are periodic functions of their atomic numbers. (Note that Mendeleev's original statement was based on atomic masses.)
proton (2.1) a positively charged particle in the nucleus of an atom.
quantization (2.3) a characteristic that energy can occur only in discrete units called quanta.
representative element (2.4) member of the group of the periodic table designated as "A".

spectroscopy (2.3) the measurement of intensity and energy of electromagnetic radiation.

speed of light (2.3) 2.99×10^8 m/s in a vacuum.

sublevel (2.5) set of equal-energy orbitals within a principal energy level.

transition element (2.4) occurs between groups IIIB (3) and IIB (12) in the long period of the periodic table.

valence electron (2.5) electron in the outermost shell (principal quantum level) of an atom.

Self Test for Chapter Two

1. How many protons and electrons are found in a magnesium atom? *12, 12*

2. How many protons and electrons are found in a chloride ion, Cl^-? *17, 18*

3. Give the number of protons and electrons found in a lithium atom and a lithium ion, Li^+. Write the reaction showing how a lithium atom can form a lithium ion.

4. What is the dense positive center of an atom? *nucleus*

5. Which particle of an atom has the least mass? *electron*

6. What number is equal to the number of protons in an atom? *atomic number*

7. The mass number minus the number of protons equals the number of what particle? *neutrons*

8. If an atom loses an electron, what charge does the resulting ion have? *positive*

9. Symbolically represent the three isotopes of the element hydrogen. *^1H ^2H ^3H*

10. What do we call the outermost electrons of an element, which are involved in chemical bonding?

11. How many valence electrons are in sulfur? *2 8 6 0*

12. Name the only noble gas with two valence electrons. *He*

13. Name the group that contains elements frequently isoelectronic with stable cations and anions.

14. In chemical reactions, elements will gain, lose, or share the minimum number of electrons necessary to achieve what kind of energy state?

15. What do we call an ion that has the same electronic arrangement as its nearest noble gas neighbor?

16. What are the Group A elements often called? *representative*

17. What are the Group B elements often called? *transition*

18. List the values of the principal (main) energy levels of an atom.

19. How many sublevels are present in the second principal energy level?

20. What is the name of a specific region of space of a sublevel where a maximum of two electrons may be found?

21. What orbital has a spherical shape?

22. What is the relationship of the spins of two electrons in the same orbital?

23. What term refers to the arrangement of electrons in atomic orbitals?

24. Write the electron configuration of carbon.

25. Which elements do not readily bond to other elements?
26. All of the ions in this group always form a + 1 ion.
27. What do we call the ability of an atom to attract electrons to itself?

Vocabulary Quiz for Chapter Two

1. The sum of the number of protons and neutrons in an atom is the _____.
2. A(n)_____ is a positively charged atom (or group of atoms).
3. A(n)_____ is an integer used to describe the orbitals of an atom.
4. A(n)_____ is a negatively charged particle outside of the nucleus of an atom.
5. A(n)_____ is a substance that cannot be decomposed into simpler substances by chemical or physical means.
6. According the Bohr theory, a region of the atom where electrons may be found is called a(n)_____.
7. A neutral atom that gains one electron has a _____ charge and is called a(n) _____.
8. Atoms of the same element that differ in mass because they contain different numbers of neutrons are called _____.
9. A(n) _____ is an uncharged particle in the nucleus of an atom that has the same mass as the proton.
10. A positively charged particle in the nucleus of an atom is called a (n)_____.
11. The _____ occupy group IA of the periodic table.
12. The _____ is an organized "map" of the elements.
13. A horizontal row of elements in the periodic table is referred to as a _____.
14. Columns in the period table are also called _____.
15. The outermost electrons in an atom are _____ electrons.
16. Neon, argon, krypton, xenon, and radon are termed _____.
17. Energy released when an electron is added to an atom is _____.
18. Energy required to remove an electron from an atom is _____.

3 *Structure and Properties of Ionic and Covalent Compounds*

Learning Goals

1 Classify compounds as having ionic, covalent, or polar covalent bonds.
2 Write the formulas of compounds when provided with the name of the compound.
3 Name common inorganic compounds using standard naming conventions, and recognize the common names of frequently used substances.
4 Predict differences in physical state, melting and boiling points, solid-state structure, and solution chemistry that result from differences in bonding.
5 Draw Lewis structures for covalent compounds and polyatomic ions.
6 Describe the relationship between stability and bond energy.
7 Predict the geometry of molecules and ions using the octet rule and Lewis structure.
8 Understand the role that molecular geometry plays in determining the solubility and melting and boiling points of compounds.
9 Use the principles of VSEPR theory and molecular geometry to predict relative melting points, boiling points, and solubilities of compounds.

Introduction

This chapter describes the role of valence electrons in bond formation between atoms. Systems of naming the resultant compounds are discussed, as well as the procedure for writing formulas based on the names of the compounds. The chemical and physical properties of these compounds are related to structure and bonding.

3.1 Chemical Bonding

When two atoms are joined together to make a chemical compound, the force of attraction between the two species is referred to as a **chemical bond**. Interactions involving valence electrons are responsible for the chemical bond.

Lewis Symbols

The **Lewis symbol** is a convenient way of representing atoms singly or in combination. Its principal advantage is that only valence electrons are shown. This results in simpler structures and greater clarity. The chemical symbol of the atom is written; this symbol represents the nucleus and all of the lower-energy nonvalence electrons, which do not directly participate in bonding. The valence electrons are indicated by dots (•) or crosses (x) arranged around the atomic symbol.

Principal Types of Chemical Bonds: Ionic and Covalent

Learning Goal 1:	Classify compounds as having ionic, covalent, or polar covalent bonds.

Ionic bonding is characterized by an electron transfer process occurring prior to bond formation. In **covalent bonding,** electrons are shared between atoms in the bonding process.

The essential features of ionic bonding are as follows:

- Elements with low ionization energy and low electron affinity tend to form positive ions.

- Elements with high ionization energy and high electron affinity tend to form negative ions.

- Ion formation takes place by an electron transfer process.

- The resulting positive and negative ions are held together by the electrostatic force between ions of opposite charge in an ionic bond.

- Reactions between metals and nonmetals (elements far to the left and right, respectively, on the periodic table) tend to result in ionic bonds.

When electrons are shared rather than transferred, the shared electron pair is referred to as a covalent bond. Compounds characterized by covalent bonding are called covalent compounds. Covalent bonds tend to form among atoms with similar tendencies to gain or lose electrons. The most obvious examples are the diatomic molecules H_2 as well as N_2, O_2, F_2, Cl_2, I_2, and Br_2. Bonding in these molecules is totally covalent because there is no net tendency for electron transfer between identical atoms.

Two atoms do not have to be identical in order to form a covalent bond. Compounds such as hydrogen fluoride, water, methane, and ammonia are common examples.

Polar Covalent Bonding and Electronegativity

Polar covalent bonding, like covalent bonding, is based on the concept of electron sharing; however, the sharing is unequal and based on the electronegativity difference between joined atoms. **Electronegativity** is a measure of the tendency of an atom in a molecule to attract shared electrons.

Electronegativity is represented by a scale derived from the measurement of energies of chemical bonds. The most electronegative element is fluorine, F, with a value of 4.0, and the least electronegative elements are Cs and Fr, each of which has a value of 0.7. The periodic trends for electronegativity, which increases from left to right and decreases from top to bottom, are similar to both ionization energy and electron affinity.

3.2 Naming Compounds and Writing Formulas of Compounds

Learning Goal 2:	Write the formulas of compounds when provided with the name of the compound.

Learning Goal 3:	Name common inorganic compounds using standard naming conventions and recognize the common names of frequently used substances.

Proper use of **nomenclature**, the assignment of a correct and unambiguous name to every chemical compound, is fundamental to the study of chemistry. The student must be able to write the name corresponding to a compound and the formula, when provided only with the name.

Ionic Compounds

The "shorthand" symbol for a compound is its **formula**. Examples are: $NaCl$, $MgBr_2$, and $NaIO_3$. The formula identifies the number and type of the various atoms that compose the compounds. The number of like atoms is denoted by a subscript. The presence of one atom is implied when no subscript is present.

Remember that positive ions are formed from elements that
1. are located to the left on the periodic table.
2. are referred to as metals.
3. have low ionization energies and low electron affinities, and hence easily lose electrons.

Elements that form negative ions, on the other hand,
1. are located to the right on the periodic table (but exclude the noble gases).
2. are referred to as nonmetals.
3. have high ionization energies and high electron affinities, and hence easily gain electrons.

Metals and nonmetals react to produce ionic compounds as a result of electron transfer. Although we refer to ionic compounds as **ion pairs**, in the solid state these ion pairs do not actually exist as individual units. Positive and negative ions arrange themselves in a regular, three-dimensional, repeating array known as a **crystal lattice**.

The names given to ionic compounds are based upon their formulas, with the name of the cation appearing first, followed by the anion name. The positive ion is simply the name of the element, while the negative ion is named as the stem of the element's name joined to the suffix *-ide*.

Example 1

Give the correct name for each of the following ionic compounds:

1. NaCl   3. $AlCl_3$ 5. Ba_3N_2
2. Li_2S 4. CaO   6. MgO

Answer

1. sodium chloride   3. aluminum chloride   5. barium nitride
2. lithium sulfide   4. calcium oxide   6. magnesium oxide

    If the cation and anion exist in only one common charged form, there is no ambiguity between formula and name. Sodium chloride must be NaCl, and lithium sulfide must be Li_2S, so that the sum of positive and negative charges is zero. With many elements, such as the transition metals, several ions of different charge may exist. Fe^{2+}, Fe^{3+} and Cu^{+}, Cu^{2+} are a few common examples. Clearly, an ambiguity exists if we use the name iron for both Fe^{2+} and Fe^{3+} or copper for Cu^{+} and Cu^{2+}. Two systems have been developed that avoid this problem: the stock system and the common nomenclature system.

    In the stock system for naming an ion (the systematic name), a Roman numeral indicates the magnitude of charge of the cation. In the older common nomenclature, the suffix *-ous* indicates the lower of the ionic charges, and the suffix *-ic* indicates the higher ionic charge. Systematic names are preferred; they are easier and less ambiguous.

Example 2

Give the correct names for each of the following ionic compounds that contain transition metal ions in their structures.

            1. $FeCl_3$ 2. $FeCl_2$

Work

When a transition metal is found in an ionic compound, we must first find the charge on the transition metal before we can name the compound.

1. $FeCl_3$
   a. First use the formula: we find that one Fe ion and three Cl ions are present.
      Fe + 3 Cl
   b. From the periodic table, we find that the Cl is in family VIIA. These elements always form a –1 ion.
   c. $Fe^{?+}$ + 3 Cl^{-}
      We now know that there is a total charge of –3 in the compound. Thus, the iron ion must be Fe^{3+} so it can balance the –3 charge.

Note: The total + charges must always equal the total – charges in any ionic compound!

Formula = $FeCl_3$ Ion-pair form = Fe^{3+} + 3 Cl^{-} Correct name: iron(III) chloride

Example 2 continued

2. $FeCl_2$
   a. $FeCl_2$ means $Fe^{2+} + 2\ Cl^-$
      Since there are a total of -2, then the Fe^{2+} must be an Fe^{2+} to balance the charges: $+2$ balances -2
   b. Correct name: iron(II) chloride

Answer
1. iron(III) chloride   2. iron(II) chloride

Ions consisting of only a single atom are said to be **monatomic**. In contrast, **polyatomic ions**, such as the hydroxide ion, OH^-, are composed of two or more atoms bonded together. The polyatomic ion has an overall positive or negative charge. Some common polyatomic ions are listed in Table 3.3 of the textbook. Your instructor may suggest that several of the most common polyatomic ions be committed to memory.

It is equally important to write the correct formula when given the compound name. It is essential to be able to predict the charge of monatomic ions and the charge and formula of polyatomic ions. Remember, the relative number of positive and negative ions in the unit must result in a unit (compound) charge of zero.

Example 3
Name the following ionic compounds. Use Roman numerals when needed.

1. $Ca_3(PO_4)_2$ 3. $CsOH$ 5. Na_2SO_4
2. $Mg(C_2H_3O_2)_2$ 4. $FeSO_4$ 6. $Fe_2(CO_3)_3$

Answer
1. calcium phosphate
2. magnesium acetate
3. cesium hydroxide
4. $FeSO_4 = Fe^{2+} + SO_4^{2-}$; iron (II) sulfate
5. sodium sulfate
6. $Fe_2(CO_3)_3 = 2\ Fe + 3\ CO_3^{2-} = 2\ Fe^{3+} + 3\ CO_3^{2-}$; iron (III) carbonate

---

Example 4
Write the formulas of the following ionic compounds:
1. sodium chloride   3. iron (III) hydroxide
2. magnesium phosphate   4. ammonium carbonate

Answer
1. sodium chloride
    a. $Na^+ + Cl^-$ IA family $= +1$ charge; VIIA $= -1$ charge
    b. formula $= NaCl$

Example 4 continued

Answer

2. magnesium phosphate
   a. $Mg^{2+} + PO_4^{3-}$ IIA = +2 charge; phosphate ion = −3
   b. need 3 Mg^{2+} to balance 2 PO_4^{3-}
      +6 balances −6
   c. formula = $Mg_3(PO_4)_2$

3. iron (III) hydroxide
   a. $Fe^{3+} + OH^-$ (Charges are not balanced.)
      need 3 OH^- to balance the +3 charge
      $Fe^{3+} + 3\ OH^-$
      +3 balances −3
   b. formula = $Fe(OH)_3$

4. ammonium carbonate.
   a. $NH_4+ + CO_3^{2-}$ (Charges are not balanced.)
   b. formula = $(NH_4)_2\ CO_3$

Covalent Compounds

Covalent compounds are formed by the reaction of nonmetals. Covalently bonded compounds are **molecules**. The existence of this compound unit is a major distinctive feature of covalently bonded substances. The convention used for naming covalent compounds is as follows:

1. The names of the elements are written in the order in which they appear in the formula.

2. A prefix indicating the number of each kind of atom found in the unit is placed before the name of the element.

3. If only one atom of a particular kind is present in the molecule, the prefix *mono-* is usually omitted.

4. The stem of the name of the last element is used with the suffix *-ide*.

Common names are often used. For example, H_2O is water, NH_3 is ammonia, C_2H_5OH (ethanol) is alcohol, and $C_6H_{12}O_6$ is glucose. It is useful to be able to correlate both systematic and common names with the corresponding molecular formula and vice versa.

Writing formulas of covalent compounds can only be done from memory when common names are used. You must remember that water is H_2O, ammonia is NH_3, and so forth. This is the major disadvantage of common names; however, they cannot be avoided.

3.3 Properties of Ionic and Covalent Compounds

Learning Goal 4:   Predict differences in physical state, melting and boiling points, solid-state structure, and solution chemistry that result from differences in bonding.

The differences in ionic and covalent bonding account for the different properties of ionic and covalent compounds. Covalently bonded molecules are discrete units, and they have less tendency to form an extended structure in the solid state. Ionic compounds do not have definable units, but form a **crystal lattice** composed of hundreds of billions of positive and negative ions in an extended three-dimensional network.

Major differences in the properties of these compounds are summarized below.

Physical State

All ionic compounds are solids at room temperature, while covalent compounds may be solids (sugar, silicon dioxide), liquids (water, ethyl alcohol), or gases (carbon monoxide, carbon dioxide).

Melting and Boiling Points

The **melting point** is the temperature at which a solid is converted to a liquid, and the **boiling point** is the temperature at which a liquid is converted to a gas. Considerable energy is needed to break apart an ionic crystal lattice and convert an ionic solid to a liquid or a gas. Therefore, the melting and boiling temperatures for ionic compounds are generally higher than those of covalent compounds, which interact less strongly in the solid state.

Structure of Compounds in the Solid State

Ionic solids are crystalline, characterized by a regular structure, whereas covalent solids may be either crystalline or amorphous (no regular structure).

Solutions of Ionic and Covalent Compounds

Many ionic solids dissolve in solvents, such as water. If soluble, an ionic solid will dissociate in solution to form positive and negative ions (ionization). Because these ions are capable of conducting electric current, these compounds are **electrolytes**, and the solution is termed an **electrolytic solution**. Covalent solids in solution are neutral and are **nonelectrolytes**. The solution is not an electrical conductor.

3.4 Drawing Lewis Structures of Molecules and Polyatomic Ions

Learning Goal 5:	Draw Lewis structures for covalent compounds and polyatomic ions.

Lewis Structures of Molecules

Rule 1: Use chemical symbols for the various elements to write the skeletal structure of the compound.

Rule 2: Determine the number of valence electrons associated with each atom; combine them to determine the total number of valence electrons in the compound.

Rule 3: Connect the central atom to each of the surrounding atoms using electron pairs.

Rule 4: If the octet rule is not satisfied for the central atom, move one or more electron pairs from the surrounding atoms to create double or triple bonds until all atoms have an octet.

Rule 5: After you are satisfied with the Lewis structure that you have constructed, perform a final electron count verifying that the total number of electrons and the number around each atom are correct.

Example 6

Draw the Lewis dot and line structures of carbon dioxide. The carbon is in the middle of two oxygen atoms. The carbon atom is the central atom.

Example 6 continued

Work
1. O C O (arrangement as in nature)
2. O = 6 valence e$^-$
   C = 4 valence e$^-$
   O = 6 valence e$^-$
       16 total valence e$^-$ in CO_2

3. Dot structure: $:\ddot{O}::C::\ddot{O}:$
   (All atoms must have 8 valence e$^-$ in their final valence shell.)

4. Line structure: **O=C=O**

   This structure is also acceptable: $|\overline{O}=C=\overline{O}|$

Example 7
Draw the Lewis dot and line structures of SO_3. Sulfur is the central atom in the compound.

Work
1.   O
     O S O

2. 3 oxygen atoms =   18 valence e
   1 sulfur atom –   6 valence e$^-$
             24 total valence e$^-$ in SO_3

3. Dot structure (Any one of the following is correct.)

   $:\ddot{O}:$ $:\ddot{O}:$ $:\ddot{O}:$
   $:\ddot{O}:S::\ddot{O}:$ or $:\ddot{O}::S:\ddot{O}:$ or $:\ddot{O}:S:\ddot{O}:$

4. Line structure (Any one of the following is correct.)

     O   **O**   **O**
     | **|**   **|**
   O–S=O   or   **O–S=O**   or   **O–S=O**

Lewis Structures of Polyatomic Ions

The charge on the ion must be accounted for while computing the total number of valence electrons when writing Lewis structures of ions. For negative ions, one valence electron is added for each unit of negative charge and we subtract one valence electron for each unit of positive charge in a positive ion.

Lewis Structure, Stability, Multiple Bonds, and Bond Energies

Learning Goal 6: Describe the relationship between stability and bond energy.

The order of stability often parallels the bond order (the number of bonds between adjacent atoms). Stability is also related to the bond energy, where the **bond energy** is defined as the amount of energy in kilocalories required to break a bond holding two atoms together. The magnitude of the bond energy decreases in the order triple bond > double bond > single bond. The bond length decreases in the order single bond > double bond > triple bond.

Lewis Structures and Resonance

More than one Lewis structure may satisfy the octet rule for a particular compound. When a compound has two or more Lewis structures that contribute to the real structure, the compound displays **resonance**. The contributing Lewis structures are **resonance forms**. The true structure, a hybrid or mixture of the resonance forms, is known as a **resonance hybrid**. The answers for Examples 6 and 7 (above) show a variety of possible structures for SO_3 and CO_3^{2-}. These are resonance forms.

Example 8
Draw the Lewis dot and line structures of the ammonium ion.
The nitrogen atom is the central atom in the ion.

Work
1.

$$\begin{bmatrix} & H & \\ H & N & H \\ & H & \end{bmatrix}^{+}$$

2. N = 5 valence e⁻
    H = 1 valence e⁻
    H = 1 valence e⁻
    H = 1 valence e⁻
    H = <u>1 valence e⁻</u>
        9 valence e⁻ total

Example 8 continued

Remember however, that since the ion is a +1, one valence electron has been removed. Thus, *only eight valence electrons* are found in the ammonium ion.

3. The dot structure is:

$$\left[\begin{array}{c} \text{H} \\ \overset{\displaystyle\cdot\cdot}{\text{H : N : H}} \\ \text{H} \end{array} \right]^{+}$$

4. The line structure is:

$$\left[\begin{array}{c} \text{H} \\ | \\ \text{H---N---H} \\ | \\ \text{H} \end{array} \right]^{+}$$

Lewis Structures and Exceptions to the Octet Rule

Molecules with fewer than four and as many as five or six electron pairs around the central atom also exist. They are exceptions to the octet rule. In general, if one atom in a molecule (or polyatomic ion) must be an exception to the octet rule, it will be the central atom.

Example 9
Draw the Lewis structure for BF_3. Boron is the central atom in this molecule.

Work
Boron has 3 valence electrons, and each fluorine has 7 valence electrons, for a total of: $3 + (3 \times 7) = 3 + 21 = 24$ valence electrons, which are not enough for each atom to have a full octet. One atom – the central atom, boron – must be an exception to the octet rule.

Note: This is why it's important to count valence electrons and then to double-check your total. Drawing a full octet around each atom may *look* correct, but for BF_3, that would be incorrect.

Answer

Lewis Structures and Molecular Geometry; VSEPR Theory

Learning Goal 7:	Predict the geometry of molecules and ions using the octet rule and Lewis structure.

The shape of a molecule contributes to its properties and reactivity. We may predict the shapes of various molecules by inspecting their Lewis structures for the orientation of their electron pairs. The covalent bond, in which bonding electrons are localized between the atoms, is directional: the bond has a specific orientation in space among the bonded atoms. Electrostatic forces in ionic bonds, in contrast, are nondirectional: they have no specific orientation in space. As a result, the specific orientation of electron pairs in covalent molecules imparts a characteristic shape to the molecules.

Electron pairs around the central atom of the molecule arrange themselves to minimize repulsion; the electron pairs are as far as possible from each other. This is termed the valence shell electron pair repulsion theory (**VSEPR Theory**). Two electron pairs around the central atom lead to a **linear** arrangement of the attached atoms; three indicate a **trigonal planar** arrangement; and four result in a **tetrahedral** geometry. Molecules with five and six electron pairs also exist. They may have structures that are trigonal bipyramidal or octahedral.

Compounds containing the same central atom will have structures with similar geometries. This is an approximation; it is not always true, but useful in writing reasonable, geometrically accurate structures for a large number of compounds. Consider the following:

BeH_2 (linear)
   BeI_2, $BeCl_2$, $Be(OH)_2$ are also linear.

BF_3 (trigonal planar)
   BCl_3, BBr_3, BI_3 are also trigonal planar.

CH_4 (tetrahedral)
   CF_4, CCl_4, CBr_4 are also tetrahedral.

Furthermore, the periodic similarity of group members is useful in predictions involving bonding. Consider the following:

H_2O (angular)
   H_2S, H_2Se are also angular because O, S, and Se are in the same group, VI, 6 valence electrons.

Lewis Structures and Polarity

Learning Goal 8:	Understand the role that molecular geometry plays in determining the solubility and melting and boiling points of compounds.

A molecule is **polar** if its centers of positive and negative charges do not coincide. On the other hand, the positive and negative centers are coincident in nonpolar molecules. No dipole exists, and the dipole moment is equal to zero.

This property may also be described by considering the equality of electron sharing between the atoms being bonded. The degree of electron sharing can be described using the property of electronegativity (which was introduced in section 4.1). The atoms of H_2

are identical; their electronegativity is the same. The electrons remain, on average, at the center of the molecule, and the bond is nonpolar. The hydrogen molecule itself is nonpolar as well, with a dipole moment of zero. O_2, N_2, Cl_2, and F_2 are also nonpolar molecules with nonpolar bonds.

Bonds formed from atoms of unequal electronegativity are polar. The more electronegative end of the bond is designated δ^- (partial negative) and the less electronegative end δ^+ (partial positive). A polar covalent molecule is characterized by **polar covalent bonding**. This implies that the electrons are shared unequally.

A molecule containing all nonpolar bonds must itself be nonpolar. A molecule containing polar bonds may be either polar or nonpolar depending on the relative orientation of the bonds.

3.5 Properties Based on Electronic Structure and Molecular Geometry

Learning Goal 9:	**Use the principles of VSEPR theory and molecular geometry to predict relative melting points, boiling points, and solubilities of compounds.**

The effects of molecular polarity result from the strength of attractive forces among individual molecules of a compound. These attractions are **intermolecular forces**. The student should not confuse intermolecular forces with intramolecular forces. **Intramolecular forces** are the attractive forces within molecules. It is the intermolecular forces that determine such properties as solubility, melting point, and boiling point.

Solubility

Solubility is defined as the maximum amount of solute that dissolves in a given amount of solvent. (The solute is present in lesser quantity, and the solvent is present in the greater amount.) Polar molecules are most soluble in polar solvents, and nonpolar molecules are most soluble in nonpolar solvents. This is the rule of "like dissolves like."

Boiling Points of Liquids and Melting Points of Solids

Boiling is the conversion of a liquid to a vapor. Such a process requires energy, overcoming the intermolecular attractive forces in the liquid. The energy required is related to the magnitude of the boiling point and depends upon the strength of the intermolecular attractive forces in the liquid. These attractive forces are directly related to polarity. Molecular size is also an important consideration. The larger or heavier the molecule, and the more polar the molecule, the more difficult it becomes to convert to the gas phase.

The melting points of solids may be described on the basis of intermolecular forces as well. As a general rule, polar compounds have strong attractive (intermolecular) force, and their boiling and melting points tend to be higher than for nonpolar substances.

Glossary of Key Terms in Chapter Three

angular structure (3.4) planar molecule with bond angles less than 180°.

boiling point (3.3) the temperature at which the vapor pressure of a liquid is equal to the atmospheric pressure.

bond energy (3.4) the amount of energy necessary to break a chemical bond.

chemical bond (3.1) the attractive force holding two atomic nuclei together in a chemical compound.

covalent bond (3.1) a pair of electrons shared between two atoms.

crystal lattice (3.2) a unit of a solid characterized by a regular arrangement of components.

dissociation (3.3) production of positive and negative ions when an ionic compound dissolves in water.

double bond (3.4) a bond in which two pairs of electrons are shared by two atoms.

electrolyte (3.3) a material that dissolves in water to produce a solution that conducts an electrical current.

electrolytic solution (3.3) a solvent composed of an electrolytic solute dissolved in water.

electronegativity (3.1) a measure of the tendency of an atom in a molecule to attract shared electrons.

formula (3.2) the representation of the fundamental compound unit using chemical symbols and numerical subscripts.

intermolecular force (3.5) attractive force that occurs between molecules.

intramolecular force (3.5) attractive force that occurs within molecules.

ionic bonding (3.1) an electrostatic attractive force between ions resulting from electron transfer.

ion pair (3.1) the simplest formula unit for an ionic compound.

Lewis symbol (3.1) representation of an atom or ion using the atomic symbol (for the nucleus and core electrons) and dots to represent valence electrons.

linear structure (3.4) the structure of a molecule in which the bond angle(s) about the central atom(s) is (are) 180°.

lone pair (3.4) an electron pair that is not involved in bonding.

melting point (3.3) the temperature at which a solid converts to a liquid.

molecule (3.2) a collection of two or more atoms held together by covalent bonds.

monatomic ion (3.2) an ion formed by electron gain or loss from a single atom.

nomenclature (3.2) a system for naming chemical compounds.

nonelectrolyte (3.3) a substance that, when dissolved in water, produces a solution that does not conduct an electrical current.

polar covalent bonding (3.4) a covalent bond in which the electrons are not equally shared.

polar covalent molecule (3.4) a molecule that has permanent electric dipole moment resulting from an unsymmetrical electron distribution; a dipolar molecule.

polyatomic ion (3.2) an ion containing a number of atoms.

resonance (3.4) a condition that occurs when more than one valid Lewis structure can be written for a particular molecule.

resonance form (3.4) one of a number of valid Lewis structures for a particular molecule.

resonance hybrid (3.4) a description of bonding in a molecule resulting from a superimposition of all valid Lewis structures (resonance forms).

single bond (3.4) a bond in which one pair of electrons is shared by two atoms.

solubility (3.5) the amount of a substance that will dissolve in a given volume of solvent at a specified temperature.

tetrahedral structure (3.4) a molecule consisting of four groups attached to a central atom which occupy the four corners of an imagined tetrahedron.

trigonal pyramidal molecule (3.4) a molecular structure resulting from three groups bonded to a central atom; all groups are equidistant from the central atom.

triple bond (3.4) a bond in which three pairs of electrons are shared by two atoms.

valence shell electron pair repulsion theory (3.4) a model that predicts molecular geometry using the premise that electron pairs will arrange themselves as far apart as possible, to minimize electron repulsion.

Self Test for Chapter Three

1. The transfer of electrons from one atom to another forms which type of chemical bond?
2. The sharing of electrons between two atoms forms which type of chemical bond?
3. What particles are found in ionic compounds?
4. How many sodium ions are in Na_2S?
5. On what side of the periodic table are elements found that usually form positive ions?
6. On what side of the periodic table are elements found that usually form negative ions?
7. The name of an ionic compound always contains the name of which ion first?
8. Provide the best name of Na_2O.
9. Provide the best name of Li_2S.
10. Provide the best name of $AlBr_3$.
11. What is the common name of the iron (II) ion?
12. What is the best name of Fe^{3+}?
13. What does the ending "ic" on the common names of ions mean?
14. Give the common name of $FeSO_4$.
15. Give the systematic name of CuO.
16. Write the formula of sodium sulfate.
17. How many bonding electrons are in H_2O?
18. How many nonbonding electrons are in CO_2?
19. Which end of HCl is slightly more negative?
20. Why is HF a polar molecule?
21. When H_2O is compared to C_2H_4, which has the higher boiling point?
22. When Cl_2 is compared to ICl, which has the higher boiling point?
23. Draw the dot and line structures of ammonia.
24. Draw the dot and line structures of water.
25. Draw the dot and line structures of a hydroxide ion.
26. Draw the dot and line structures of a hydronium ion. The H_3O^+ ion has oxygen at the center of the ion.
27. Which electrons of an atom are involved in chemical bond formation?
28. What type of bond is formed by the sharing of valence electrons?
29. Write the formula of the ions that result when sodium atoms react with chlorine atoms.
30. Write the formula of the ions that result when calcium atoms react with sulfur atoms.
31. What process results in ion formation in ionic reactions?
32. How many electrons are jointly shared in the hydrogen molecule?
33. How many electrons are shared in H_2O?

Vocabulary Quiz for Chapter Three

1. A(n) _____ force is a attraction between molecules.

2. A(n) _____ is a unit of solid characterized by regular arrangement of components.

3. A material that dissolves in water to produce a solution that conducts an electrical current is a(n)_____.

4. _____ are attractive forces that occur between molecules.

5. A(n) _____ is a diagram of an atom, ion, or molecule showing valence electron arrangement.

6. A(n) _____ is a electron pair that is not involved in bonding.

7. A covalent bond in which the electrons are not equally shared is referred to as a(n) _____ bond.

8. A(n) _____ is an ion containing a number of atoms.

9. A molecule consisting of four groups attached to a central atom usually has a(n) _____ structure.

10. A bond in which three pairs of electrons are shared by two atoms is called a(n) _____.

4 Calculations and the Chemical Equation

Learning Goals

1 Know the relationship between the mole and Avogadro's number, and the usefulness of these quantities.
2 Perform calculations using Avogadro's number and the mole.
3 Write chemical formulas for common inorganic substances.
4 Calculate the formula weight and molar mass of a compound.
5 Know the major function served by the chemical equation, the basis for chemical calculations.
6 Classify chemical reactions by type: combination, decomposition, or replacement.
7 Recognize the various classes of chemical reactions: precipitation, reactions with oxygen, acid-base, and oxidation-reduction.
8 Balance chemical equations given the identity of products and reactants.
9 Calculate the number of moles or grams of product resulting from a given number of moles or grams of reactants or the number of moles or grams of reactant needed to produce a certain number of moles or grams of product.
10 Calculate theoretical and percent yield.

Introduction

You should be able to predict the quantity of a product produced from the reaction of a given amount of reactant. You should also be able to calculate how much of a reactant, or group of reactants, would be required to produce a desired amount of product.

These calculations are based on the **chemical equation**. The chemical equation provides all of the needed information: the combining ratio of elements or compounds (reactants) that are necessary to produce a particular product or products as well as the relationship between the amounts of reactants and products formed.

4.1   The Mole Concept and Atoms

The Mole and Avogadro's Number

Learning Goal 1:	Know the relationship between the mole and Avogadro's number, and the usefulness of these quantities.

Atomic masses have been experimentally determined for each of the elements. Their unit of measurement is the **atomic mass unit**, abbreviated **amu**.

$$1 \text{ amu} = 1.661 \times 10^{-24} \text{ grams}$$

A more practical unit for defining a "collection" of atoms is the **mole**.

$$1 \text{ mole of atoms} = 6.022 \times 10^{23} \text{ atoms of an element}$$

This number is **Avogadro's number**.

The mole and the amu are related. The atomic mass of a given element corresponds to the average mass of a single atom in amu *and* the mass of a mole of atoms in grams. For example, the average mass of one copper atom is 63.54 amu, while the mass of one *mole* of copper atoms is 63.54 grams. One mole of atoms of *any element* contains the same number, Avogadro's number, of atoms.

Calculating Atoms, Moles, and Mass

Learning Goal 2: Perform calculations using Avogadro's number and the mole.

Calculations based on the chemical equation (Section 4.6 of the textbook) utilize the relationship of the number of atoms, number of moles, and mass in grams. Conversion factors are used to proceed from the information *provided* in the problem to the information *requested* by the problem.

Before you begin any calculations, map out a pattern for the required conversion. You may be given the number of grams and need the number of atoms that corresponds to that mass. Begin by tracing a path to the answer:

$$\text{grams} \xrightarrow{\text{Step 1}} \text{moles} \xrightarrow{\text{Step 2}} \text{atoms}$$

Two transformations, or conversions are required:

Step 1:   grams to moles

Step 2:   moles to atoms

The conversion between the three principal measures of quantity of matter, the number of grams (mass), the number of moles, and the number of individual particles (atoms, ions, or molecules) is depicted in Figure 5.2 of the textbook.

Example 1
Calculate the mass of one mole of iron atoms.

Work
From the periodic table, we find that the average mass of one iron atom is 55.85 atomic mass units. Thus, one mole of iron atoms has a mass of 55.85 grams.

4.2   The Chemical Formula, Formula Weight, and Molar Mass

The Chemical Formula

Learning Goal 3:   Write chemical formulas for common inorganic substances.

Chemical compounds are represented by their **chemical formulas**, a combination of symbols of the various elements that make up the compounds. The chemical formula is based upon the **formula unit**, the smallest collection of atoms that provides the following information:

1. the identity of the atoms present in the compound and
2. the relative numbers of each type of atom.

The term *formula* may be used in reference to any ionic or covalent compound. In Chapter 3 compounds were classified as nonpolar covalent, polar covalent, or ionic, depending upon the type of bonding that holds the unit together.

A covalent compound is composed of discrete molecules, and the term *molecular formula* is most appropriate. Ionic compounds, on the other hand, are composed of ion pairs, not molecules, and *cannot* be described as having a molecular formula.

Formula Weight and Molecular Mass

Learning Goal 4:   Calculate the formula weight and molar mass of a compound.

A mole of a compound is based upon the formula mass or **formula weight**. The formula weight is calculated by adding the masses of all the atoms of which the unit is composed. To calculate the formula weight, the formula unit must be known.

Example 2

Calculate the mass of one mole of carbon dioxide (the formula weight of carbon dioxide).

Work

From the periodic table, we find that one atom of carbon has a mass of 12.01 amu, and one atom of oxygen has a mass of 16.00 amu.

CO_2 = 1 carbon atom + 2 oxygen atoms
CO_2 = 1 C + 2 O
CO_2 = 1(12.01 amu) + 2(16.00 amu) = 44.01 amu

Answer

Thus, one molecule of CO_2 has a mass of 44.01 amu, and one mole of carbon dioxide molecules weighs 44.01 grams.

Example 3
How many grams of carbon dioxide are contained in 10.00 moles of carbon dioxide?

Work
Since 1 mole of CO_2 = 44.01 grams, we can use the factor-label method to convert grams into mole units.

$$10.00 \text{ mol } CO_2 \times \frac{44.01 \text{ g } CO_2}{1 \text{ mol } CO_2} = 4.401 \times 10^2 \text{ g } CO_2$$

Answer
10.00 moles of CO_2 = 4.401×10^2 grams of CO_2

Example 4
Calculate the formula weight of water.

Work
Water (H_2O) is composed of two hydrogen atoms plus one oxygen atom. From the periodic table, we find that O = 16.00 amu, and H = 1.008 amu.

H_2O = 2 hydrogen atoms + 1 oxygen atom
H_2O = 2 H + 1 O
H_2O = 2(1.008 amu) + 1(16.00 amu) = 18.01 amu

Answer
The formula weight of water equals 18.01 amu. Notice that amu are used for formula weights. A mole of water, however, is equal to 18.01 grams.

4.3   The Chemical Equation and the Information It Conveys

A Recipe for Chemical Change

Learning Goal 5:   Know the major function served by the chemical equation, the basis for chemical calculations.

The chemical equation describes all of the substances that react to produce the product(s). The chemical equation also describes the physical state of the reactants and products as solid, liquid, or vapor. It tells us whether the reaction occurs, and identifies the solvent and experimental conditions. Most important, the relative number of moles of reactants and products appears in the equation. According to the law of conservation of mass, matter cannot be either gained or lost in the process of a chemical reaction. The law of conservation of mass states that we must have a balanced chemical equation.

Features of a Chemical Equation

1. The identity of products and reactants must be specified.
2. Reactants are written to the left of the reaction arrow and products to the right.
3. The physical state of reactants and products is shown in parentheses.
4. The symbol Δ over the reaction arrow means that heat energy is necessary for the reaction to occur.
5. The equation must be balanced.

The Experimental Basis of a Chemical Equation

The chemical equation represents a real chemical transformation. Evidence for the reaction may be based on observations such as the formation of a gas, a solid precipitate, the evolution of heat, or a color change in the solution.

Writing Chemical Reactions

Learning Goal 6:	Classify chemical reactions by type: combination, decomposition, or replacement.

Spontaneous chemical reactions occur for a variety of reasons, linked by the tendency to achieve the lowest (most stable) electronic energy state. Strong electrolytes will react to form weak (less dissociated) electrolytes, if possible. Reactions forming gaseous products are favored. Formation of insoluble solid products favors a chemical reaction.

Chemical reactions involve the combination of reactants to produce products, the decomposition of reactant(s) into products, or the replacement of one or more elements in a compound to yield products.

Combination reactions: Combination reactions involve the joining of two or more atoms or compounds, producing a product of different composition.

Decomposition reactions: Decomposition reactions produce two or more products from a single reactant.

Replacement reactions: Replacement reactions are subcategorized as either single-replacement or double-replacement. Single-replacement reactions occur between atoms and compounds; the atom replaces another in the compound, producing a new compound and the replaced atom. Double-replacement reactions, on the other hand, involve two compounds undergoing a "change of partners."

Types of Chemical Reactions

Learning Goal 7:	Recognize the various classes of chemical reactions: precipitation, reactions with oxygen, acid-base, and oxidation-reduction.

Precipitation reactions: Precipitation reactions involve the conversion of soluble reactants into one or more insoluble products. Information contained in solubility tables enables prediction of possible insoluble products.

Reactions with oxygen: Reactions involving combination of metals or nonmetals with oxygen are generally exothermic; they are often used specifically for their heat-generating properties. Examples include the reaction of oxygen with carbon/hydrogen-containing compounds (natural gas, gasoline, and oil).

Acid-base and oxidation-reduction reactions: Another approach to the classification of chemical reactions is based upon a consideration of charge transfer. Acid-base reactions involve the transfer of a hydrogen ion, H^+, from one reactant to another. Another important reaction type, oxidation-reduction, takes place because of the transfer of one or more electrons from one reactant to another.

4.4   Balancing Chemical Equations

Learning Goal 8:	**Balance chemical equations given the identity of products and reactants.**

A chemical equation tells us both the identity and state of the reactants and products. **Reactants**, or starting materials, are all substances that undergo change in a chemical reaction, while **products** are substances produced by a chemical reaction.

The chemical equation also shows the *molar quantity* of reactants needed to produce a certain *molar quantity* of products.

The number of moles of each product and reactant is indicated by placing a whole number *coefficient* before the formula of each substance in the chemical equation.

Many equations are balanced by trial and error. The following steps provide a method for correctly balancing a chemical equation:

Step 1:   Count the number of atoms of each element on both product and reactant side.

Step 2:   Determine which atoms are not balanced.

Step 3:   Balance one atom at a time, using coefficients.

Step 4:   After you believe that you have successfully balanced the equation, check, as in Step 1, to be certain that mass conservation has been achieved.

Example 5

Write the balanced reaction of an aqueous solution of hydrochloric acid (HCl) reacting with solid calcium to produce calcium chloride plus hydrogen gas.

Work

1. First write the correct symbols and formulas needed for the reactants and products:

Example 5 continued

$$HCl \ + \ Ca \ \rightarrow CaCl_2 \ + \ H_2$$

2. Next, count the number of atoms of each element on the right and left side of the reaction arrow:

$$HCl \ + \ Ca \ \rightarrow \ CaCl_2 \ + \ H_2$$

H	Cl	Ca	‖	H	Cl	Ca
1 atom	1 atom	1 atom	‖	2 atoms	2 atoms	1 atom

Notice that the numbers of atoms on each side are not equal; the equation is **not** balanced.

3. After the correct formulas of all reactants and products are given, you can now use coefficients alone to balance the equation.

4. $$2\,HCl \ + \ Ca \ \rightarrow \ CaCl_2 \ + \ H_2$$

H	Cl	Ca	‖	H	Cl	Ca
2	2	1	‖	2	2	1

The equation is now balanced.

Answer
$$2\,HCl \ + Ca \ \rightarrow \ CaCl_2 \ + \ H_2$$

Note that the equation becomes more complete when the states of each reactant and product are added:
$$2\,HCl(aq) + Ca(s) \rightarrow \ CaCl_2(aq) \ + \ H_2(g)$$

Example 6
Balance the following reaction: $FeCl_3(aq) + Cu(s) \rightarrow CuCl_2(aq) \ + \ Fe(s)$

Work
1. First, find the number of atoms of each element:

$$FeCl_3 \ + \ Cu \ \rightarrow \ CuCl_2 \ + \ \ Fe$$

Fe	Cl	Cu	‖	Fe	Cl	Cu
1	3	1	‖	1	2	1

The equation is **not** balanced.

Example 6 continued

2. Next, use only coefficients to balance the equation:

$$2\ FeCl_3\ +\ 3\ Cu\ \rightarrow\ 3\ CuCl_2\ +\ 2\ Fe$$

Fe	Cl	Cu	‖	Fe	Cl	Cu
2	6	3	‖	2	6	3

The equation is now balanced.

Answer

$$2\ FeCl_3(aq)\ +3\ Cu(s)\ \rightarrow\ 3\ CuCl_2(aq)\ +\ 2\ Fe(s)$$

4.5 Calculations Using the Chemical Equation

Learning Goal 9: Calculate the number of moles or grams of product resulting from a given number of moles or grams of reactant or the number of moles or grams of reactant needed to produce a certain number of moles or grams of product.

General Principles

In doing calculations involving chemical reactions, we apply the following rules:

1. The basis for the calculations is a balanced equation.

2. The calculations are performed in terms of moles.

3. The conservation of mass must be obeyed.

The mole is the basis for calculations. However, masses are generally measured in grams (or kilograms). A facility for interconversion of moles and grams is therefore necessary in calculations involving chemical reactions.

Example 7

In the following balanced equation, how many moles of each element or compound are involved?

$$2\ C_8H_{18}\ +\ 25\ O_2\ \rightarrow\ 16\ CO_2\ +\ 18\ H_2O$$

Work

The coefficients in a balanced equation tell us the number of moles of each reactant and product.

Example 7 continued

Answer
Two moles of octane (C_8H_{18} is found in gasoline) react with 25 moles of oxygen (always O_2 in nature) to form 16 moles of carbon dioxide plus 18 moles of water.

Use of Conversion Factors

1. *Conversion between moles and grams.* Conversion from moles to grams and vice versa requires only the formula weight of the compound of interest.

2. *Conversion of moles of reactants and products.* Conversion factors, based on the chemical equation, permit us to perform a variety of calculations.

A general problem-solving strategy is summarized in Figure 4.7 (p. 143) of the textbook.

Example 8
The reaction of propane (C_3H_8 is found in bottled gas) with oxygen produces carbon dioxide and water. The balanced equation is given below:
$$C_3H_8 + 5 O_2 \rightarrow 3 CO_2 + 4 H_2O$$
Calculate the number of grams of oxygen (O_2, as in nature) needed to completely react with 1 mole of propane.

Work
1. From the balanced equation, we see that 1 mole of propane reacts with 5 moles of O_2 to form 3 moles of CO_2 plus 4 moles of water.
2. One mole of O_2 equals 32.00 grams.
3. Since 5 moles of O_2 are needed to react with 1 mole of propane, then 160.0 grams of oxygen (O_2) are needed.

$$1 \text{ mol } C_3H_8 \times \frac{5 \text{ mol } O_2}{1 \text{ mol } C_3H_8} \times \frac{32.00 \text{ g } O_2}{1 \text{ mol } O_2} = 160.0 \text{ g } O_2$$

Answer
One mole of propane will react with 160.0 grams of oxygen (O_2).

Learning Goal 10:   Calculate theoretical and percent yield.

Theoretical and Percent Yield

The **theoretical yield** is the *maximum* amount of product that can be produced (in an ideal world). In the real world it is difficult to produce the amount calculated as the theoretical yield.

A percent yield, the ratio of the actual and theoretical yields multiplied by 100%, is often used to show the relationship between predicted and experimental quantities. Thus

$$\% \text{ yield} = \frac{\text{actual yield}}{\text{theoretical yield}} \times 100\%$$

The percent yield for most chemical reactions is less than 100%. This is due, in part, to experimental limitations in isolation, transfer, and recovery of the product. Many reactants do not completely convert to products. We will study these reactions in detail in Chapter 8.

Glossary of Key Terms for Chapter Four

acid-base reaction (4.3) reactions that involve the transfer of a hydrogen ion (H^+) from one reactant to another.

atomic mass unit (4.1) the unit of measure of atomic mass, one-twelfth of the mass of the 12C atom, equivalent to 1.6606×10^{-24} gram.

Avogadro's number (4.1) 6.022×10^{23} particles of matter contained in one mole substance.

chemical equation (4.3) a record of chemical change, showing the conversion of reactants to products.

chemical formula (4.2) the representation of a compound or ion in which elemental symbols represent types of atoms and subscripts show the relative numbers of atoms.

combination reaction (4.3) a reaction in which two substances join to form another substance.

decomposition reaction (4.3) the breakdown of a substance into two or more substances.

double-replacement reaction (4.3) a chemical change in which cations and anions "exchange partners."

formula unit (4.2) the smallest collection of atoms from which the formula of a compound can be established.

formula weight (4.2) the mass of a formula unit of a compound relative to a standard (carbon-12).

hydrate (4.2) any substance that has water molecules incorporated in its structure.

law of conservation of mass (4.3) a law stating that, in chemical change, matter cannot be created or destroyed.

molar mass (4.1) the mass in grams of one mole of a substance.

mole (4.1) the amount of substance containing Avogadro's number of particles.

oxidation-reduction reaction (4.3) also called redox reaction, a reaction involving the transfer of one or more electrons from one reactant to another.

percent yield (4.5) the ratio of the actual and theoretical yields of a chemical reaction multiplied by 100%.

product (4.3) the chemical species that results from a chemical reaction and that appears on the right side of the chemical equation.

reactant (4.3) starting material for a chemical reaction that appears on the left side of the chemical equation.

single-replacement reaction (4.3) also called substitution reaction, one in which one atom in a molecule is displaced by another.

theoretical yield (4.5) the maximum amount of product that can be produced from a given amount of reactant.

Self Test for Chapter Four

1. 6.02×10^{23} molecules of a covalent compound equal how many moles of that compound?

2.   One mole of hydrogen molecules weighs how many grams? Give answer to the nearest tenth of a gram.

3.   How many iron atoms are present in 1 mole of iron?

4. How many grams of sulfur are found in 0.150 mole of sulfur? Use S = 32.06 amu.

5. Write the formula of the smallest unit of nitrogen as it is normally found in nature.

6. The term *molecular formula* is used for what type of compound?

7. What units are used for the formula weights of atoms and individual molecules?

8. What is the formula weight of one molecule of H_2O (nearest hundredth)?

   Use H = 1.01; O = 16.00 amu.

9. The formula weight of a covalent compound may also be called what weight?

10. What term means the starting materials in a chemical reaction?

11. How many moles of hydrogen are needed to react with oxygen to form 2 moles of water?

    $$2 H_2 + O_2 \rightarrow 2 H_2O$$

12. How many moles of oxygen are needed to react with hydrogen to form 1 mole of water?

    $$2 H_2 + O_2 \rightarrow 2 H_2O$$

13. The number of moles of reactants and products in a balanced chemical equation is given by what kind of numbers?

14. What number will be found in front of HCl when the following equation is balanced?

    $$Ca + HCl \rightarrow CaCl_2 + H_2$$

15. What number will be found in front of H_2O when the following equation is balanced?

    $$Mg(OH)_2 + HCl \rightarrow MgCl_2 + H_2O$$

16. What number will be found in front of Cl_2 when the following equation is balanced?

    $$Na + Cl_2 \rightarrow NaCl$$

17. How many grams of sodium hydroxide will react with 73.00 grams of hydrochloric acid?

    Use NaOH = 40.00; HCl = 36.45 amu.

    $$NaOH + HCl \rightarrow NaCl + H_2O$$

18. Calculate the number of grams of oxygen that must react with 46.85 grams of C_3H_8 to produce carbon dioxide and water.

    Use C = 12.01, H = 1.01, and O = 16.00 amu.

19. Iron reacts with oxygen to form iron (III) oxide (Fe_2O_3). How many grams of product will be formed from 5.00 grams of Fe?

    Use Fe = 55.85, and 0 = 16.00 amu.

20. How many moles of sulfur are found in 1.81×10^{24} atoms of sulfur?

    Use S = 32.06 amu.

21.   What is the name given to the number of atoms in a mole of iron?

22.   Write the formula of the smallest unit of oxygen as it is normally found in nature.

Vocabulary Quiz for Chapter Four

1.   _____ is 6.022×10^{23} particles of matter contained in one mole of a substance.

2.   A record of chemical change showing the conversion of reactants to products is called a(n) _____.

3.   A(n) _____ is the representation of a compound in which elemental symbols represent types of atoms and subscripts show the relative numbers of atoms.

4.   The smallest collection of atoms from which the formula of a compound can be established is called _____.

5.   The mass of a formula unit of a compound relative to a standard (carbon-12) is called _____.

6.   In chemical change, "matter cannot be created or destroyed" is a statement of the

      _____.

7.   A(n) _____ is the amount of substance containing Avogadro's number of particles.

8.   The _____ is the mass of a molecule relative to a standard (carbon-12).

9.   _____ result from a chemical reaction and appear on the right side of a chemical equation.

10.   _____ are starting materials for a chemical reaction, appearing on the left side of a chemical equation.

5 States of Matter: Gases, Liquids, and Solids

Learning Goals

1. Describe the major points of the kinetic molecular theory of gases.
2. Explain the relationship between the kinetic molecular theory and the physical properties of macroscopic quantities of gases.
3. Describe the behavior of gases expressed by the gas laws: Boyle's law, Charles's law, combined gas law, Avogadro's law, the ideal gas law, and Dalton's law.
4. Use gas law equations to calculate conditions and changes in conditions of gases.
5. Describe the properties of the liquid state in terms of the properties of the individual molecules that comprise the liquid.
6. Describe the processes of melting, boiling, evaporation, and condensation.
7. Describe the dipolar attractions known collectively as van der Waals forces.
8. Describe hydrogen bonding and its relationship to boiling and melting temperatures.
9. Relate the properties of the various classes of solids (ionic, covalent, molecular, and metallic) to the structure of these solids.

Introduction

The major differences between solid, liquid, and gaseous substances lie in the following properties:

1. The average distance of separation of particles in each state
2. The strength of the attractive forces between the particles
3. The degree of organization of particles

The behavior of the states of matter, their properties, and their interconversion are considered in this chapter.

5.1 The Gaseous State

Ideal Gas Concept

An ideal gas is a model that describes how molecules or atoms behave at the microscopic level. The model works well for most real gases that are *not* at high concentration, high pressure, or close to the temperature at which they would condense, that is, for gases that are not about to become a liquid! (Also, attractive forces between gas particles can cause deviation from the ideal gas law. This is addressed later in the chapter.)

The ideal gas model has been used to determine the interrelationship of temperature, volume, pressure, and quantity (mass or number of moles) in gaseous systems. These relationships are summarized in the ideal gas law and are described in this chapter.

Measurement of Gases

Measuring the temperature and volume of a gas is as familiar as using a thermometer and looking at a one-liter container, for example. (More precise measurements of volume can be taken by using glassware that is carefully marked, such as a graduated cylinder.)

Pressure can be measured with a mercury barometer, in which a long tube filled with mercury (Hg) is inverted into a dish of mercury. The atmospheric pressure pushing on the surface of the mercury in the dish supports the column of mercury in the tube. So, the height of the mercury in the tube is proportional to the atmospheric pressure, and "mm Hg" can be used as a unit of pressure.

In addition to *mm Hg (millimeters of mercury)*, other common units of pressure are: *atmosphere (atm)* and *torr*. Other units of pressure include the *pascal (Pa)* or *kilopascal (kPa)*, *lb/in² (pounds per square inch,* or *psi)*, and *in Hg (inches of mercury)*. These units are related as follows:

$$1 \text{ atm} = 760 \text{ mm Hg} = 760 \text{ torr} = 14.7 \text{ lb/in}^2 = 29.9 \text{ in Hg} =$$
$$1.01 \times 10^5 \text{ Pa} = 101 \text{ kPa}$$

Kinetic Molecular Theory of Gases

Learning Goal 1:	Describe the major points of the kinetic molecular theory of gases.

Learning Goal 2:	Explain the relationship between the kinetic molecular theory and the physical properties of macroscopic quantities of gases.

The fundamental model of particle behavior in the gas phase is the kinetic-molecular theory. This theory describes an ideal gas, in which gas particles exhibit no interactive or repulsive forces, and the volumes of the individual gas particles are assumed to be negligible. The theory may be summarized as follows:

1. A gas consists of particles that are far apart. The volume of the individual particles is assumed to be small in comparison to the average distance of separation of the particles.
2. The gas particles are in continuous, rapid, random motion. Particles change direction only as a result of collisions with other particles or with the wall of a container.
3. Upon collision, there is no net loss of energy; energy may only be transferred from one particle to another.

4. The speed of the particles is directly proportional to the absolute (Kelvin) temperature; as a result, the average kinetic energy is directly proportional to the absolute temperature.

Learning Goal 3:	Describe the behavior of gases expressed by the gas laws: Boyle's law, Charles's law, combined gas law, Avogadro's law, the ideal gas law, and Dalton's law.

Learning Goal 4:	Use gas law equations to calculate conditions and change in conditions of gases.

Boyle's Law

Robert Boyle found that the volume of a gas varies inversely with the pressure exerted by the gas, if the number of moles and the temperature of the gas are held constant. This relationship is known as Boyle's law.

Mathematically, the product of pressure (P) and volume (V) is a constant:

$$PV = k_1$$

Boyle's law is often used to calculate the volume resulting from a pressure change or vice versa. We consider $P_iV_i = k_1$ the initial condition, and $P_fV_f = k_1$ the final condition. Since (PV), initial or final, is constant and is equal to k_1,

$$P_iV_i = P_fV_f$$

Note that the calculation can be done with volume units of mL or L. It is only important that the units be the same on both sides of the equation. This is also true of pressure; however, the most commonly used unit of pressure measurement is the atmosphere. The relationship between the various pressure units follows: 1 atmosphere (atm) = 76 cm Hg = 760 mm Hg = 760 torr = 30.0 inches Hg = 14.7 psi (pounds per square inch).

Example 1
Convert each of the following into atmospheres.

1. 77.0 mm of Hg
2. 30.2 inches of Hg
3. 16.8 psi (The abbreviation psi means pounds per square inch.)
4. 800.00 torr

Work
One atmosphere of pressure equals 760 mm of Hg pressure, which also equals 760 torr of pressure, which also equals 14.7 psi, or

Example 1 continued

$$1 \text{ atmosphere } = 760 \text{ mm Hg } = 760 \text{ torr } = 14.7 \frac{\text{lb}}{\text{in}^2}$$

1. $77.0 \text{ mm Hg} \times \dfrac{1 \text{ atm}}{760 \text{ mm Hg}} = 1.01 \times 10^{-1} \text{ atm}$

2. $30.2 \text{ in Hg} \times \dfrac{2.54 \text{ cm}}{1 \text{ inch}} \times \dfrac{1 \text{ meter}}{100 \text{ cm}} \times \dfrac{1000 \text{ mm}}{1 \text{ meter}} \times \dfrac{1 \text{ atm}}{760 \text{ mm Hg}} = 1.01 \text{ atm}$

3. $16.8 \text{ psi} \times \dfrac{1 \text{ atm}}{14.7 \text{ psi}} = 1.14 \text{ atm}$

4. $800.0 \text{ torr} \times \dfrac{1 \text{ atm}}{760 \text{ torr}} = 1.053 \text{ atm}$

Answer
The pressure is 1.053 atm.

Example 2
A given mass of carbon dioxide at 25°C occupies a volume of 500.0 mL at 2.00 atmospheres of pressure. What pressure must be applied to compress the gas to a volume of 50.0 mL, assuming no temperature change?

Work

1.   Use Boyle's law: $P_i V_i = P_f V_f$
     P_i represents the starting pressure; P_f represents the final pressure.
     V_i represents the starting volume; V_f represents the final volume.

2.   From the given information, we have the following:

     $P_i = 2.00$ atmospheres $V_i = 500.0$ mL (the initial volume)

     $P_f = ?$ (not known yet) $V_f = 50.0$ mL (the final volume)

3.   Rearrange the equation as follows: $P_i V_i = P_f V_f$
                                          $P_f V_f = P_i V_i$

$$P_f = \frac{P_i V_i}{V_f}$$

Example 2 continued

We can thus find the pressure needed to compress the 500.0 mL of gas to 50.0 mL.

4. $P_f = \dfrac{P_i V_i}{V_f}$

$P_f = \dfrac{(2.00\ atm)(500.0\ mL)}{50.0\ mL}$

$P_{final} = 20.0\ atm$

Answer

20.0 atmospheres of pressure will be needed.

Charles's law

Jacques Charles found that the volume of a gas varies directly with the absolute temperature (K) if pressure and number of moles of gas are constant. This relationship is Charles's law.

Mathematically the ratio of volume (V) and temperature (T) is a constant:

$$V/T = k_2$$

In a way that is analogous to Boyle's law, we may use this expression to solve some practical problems.

Example 3

A balloon filled with helium has a volume of 5.00 liters at 0.00°C. What would be the balloon's volume at 25.00°C if the pressure surrounding the balloon remained constant?

Work

1. Use Charles's law: $\dfrac{V_i}{T_i} = \dfrac{V_f}{T_f}$

V_i represents starting volume; V_f means final volume.
T_i represents starting temperature in Kelvin units.
T_f represents final temperature in Kelvin units.

2. Change the °C values into Kelvin values. Use the following equation:

$K = °C + 273$

$T_i = 0.00°C$;   in K, $T_i = 0.00°C + 273 = 273.00\ K$

Example 3 continued

$$T_f = 25.00°C; \quad \text{in K, } T_f = 25.00°C + 273 = 298.00 \text{ K}$$

3.   Rearrange the equation: $\dfrac{V_i}{T_i} = \dfrac{V_f}{T_f}$

$$(V_f)(T_i) = (V_i)(T_f)$$

$$V_f = \dfrac{(V_i)(T_f)}{T_i}$$

4. $V_{final} = \dfrac{(5.00 \text{ liters})(298 \text{ K})}{273 \text{ K}} = 5.46 \text{ L}$

Answer

5.46 liters

Combined Gas Law

Often a sample of gas (a fixed number of moles of gas) undergoes change involving volume, pressure, and temperature simultaneously. It would be useful to have one equation that describes such processes.

A **combined gas law** is such an equation. It can be derived from Boyle's law and Charles's law and takes the form:

$$\dfrac{P_i V_i}{T_i} = \dfrac{P_f V_f}{T_f}$$

Avogadro's Law

The relationship between the volume and number of moles of a gas is known as Avogadro's law. This law states that equal volumes of a gas contain the same number of moles, if measured under the same conditions of temperature and pressure.

Mathematically, the ratio of volume (V) and number of moles (n) is a constant:

$$V/n = k_3$$

A relationship comparing initial and final conditions, similar to Boyle's and Charles's laws, may be derived:

$$\dfrac{V_i}{n_i} = \dfrac{V_f}{n_f}$$

Molar Volume of a Gas

The volume occupied by 1 mole of any gas is referred to as its molar volume. The basis for this relationship is Avogadro's law. At **standard temperature and pressure** (STP), the molar volume of any gas is 22.4 L. STP conditions are defined as follows:

$$T = 273 \text{ K (or } 0°\text{C)} \qquad P = 1 \text{ atm}$$

Thus, 1 mole of N_2 (28g), O_2 (32g), H_2 (2g), and He (4g) all occupy the same volume, 22.4 L at STP.

Gas Densities

It is also possible to compute the density of various gases at STP if one recalls that density is the mass/unit volume:

$$d = m/V$$

The Ideal Gas Law

Boyle's law (relating volume and pressure), Charles' law (relating volume and temperature) and Avogadro's law (relating volume to the number of moles) may be combined into a single expression relating all four terms. This expression is the **ideal gas law**:

$$\textbf{PV = nRT}$$

where R is the ideal gas constant.

$$R = 0.0821 \text{ L•atmK}^{-1}\text{mole}^{-1}$$

if the units P (atmospheres), V (liters), n (number of moles), and T (Kelvin) are used.

Example 4
Calculate the number of grams of helium in a 1.00-liter balloon at 27.0°C and under 1.00 atmosphere of pressure.

Work
1.   Use the ideal gas law:
       PV = nRT
       P is the pressure in atmospheres.
       V is the volume of the gas in liters.

       R is the ideal gas constant, which is equal to $\dfrac{0.0821 \text{ (L)(atm)}}{\text{(K)(mole)}}$

       T is the temperature given in Kelvin units.
       n is the number of moles of gas present.

2.   Convert 27.0°C into Kelvin units:
       K = °C + 273
       K = 27.0 + 273 = 300 K

Example 4 continued

3.   Rearrange PV = nRT so we can find the number of moles (n in the equation) of gas present:

$$n = \frac{PV}{RT}$$

4. $$n_{helium} = \frac{(1.00 \text{ atm})(1.00 \text{ L})}{\left(\dfrac{0.0821 \text{ (L)(atm)}}{\text{(K)(mole)}}\right)(300 \text{ K})}$$

$n_{helium} = 4.06 \times 10^{-2}$ moles of helium are present.

5.   Convert moles of helium to grams of helium using its molar mass, 4.00 g/mole. $(4.06 \times 10^{-2}$ moles$)(4.00$ g/mole$) = 0.162$ grams (or, 1.62×10^{-1} grams)

Answer
0.162 grams of helium are present.

Dalton's Law of Partial Pressures

A mixture of gases exerts a pressure that is the sum of the pressure that each gas would exert if it were present alone under similar conditions. This is known as Dalton's law of partial pressures.

$$P_t = P_1 + P_2 + P_3 + \bullet \bullet \bullet \bullet + P_n$$

where P_t = total pressure and P_1, P_2, P_3, and so on are the partial pressures of the component gases.

Ideal Gases vs. Real Gases

We have assumed so far, both in theory and calculations, that all gases behave as ideal gases. However, in reality, there is no such thing as an ideal gas. Interactive forces, even between the widely spaced particles of gas, are not totally absent in any sample of gas.

Attractive forces are particularly significant in gases composed of polar molecules. Calculations involving polar gases such as HF, NO, and SO_2, using ideal gas equations (which presume no such interactions), are approximate at best.

Nonpolar molecules exhibit a temporary dipole-dipole interaction known as a London force. The temporary dipole-dipole interaction is weak, and London forces are much weaker than permanent dipole interactions. As a result, the behavior of nonpolar molecules, such as N_2, O_2, and H_2, is explained rather well by the ideal-gas equations.

5.2   The Liquid State

Molecules in the liquid state are close to one another. Attractive forces are large enough to keep the molecules together, in contrast to gases, in which cohesive forces are so low that a gas expands to fill any volume. These attractive forces in a liquid are, however, not large enough to restrict movement, as in solids.

Learning Goal 5:	Describe the properties of the liquid state in terms of the properties of the individual molecules that comprise the liquid.

Compressibility

Liquids are almost incompressible. In fact, the spacing between molecules is so small that even the application of many atmospheres of pressure does not significantly decrease the volume of a liquid.

Viscosity

The **viscosity** of a liquid is a measure of its resistance to flow. Viscosity is a function of the attractive forces present between molecules as well as the molecular geometry. Complex molecules, which do not "slide" smoothly past each other, as well as polar molecules, tend to have higher viscosity than less structurally complex, less polar liquids.

Surface Tension

The surface tension of a liquid is a measure of the attractive forces at its surface. Intermolecular attraction is stronger at the surface of a liquid. This increased surface force is responsible for the spherical shape of drops of liquid. Substances known as surfactants may be added to a liquid to decrease surface tension.

Vapor Pressure of a Liquid

Learning Goal 6:	Describe the processes of melting, boiling, evaporation, and condensation.

According to the kinetic theory, the molecules of a liquid are in continuous motion, with their average kinetic energy proportional to the temperature. Although the average kinetic energy is too small to allow molecules to "escape" from liquid to vapor, a few high-energy molecules possess sufficient energy to escape from the bulk liquid.

At the same time, a fraction of these vaporized molecules lose energy (perhaps by collision with the walls of the container) and return to the liquid state. The process of conversion of liquid to vapor, at a temperature too low to boil, is **evaporation**. The reverse process, conversion of the gas to the liquid state, is **condensation**. After some period of time, the rates of evaporation and condensation become equal, and this constitutes a dynamic equilibrium between liquid and vapor states. The vapor pressure of the liquid is therefore defined as the pressure exerted by the vapor at equilibrium.

The boiling point of a liquid is defined as the temperature at which the vapor pressure of the liquid becomes equal to the prevailing atmospheric pressure. The "normal" atmospheric pressure is 760 torr, or one atmosphere, and the **normal boiling point** is the temperature at which the vapor pressure of the liquid is equal to one atmosphere.

van der Waals Forces

Learning Goal 7:	Describe the dipolar attractions known collectively as van der Waals forces.

In 1930 Fritz London demonstrated that he could account for a weak attractive force between any two molecules, whether polar or nonpolar. He postulated that the electron distribution in molecules is not fixed; electrons are in continuous motion, relative to the nucleus. So, for a short period of time a nonpolar molecule could experience an instantaneous dipole, a short-lived polarity due to a temporary dislocation of the electron cloud. These temporary dipoles could interact with other temporary dipoles, just permanent dipoles interact in polar molecules. We now call these intermolecular forces **London forces**.

London forces and dipole-dipole interactions are collectively known as **van der Waals forces**. London forces exist among polar and nonpolar molecules because electrons are in constant motion in all molecules. Dipole-dipole attractions occur only among polar molecules. In the next section we will see a third type of intermolecular force, the **hydrogen bond**.

Hydrogen Bonding

Learning Goal 8:	Describe hydrogen bonding and its relationship to boiling and melting temperatures.

Molecules in which a hydrogen atom is bonded to a highly electronegative atom such as nitrogen, oxygen, or fluorine exhibit **hydrogen bonding**. This arrangement of atoms produces a very polar bond, often resulting in a polar molecule with strong intermolecular attractive forces.

Example 5

Illustrate the hydrogen bonding found between a water molecule and its closest three neighboring water molecules.

Work

1. The polar nature of water is caused by the large electronegativity difference between oxygen and hydrogen atoms.
2. One water molecule has the following polar structure:

Example 5 continued

$$\delta^-$$
$$\delta^+ \ \text{H}\overset{\text{O}}{\frown}\text{H} \ \delta^+$$

Notice that the oxygen end is slightly negative, and the two hydrogen ends are slightly positive.

3. The hydrogen bonding in liquid water is due to the attraction of the slightly negative oxygen atom to a slightly positive hydrogen in another water molecule. In similar fashion each slightly positive hydrogen atom attracts an oxygen atom in two other water molecules.
4. Thus, one water molecule has three hydrogen bonds with three other water molecules.

Answer

Example 6
Show the hydrogen bonding that will occur between one molecule of methyl alcohol (CH₃OH) and water molecules. The methyl alcohol has the following polar structure:

$$\delta^-$$
$$\delta^+ \ \text{H}\overset{\text{O}}{\frown}\text{CH}_3$$

Answer

Example 7

Show the hydrogen bonding that will occur between water and the amide structure that is found in proteins.

$$\begin{array}{c} O \\ \| \\ -C-N- \\ | \\ H \end{array}$$

Work

The amide structure has the following polar arrangement:

$$\begin{array}{c} \delta^- \\ O \quad \delta^- \\ \| \\ -C-N- \\ \delta^+ \quad | \\ H \delta^+ \end{array}$$

Answer

5.3   The Solid State

The close packing of the particles of a solid results from attractive forces that are strong enough to restrict motion. The particles are "locked" together in a defined and highly organized fashion. This results in a fixed shape and volume (recall that gases have no fixed shape or volume, and liquids have a fixed volume but no fixed shape).

Properties of Solids

Learning Goal 9:	Relate the properties of the various classes of solids (ionic, covalent, molecular, and metallic) to the structure of these solids.

Solids are incompressible, due to the small distance between particles. Solids may be **crystalline**, having a regular repeating structure, or **amorphous**, having no organized structure.

Types of Crystalline Solids

Crystalline solids may exist in one of four general groups:

1. **Ionic solids.** The units that make up an **ionic solid** are positive and negative ions. Electrostatic forces hold the crystal together.
2. **Covalent solids.** The units that make up a **covalent solid** are atoms held together by covalent bonds.
3. **Molecular solids.** The units composing **molecular solids** are molecules held together by intermolecular forces (dipole-dipole and London).
4. **Metallic solids**. **Metallic solids** are composed of metal atoms held together by metallic bonds. Metallic bonds are formed by the overlap of orbitals of metal atoms, resulting in regions of high electron density surrounding the positive metal centers. Electrons in these regions are extremely mobile, resulting in the high conductivity (ability to carry electrical current) exhibited by many metallic solids.

Glossary of Key Terms in Chapter Five

amorphous solid (5.3) solids that have no organized, regular structure.
Avogadro's law (5.1) states that the volume is directly proportional to the number of moles of gas particles if the pressure and temperature are constant.
barometer (5.1) a device for measuring pressure.
Boyle's law (5.1) states that the volume of a gas varies inversely with pressure exerted if the temperature and number of moles are constant.
Charles's law (5.1) states that the volume of a gas is directly proportional to the temperature of the gas if the pressure and number of moles of the gas are constant.
combined gas law (5.1) an equation that describes the behavior of a gas when volume, pressure, and temperature may change simultaneously.
condensation (5.2) the conversion of gas to a liquid.
covalent solid (5.3) a collection of atoms held together by covalent bonds.
crystalline solid (5.3) a solid having a regular repeating structure.
Dalton's law (5.1) the law of partial pressure, states that the total pressure exerted by a gas mixture is the sum of the partial pressures of the component gases.
dipole-dipole interactions (5.2) attractive forces between polar molecules.
evaporation (5.2) the conversion of a liquid to a gas below the boiling point of the liquid.
hydrogen bonding (5.2) the attractive force between a hydrogen atom covalently bonded to a small, highly electronegative atom and another atom containing an unshared pair of electrons.
ideal gas (5.1) a gas in which the particles do not interact and the volume of the individual gas particles is assumed to be negligible.
ideal gas law (5.1) states that for an ideal gas the product of pressure and volume is proportional to the product of the number of moles of the gas and its temperature. The proportionality constant for an ideal gas is symbolized as R.
ionic solid (5.3) solids composed of positive and negative ions in a regular three-dimensional crystalline arrangement.
kinetic-molecular theory (5.1) the fundamental model of particle behavior in the gas phase.
London forces (5.2) weak attractive forces between molecules that result from short-lived dipoles that occur because of continuous motion of electrons in the molecules.
melting point (5.3) the temperature at which a solid converts to a liquid.
metallic bond (5.3) bond that results from the orbital overlap of metal atoms.
metallic solid (5.3) solid composed of metal atoms held together by metallic bonds.
molar volume (5.1) the volume occupied by 1 mol of a substance.
molecular solid (5.3) solid in which the molecules are held together by dipole-dipole and London forces.
normal boiling point (5.2) the temperature at which a substance will boil at one atmosphere of pressure.
partial pressure (5.1) the pressure exerted by one component of a gas mixture.

pressure (5.1) a force per unit area.

standard temperature and pressure (STP) (5.1) 273 K and one atmosphere pressure.

surface tension (5.2) a measure of the strength of the attractive force at a liquid's surface.

surfactant (5.2) a substance that decreases the surface tension of a liquid.

van der Waals forces (5.2) a general term for intermolecular forces that include dipole-dipole and London forces.

vapor pressure of a liquid (5.2) the pressure exerted by the vapor at the surface of a liquid at equilibrium.

viscosity (5.2) a measure of the resistance to flow of a substance at constant temperature.

Self Test for Chapter Five

1. _____ is the conversion of a liquid to a gas below the boiling point of the liquid.

2. A(n) _____ gas is one in which gas particles do not interact and the volume of the individual gas particles is assumed to be negligible.

3. Convert 29.96 inches of mercury pressure into mm of Hg pressure.

4. A given mass of gas at 20.0°C occupies a volume of 250.0 mL at 1.00 atmosphere of pressure. What pressure must be applied to compress the gas to a volume of 50.0 mL, assuming no temperature change?

5. A balloon filled with air has a volume of 2.00 liters at 24.0°C. What would be the balloon's volume at 37.0°C if the pressure surrounding the balloon remained constant?

6. Calculate the number of moles of helium in a 0.500-liter balloon at 25.0°C and under 0.960 atmospheres of pressure.

7. Show the hydrogen bonds between hydrogen fluoride molecules. The polar structure of HF is: $\delta^+ H - F \delta^-$.

8. The total pressure of our atmosphere is primarily equal to the sum of the pressures of what two gases?

9. What law explains why oxygen is distributed to the cells and the waste product, carbon dioxide, is expelled by the lungs?

10. What is defined as a measure of the attractive forces at the surface of a liquid?

11. What type of substances may be added to a liquid to decrease its surface tension?

12. What is defined as the pressure exerted by the vapor of a liquid at equilibrium?

13. What is defined as the temperature at which a solid is converted into a liquid?

14. Ionic solids have what kind of melting points?

15. If the temperature of 10.0 liters of gas is increased from 273 K to 546 K, what will be the new volume of that gas? Assume that all other conditions are held constant.

16. _____ is the conversion of a gas to a liquid.

17. _____ solids have no organized, regular structure.

18. A(n) _____ is a device for measuring gas pressure.

19. A(n) _____ is a device for measuring heat.
20. Ice is a _____ solid.

Vocabulary Quiz for Chapter Five

1. The _____ theory explains the behavior of an ideal gas.
2. The inverse relationship of pressure and volume in a gas is a statement of _____ law.
3. The direct proportionality of volume and temperature of a gas was first recognized by _____.
4. _____ decrease surface tension of a liquid.
5. An _____ solid gas has no regular, repeating pattern.
6. A _____ measures gas pressure.
7. _____ is a force per unit area.
8. The standard pressure of gas is _____ atm.
9. _____ solids are good electrical conductors.
10. A(n) _____ gas is one in which gas particles do not interact and the volume of the individual gas particles is assumed to be negligible.

6 *Reactions and Solutions*

Learning Goals

1 Distinguish among the terms *solution*, *solute*, and *solvent*.
2 Describe the various kinds of solutions, and give examples of each.
3 Describe the relationship between solubility and equilibrium.
4 Calculate solution concentration in weight/volume percent, weight/weight percent, parts per thousand, and parts per million.
5 Calculate solution concentration using molarity.
6 Perform dilution calculations.
7 Interconvert molar concentration of ions and milliequivalents/liter.
8 Describe and explain concentration-dependent solution properties.
9 Describe why the chemical and physical properties of water make it a truly unique solvent.
10 Explain the role of electrolytes in blood and their relationship to the process of dialysis.

Introduction

A **solution** is a homogeneous mixture of two or more substances. A solution is composed of one or more **solutes** dissolved in a **solvent**. When the solvent is water we refer to the homogeneous mixture as an **aqueous solution**.

6.1 Properties of Solutions

Learning Goal 1:   Distinguish among the terms *solution, solute*, and *solvent*.

General Properties of Liquid Solutions

A **solution** is made up of a **solvent** (the substance in excess) and one or more **solutes** that are dissolved in the solvent.

Liquid solutions are clear and transparent. They may be colored or colorless, depending upon the properties of the solute and solvent.

Learning Goal 2:   Describe the various kinds of solutions, and give examples of each.

Solutions may be classified as **electrolytic** or **nonelectrolytic**. Electrolytic solutions are formed from ionic compounds that dissociate in solution to produce ions. Electrolytic solutions are good conductors of electricity. Nonelectrolytic solutions are formed from nondissociating molecular solutes (nonelectrolytes), and these solutions are nonconducting. Aqueous sodium chloride solution is an example of an electrolytic solution; sugar water is a common nonelectrolytic solution.

Solutions and Colloids

A **colloidal suspension** is not a true solution; the colloid particles are not identical in size nor homogeneously distributed throughout the solution. Particles with diameters between one nanometer and two hundred nanometers are colloids. Smaller particles form a true solution; larger particles are precipitates.

A colloidal suspension is characterized by its light-scattering property, the Tyndall effect.

Degree of Solubility

The degree of solubility, how much solute can dissolve in a given volume of solvent, is difficult to predict, but general trends are based upon the following considerations:

1. The magnitude of difference between polarity of solute and solvent (the rule of "like dissolves like").

2. Temperature. An increase in temperature usually, but not always, increases solubility.

3. Pressure. Pressure has little effect on the solubility of solids and liquids in liquids. However, the solubility of gases in solution is extremely pressure dependent.

When a solution contains all the solute that can be dissolved at a particular temperature, it is **saturated**. Cooling a saturated solution often results in the formation of a **precipitate**. Occasionally, however, on cooling, the excess solute may remain in solution for a period of time. Such a solution is described as a **supersaturated solution**.

Solubility and Equilibrium

Learning Goal 3:   Describe the relationship between solubility and equilibrium.

When an excess of solute is brought into contact with a solvent, the resulting dissolution establishes a dynamic equilibrium.

Solubility of Gases: Henry's Law

Henry's law states that the number of moles of a gas dissolved in a liquid at a given temperature is proportional to the partial pressure of the gas. In other words, the gas solubility is directly proportional to the pressure of that gas in the atmosphere that is in contact with the liquid.

6.2 Concentration Based on Mass

Learning Goal 4:	Calculate solution concentration in weight/volume percent, weight/weight percent, parts per thousand, and parts per million.

The amount of solute dissolved in a given amount of solution is defined as the solution concentration. The concentration of a solution has a profound effect on the properties of a solution, both physical (melting and boiling points) and chemical (solution reactivity). The more widely used concentration units are considered below.

Weight/Volume Percent

$$\% \left(\frac{W}{V} \right) = \frac{\text{grams of solute}}{\text{mL of solution}} \times 10^2 \%$$

Example 1

Calculate the %(W / V) of a sodium chloride solution that was made by mixing 15.0 grams of NaCl with enough distilled water to prepare 500.0 mL of solution.

Work

1. Use the equation for % (W / V): $\quad \% \left(\frac{W}{V} \right) = \frac{\text{grams of solute}}{\text{mL of solution}} \times 10^2$

2. The solute is NaCl, so grams of solute = 15.0 grams of NaCl.

3. $\% \left(\frac{W}{V} \right) = \frac{15.0 \text{ grams}}{500.0 \text{ mL}} \times 10^2$

   % (W / V) = 3.00 %

Answer

The solution is a 3.00 % (W / V) sodium chloride solution.

Example 2

How many grams of glucose are found in 1.00 liter of 5.00% (W / V) glucose solution?

Work

1. The 5.00% (W / V) glucose solution tells us that there are 5.00 grams of glucose per 100.00 mL of total solution.

Example 2 continued

2. Use conversion factors to find the number of grams of glucose that are present.

3. $1.00 \text{ liter} \times \dfrac{1000 \text{ mL}}{1 \text{ liter}} \times \dfrac{5.00 \text{ grams glucose}}{100.00 \text{ mL solution}} = 50.0 \text{ mL}$

Answer

In 1.00 liter of 5.00% (W / V) glucose solution, there are 50.0 grams of glucose.

Example 3

If a patient is to receive 2.00 grams of a specific drug, and the drug is supplied as a 5.00% (W / V) drug solution, how many milliliters of the drug solution must the patient be given?

Work

1. The 5.00% (W / V) drug solution means that there are 5.00 grams of the drug per 100.00 mL of the drug solution.
2. Use conversion factors to find the number of mL of drug solution needed.

3. $2.00 \text{ grams} \times \dfrac{100.00 \text{ mL of drug solution}}{5.00 \text{ grams of the drug}} = 40.0 \text{ mL}$

Answer

The patient must take 40.0 mL of the 5.00% (W / V) drug solution to receive 2.00 grams of the specific drug.

Weight/Weight Percent

$$\%\left(\frac{W}{W}\right) = \frac{\text{grams of solute}}{\text{grams of solution}} \times 10^2 \%$$

Example 4

Calculate the % (W / W) of gold in a wedding band that has a mass of 12.0 grams and was found to contain 8.97 g of gold.

Work

1. Use the equation for % (W / W): $\%\left(\dfrac{W}{W}\right) = \dfrac{\text{grams of solute}}{\text{grams of solution}} \times 10^2$

2. $\%\left(\dfrac{W}{W}\right) = \dfrac{8.97 \text{ grams}}{12.0 \text{ grams}} \times 10^2$

 $\% (W / W) = 74.8\%$

Answer

The wedding band is 74.8% (W / W) gold.

Parts per Thousand and Parts Per Million

$$\text{ppt} = \frac{\text{grams of solute}}{\text{grams of solution}} \times 10^3 \text{ ppt} \quad \text{and} \quad \text{ppm} = \frac{\text{grams of solute}}{\text{grams of solution}} \times 10^6 \text{ ppm}$$

Example 5

Recalculate Example 4 in (a) ppt and (b) ppm.

Work

1. Use the equation for ppt $\qquad \text{ppt} = \dfrac{\text{grams of solute}}{\text{grams of solution}} \times 10^3 \text{ ppt}$

$$\text{ppt} = \frac{8.97 \text{ g}}{12.0 \text{ g}} \times 10^3 \text{ ppt}$$

$$\text{ppt} = 748$$

2. Use the equation for ppm $\qquad \text{ppm} = \dfrac{\text{grams of solute}}{\text{grams of solution}} \times 10^6 \text{ ppm}$

$$\text{ppm} = \frac{8.97 \text{ g}}{12.0 \text{ g}} \times 10^6 \text{ ppm}$$

$$\text{ppm} = 7.48 \times 10^5$$

Answer

The wedding band is 748 ppt and 7.48×10^5 ppm gold.

6.3 Concentration of Solutions: Moles and Equivalents

Learning Goal 5:   Calculate solution concentration using molarity.

Molarity

Molarity, symbolized **M**, is defined as the number of moles of solute per liter of solution:

$$\text{Molarity} = \text{M} = \frac{\text{moles solute}}{\text{L solution}}$$

Example 6

Calculate the molarity of a sodium chloride solution made by mixing 3.51 grams of solid NaCl in enough water to prepare 500.0 mL of total solution.

Work

1. Use the molarity equation:
$$M = \frac{\text{moles solute}}{\text{L solution}}$$

2. **When using molarity, convert all volumes to liters!**
   500.0 mL of solution is the same as 0.5000 liter.

3. The number of moles of solute (NaCl) in the solution is found by first determining the weight of 1 mole of NaCl.

$$1 \text{ mole of NaCl} = 58.5 \text{ grams}$$

4. Since we have only 3.51 grams of NaCl, we use a conversion factor to find the number of moles of NaCl that is really in our solution:

$$(3.51 \text{ grams of NaCl})\left(\frac{1 \text{ mole NaCl}}{58.5 \text{ grams of NaCl}}\right) = 0.0600 \text{ moles NaCl}$$

5. We can now use the molarity equation, since we know all the needed values:
   The number of moles of NaCl in the solution = 0.0600 moles
   The number of liters of solution = 0.5000 liter

6. $$M = \frac{\text{moles solute}}{\text{L solution}}$$

$$M = \frac{0.0600 \text{ moles}}{0.5000 \text{ L}}$$

   Molarity of our solution = 0.120 M

Answer

We have prepared a 0.120 molar NaCl solution.

Example 7

Calculate the number of grams of solid silver nitrate needed to prepare 250.0 mL of a 0.100 molar $AgNO_3$ solution.

Work

1. One mole of $AgNO_3$ is found to be 169.9 grams/mole (from the periodic table).

Example 7 continued

2. Convert 250.0 mL into liter units:

$$250.0 \text{ mL x } \frac{1 \text{ liter}}{1000 \text{ mL}} = 0.250 \text{ liter solution}$$

3. 0.100 molar $AgNO_3$ solution means that there is 0.100 mole of $AgNO_3$ per 1 liter of solution.
4. Using suitable conversion factors:

$$0.250 \text{ L x } \frac{0.100 \text{ mole of } AgNO_3}{1 \text{ liter of solution}} \text{ x } \frac{169.9 \text{ grams of } AgNO_3}{1 \text{ mole of } AgNO_3} = 4.25 \text{ grams of } AgNO_3$$

Answer
We need to add 4.25 grams of solid $AgNO_3$ to prepare 250.0 mL of a 0.100 molar $AgNO_3$ solution.

Dilution

Learning Goal 6: Perform dilution calculations.

The technique of dilution is often used to prepare less-concentrated solutions. The approach to such a calculation is outlined below:

M_1 = molarity prior to dilution.
M_2 = molarity after dilution.
V_1 = volume prior to dilution.
V_2 = volume after dilution.
$(M_1)(V_1) = (M_2)(V_2)$

Knowing any three of these terms enables one to calculate the fourth.
The dilution equation is valid with any concentration units, such as % (W / V) or % (V / V), as well as molarity. Be certain to use the same units for both initial and final concentration values. Only in this way will proper unit cancellation occur.

Example 8
How many mL of 12.0 M HCl solution do we need to add to distilled water so we can prepare 500.0 mL of 2.50 M HCl solution?

Work
1. Use $(M_1)(V_1) = (M_2)(V_2)$
   V_1 is the starting volume: we do not know this value yet.
   M_1 is the molarity of the solution we have available to us. We will dilute some of

Example 8 continued
        it to make our final solution.
        M_2 is the molarity of our final solution. It always will be the lower value.
        V_2 is the final volume of our diluted solution.

2.   From our problem:
        V_1 = ? (not known yet)
        M_1 = 12.0 M HCl solution (the one we start with)
        M_2 = 2.50 M HCl solution (the one we want to make)
        V_2 = 500.0 mL (the volume we want to prepare)

3.   Rearrange the equation to find V_1:

        $$(M_1)(V_1) = (M_2)(V_2)$$

        $$V_1 = \frac{(M_2)(V_2)}{M_1}$$

4. $$V_1 = \frac{(2.50 \text{ M})(500.0 \text{ mL})}{(12.0 \text{ M})}$$

        $$V_1 = 104 \text{ mL}$$

Answer
If we measure out 104 mL of the 12.0 M HCl solution and then add distilled water to it in another container until we have made 500.0 mL of total solution, we will have prepared 500.0 mL of a 2.50 M HCl solution by dilution.

Representation of Concentration of Ions in Solution

Learning Goal 7:	Interconvert molar concentration of ions and milliequivalents/liter.

   The concentration of ions in solution may be represented as moles per liter (molarity) and equivalents per liter.
   Molarity emphasizes the number of individual ions. A one molar solution of Na^+ contains Avogadro's number, 6.02×10^{23} Na^+. In contrast, equivalents per liter emphasizes charge; one equivalent of Na^+ contains Avogadro's number of positive charge.
   We defined 1 mole as the number of grams of an atom, molecule, or ion corresponding to Avogadro's number of particles. One equivalent of an ion is the number of grams of the ion corresponding to Avogadro's number of electrical charges.

Changing from moles/L to equivalents/L (or the reverse) can be accomplished using conversion factors.

Example 9

Calculate the number of equivalents per liter of Mg^{2+} in a solution that is 1.0×10^{-4} M $MgCl_2$.

Work

A series of conversion factors is required to

1. Convert M $MgCl_2$ to M Mg^{2+}
2. Convert moles of Mg^{2+}/L to moles of charge/L
3. Convert moles of charge/L to equivalents of Mg^{2+}/L
   The setup is as follows:

$$\frac{1.0 \times 10^{-4} \text{ mol } MgCl_2}{1 \text{ L}} \times \frac{1 \text{ mol } Mg^{2+}}{1 \text{ mol } MgCl_2} \times \frac{2 \text{ mol charge}}{1 \text{ mol } Mg^{2+}} \times \frac{1 \text{ eq } Mg^{2+}}{1 \text{ mol charge}}$$

$$= \frac{2.0 \times 10^{-4} \text{ eq } Mg^{2+}}{\text{liter}}$$

Answer

A 1.0×10^{-4} M $MgCl_2$ solution contains 2.0×10^{-4} eq/L of Mg^{2+}.

6.4   Concentration-Dependent Solution Properties

Learning Goal 8:   Describe and explain concentration-dependent solution properties.

Solution properties that are dependent upon the concentration of the solute particles, rather than the identity of the solute, are referred to as colligative properties. There are four colligative properties of solutions.

Vapor Pressure Lowering

Raoult's law states that when a solute is added to a solvent, the vapor pressure of the solvent decreases in proportion to the concentration of solute particles.

Freezing-Point Depression and Boiling-Point Elevation

When a nonvolatile solid is added to a solvent, the freezing point of the resulting solution decreases (a lower temperature is required to form the solid state), and the boiling point increases (requiring a higher temperature to form the gaseous state). The magnitude of the freezing-point depression (ΔT_f) is proportional to the solute concentration over a limited range of concentration.

$$\Delta T_f = K_f \text{ (solute concentration)}$$

The boiling point elevation is also proportional to the solute concentration.

$$\Delta T_b = K_b \text{ (solute concentration)}$$

If the value of the proportionality factor is known, the magnitude of the freezing-point depression or boiling-point elevation may be calculated for a solution of known concentration of particles.

We have already worked with one mole-based unit, molarity, and this concentration unit can be used to calculate either the freezing-point depression or boiling-point elevation.

A second mole-based concentration unit is molality, which is more commonly used in these types of situations. **Molality** (symbolized by m) is defined as the number of moles of solute per kilogram of solvent in a solution:

$$m = \frac{\text{moles solute}}{\text{kg solvent}}$$

Molality does not vary with temperature, whereas molarity is temperature dependent.

Osmotic Pressure

Osmosis is the movement of solvent from a dilute solution to a more concentrated solution through a **semipermeable membrane**. Pressure must be applied to the more concentrated solution to stop this flow, and the magnitude of the pressure required to just stop the flow is termed the **osmotic pressure**. The "driving force" for the osmotic process is the need to establish an equilibrium between the solutions on either side of the membrane.

The osmotic pressure, not unlike the pressure exerted by a gas, may be treated quantitatively. Osmotic pressure, symbolized by π, follows the same form as the ideal gas equation:

Ideal gas	**Osmotic pressure**
$PV = nRT$	$\pi V = nRT$

or	$P = n\dfrac{RT}{V}$	or	$\pi = n\dfrac{RT}{V}$

and, since $M = \dfrac{n}{V}$ 　　　　 and, since $M = \dfrac{n}{V}$

then, $P = MRT$ 　　　　 then, $\pi = MRT$

The osmotic pressure may be calculated from the solution concentration at any given temperature. Osmosis is a colligative property; it is dependent on the concentration of solute particles.

By convention, the molarity of particles in solution is termed osmolarity, abbreviated Osm, for osmotic pressure calculations.

Osmosis is important as a transport mechanism to all living organisms. A living cell contains an aqueous solution, and material movement in and out of cells is based partly upon osmosis. If the concentration of the fluid surrounding red blood cells is higher than that inside the cell (a **hypertonic solution**), water flows from the cell, causing it to collapse. This process is known as crenation. On the other hand, too low a concentration of this fluid relative to the solution within the blood cell (a **hypotonic solution**) will cause water to flow into the cell, causing cell rupture, a process known as **hemolysis**.

Two solutions are **isotonic** if they have identical osmotic pressures. In that case, the osmotic pressure differential across the cell is zero, and no cell disruption occurs.

6.5　Water as a Solvent

Learning Goal 9:	Describe why the chemical and physical properties of water make it a truly unique solvent.

The role of water in the solution process deserves special attention because of its many unique characteristics:

1. It is often referred to as the universal solvent.

2. In view of its small size, it is remarkable that it exists as a liquid at room temperature.

3. Contrary to our expectations, liquid water is more dense than its solid form, ice.

4. Water is the principal biological solvent. Approximately 60% of the adult human body is water.

6.6　Electrolyte in Body Fluids

Learning Goal 10:	Explain the role of electrolytes in blood and their relationship to the process of dialysis.

Proper cell function in biological systems is critically dependent on the concentration of electrolytes. The maintenance of proper muscle and nervous system function depends on the sodium/potassium ratio inside and outside of the cell. A stable osmotic pressure in biological fluids depends on the electrolyte concentration inside and outside of the cell.

Glossary of Key Terms in Chapter Six

aqueous solution (6.1) any solution in which the solvent is water.

colligative property (6.4) property of a solution that is dependent only on the concentration of solute particles.

colloidal suspension (6.1) a heterogeneous mixture of solute particles in a solvent; distribution of solute particles is not uniform because of the size of the particles.

concentration (6.2) a measure of the quantity of a substance contained in a specified volume of solution.

crenation (6.4) the shrinkage of red blood cells caused by water loss to the surrounding medium.

dialysis (6.6) the removal of waste material via transport across a membrane.

electrolyte (6.1) a material that dissolves in water to produce a solution that conducts an electrical current.

equivalent (6.3) the number of grams of an ion corresponding to Avogadro's number of electrical charges.

hemolysis (6.4) the rupture of red blood cells resulting from movement of water from the surrounding medium into the cell.

Henry's Law (6.1) a law stating that the number of moles of a gas dissolved in a liquid at a given temperature is proportional to the partial pressure of the gas.

hypertonic solution (6.4) the more concentrated solution of two separated by a semipermeable membrane.

hypotonic solution (6.4) the more dilute solution of two separated by a semipermeable membrane.

isotonic solution (6.4) a solution that has the same solute concentration as another solution with which it is being compared; a solution that has the same osmotic pressure as a solution existing within a cell.

molality (6.4) the number of moles of solute per kilogram of solvent.

molarity (6.3) the number of moles of solute per liter of solvent.

nonelectrolyte (6.1) a substance that, when dissolved in water, produces a solution that does not conduct an electrical current.

osmolarity (6.4) molarity of particles in solution; this value is used for osmotic pressure calculation.

osmosis (6.4) net flow of a solvent across a semipermeable membrane in response to a concentration gradient.

osmotic pressure (6.4) the net force with which water enters a solution through a semipermeable membrane; alternatively, the pressure required to stop net transfer of solvent across a semipermeable membrane.

precipitate (6.1) an insoluble substance formed and separated from a solution.

Raoult's law (6.4) a law stating that the vapor pressure of a compound is equal to its mole fraction times the vapor pressure of the pure component.

saturated solution (6.1) one in which undissolved solute is in equilibrium with the solution.

semipermeable membrane (6.4) a membrane permeable to the solvent but not the solute; a material that allows the transport of certain substance from one side of the membrane to the other.

solubility (6.1) the amount of a substance that will dissolve in a given volume of solvent at a specified temperature.

solute (6.1) a component of a solution that is present in lesser quantity than the solvent.

solution (6.1) a homogeneous (uniform) mixture of two or more substances.

solvent (6.1) the solution component that is present in the largest quantity.

supersaturated solution (6.1) a solution that is more concentrated than a saturated solution (Note that such a solution is not at equilibrium).

suspension (6.1) a heterogeneous mixture of particles; the suspended particles are larger than those found in a colloidal suspension.

weight/volume percent (% (W/V)) (6.2) the concentration of a solution expressed as a ratio of grams of solute to milliliters of solution multiplied by 100%.

weight/weight percent (% (W/W)) (6.2) the concentration of a solution expressed as a ratio of mass of solute to mass of solution multiplied by 100%.

Self Test for Chapter Six

1. What is the % (W / V) of NaOH in 150.0 mL of a solution that contains 1.65 grams of sodium hydroxide?

2. How many milliliters of a 5.000% (W / V) glucose solution would contain 40.00 grams of glucose?

3. How many liters of D-5-W solution [5.00% (W / V) glucose solution] would need to be given to a patient by I.V. so that she would receive 250.0 food-calories of energy from the given glucose? One gram of glucose produces 4.00 food-calories of energy in the human body.

4. How many moles of KCl are present in 100.0 mL of a 0.552 molar potassium chloride solution?

5. How many grams of KCl are present in 50.0 mL of a 0.125 molar potassium chloride solution?

    KCl = 74.60 grams/mole.

6. If 20.80 grams of HCl are added to enough distilled water to form 3.0 liters of solution, what is the molarity of the solution? HCl = 36.46 grams/mole.

7. Calculate the molarity of a solution that contains 5.60 grams of KNO_3 in 300.0 mL of solution.

    KNO_3 = 101.1 grams/mole.

8. How many mL of 12.0 M HCl solution are required to prepare 250.0 mL of a 2.50 M HCl solution?

9. What term defines the amount of solute in grams or moles dissolved in a given amount of solution?

10. Normal physiological saline solution is a 0.900% (W / V) NaCl solution. How many grams of solid NaCl are needed to prepare 1.50 liters of this solution?

11. What term is defined as the number of moles of solute per liter of solution?

12. What term defines the movement of solvent through a semipermeable membrane from a dilute solution to a more concentrated solution?

13. Is a concentrated solution hypertonic or hypotonic when compared to a dilute solution?

14. What is the term used to describe solutions of exactly the same particle concentration?

Vocabulary Quiz for Chapter Six

1. _____ is the shrinkage of red blood cells due to water loss to the surrounding medium.

2.   A(n) _____ is a substance that ionizes in water.

3.   A(n) _____ is the more concentrated solution of two separated by a semipermeable membrane.

4.   A(n) _____ is the more dilute solution of two separated by a semipermeable membrane.

5.   _____ is the number of moles of solute per liter of solution.

6.   _____ is the net flow of solvent though a semipermeable membrane.

7.   _____ is the pressure required to stop net transfer of solvent across a semipermeable membrane.

8.   _____ describes a solution in which undissolved solute is in equilibrium with the solution.

9.   The _____ is a component of a solution that is present in lesser quantity than the solvent.

10.   A(n) _____ is a solution that is more concentrated than a saturated solution.

7 *Energy, Rate, and Equilibrium*

Learning Goals

1 Correlate the terms *endothermic* and *exothermic* with heat flow between a system and its surroundings.

2 State the meaning of the terms *enthalpy, entropy,* and *free energy* and know their implications.

3 Describe experiments that yield thermochemical information and calculate fuel values based on experimental data.

4 Describe the concept of reaction rate and the role of kinetics in chemical and physical change.

5 Describe the importance of *activation energy* and the *activated complex* in determining reaction rate.

6 Predict the way reactant structure, concentration, temperature, and catalysis affect the rate of a chemical reaction.

7 Write rate equations for elementary processes.

8 Recognize and describe equilibrium situations.

9 Write equilibrium-constant expressions and use these expressions to calculate equilibrium constants.

10 Use LeChatelier's principle to predict changes in equilibrium position.

Introduction

Three concepts play an important role in determining the extent and speed of a chemical reaction:

1. Thermodynamics deals with energy changes associated with chemical reactions.
2. Kinetics concerns itself with the rate or speed of a chemical reaction.
3. Equilibrium considerations determine the completeness of a reaction.

A reaction may be thermodynamically favored but very slow; conversely, a fast reaction may result in very little product being formed. Thus, each is considered independently.

In a reaction that does not go to completion, a system is at equilibrium when the rates of the forward and reverse reactions are equal.

The concepts of energy, rate, and equilibrium can be applied to changes in state as well as to chemical reactions. The behavior of the states of matter (gas, liquid, and solid), their properties, and their interconversion are considered in this chapter.

7.1 Thermodynamics

Thermodynamics is the study of energy, work, and heat. Thermodynamics may be applied to chemical change or physical change. There exist three basic laws of thermodynamics; only the first two are of concern in Chapter 7.

The Chemical Reaction and Energy

Every chemical reaction involves a change in energy. The first law of thermodynamics, also known as the law of conservation of energy, states that energy cannot be created or destroyed in the course of the reaction. It may only be converted from one form to another or transferred from one component of the system to another.

Our current view of events at the molecular level is consistent with the first law. When substances react:

- molecules and atoms in a reaction mixture are in constant, random motion;
- these molecules and atoms frequently collide with each other;
- only some collisions, those with sufficient energy, will break bonds in molecules; and
- when reactant bonds are broken, new bonds may be formed and products result.

Exothermic and Endothermic Reactions

Learning Goal 1:	Correlate the terms *endothermic* and *exothermic* with heat flow between a system and its surroundings.

A reaction that absorbs energy (exhibits a net gain in energy) is termed **endothermic**. The products of the reaction possess more energy than the reactants. On the other hand, a reaction that releases energy (exhibits a net loss of energy) is termed **exothermic**. The products of the reaction possess less energy than the reactants.

Figure 8.1 in the textbook graphically represents energy changes in endothermic and exothermic reactions.

The energy released in exothermic reactions is often in the form of **heat energy**. It may also take the form of light or even electricity. A battery is a common example of a chemical reaction producing electrical energy. The energy absorbed during chemical change is stored as **chemical energy**.

Enthalpy

Learning Goal 2:	State the meaning of the terms *enthalpy, entropy,* and *free energy* and know their implications.

Enthalpy is the term used to represent heat energy and is symbolized by H°. In chemical change, we are primarily interested in the heat gained or released, and that is the enthalpy change, ΔH°. For endothermic reactions, the enthalpy change is positive; $\Delta H^\circ = +$. For exothermic reactions, the enthalpy change is negative; $\Delta H^\circ = -$. The enthalpy change is most often reported in units of joules (or kilojoules) or calories (or kilocalories).

Spontaneous and Nonspontaneous Reactions

Spontaneous reactions are just that: they occur without any external energy input. Nonspontaneous reactions must be persuaded; they need an input of energy.

It seems that all exothermic reactions should be spontaneous. After all, an external supply of energy does not appear to be necessary; in fact, energy is a product of the reaction. It also seems that all endothermic reactions should be nonspontaneous: energy is a reactant that we must provide. However, these hypotheses are not supported by experimentation.

Experimental measurement has shown that most, but not all, exothermic reactions are spontaneous; likewise, most, but not all, endothermic reactions are not spontaneous. There must be some factor in addition to enthalpy that will help us to explain the less-obvious cases of nonspontaneous exothermic reactions and spontaneous endothermic reactions. This other factor is entropy.

Entropy

The second law of thermodynamics states that a system and its surroundings spontaneously tend toward increasing disorder (randomness). A measure of the randomness of a chemical system is referred to as **entropy**. A random, or disordered system is characterized by high entropy; a well-ordered system is said to have low entropy.

Often, reactions that are exothermic and whose products are more disordered (higher in entropy) occur spontaneously, while endothermic reactions that produce products of lower entropy are not spontaneous. If they occur at all, they will require external energy input.

Disorder or randomness increases as we proceed from the solid to liquid to the vapor state. Solids often have an ordered crystalline structure, and liquids have a loose structure, while gas particles are virtually random in their distribution.

Free Energy

In chemical change, the maximum amount of energy that can be converted to a useful form is called **free energy** (G). Free energy incorporates the two factors discussed above, the energy factor and the entropy factor, into a single expression for predicting the spontaneity of chemical change:

$$\Delta G = \Delta H - T\Delta S$$

In this expression, ΔG is the difference in free energy between products and reactants, ΔH is the difference in heat energy (enthalpy) between products and reactants, and ΔS is the difference in entropy between products and reactants. The temperature of the reaction is symbolized by T.

We find that a spontaneous reaction will have a negative ΔG. A positive ΔG is indicative of a nonspontaneous reaction. In fact, a change for which ΔG is positive will occur only if energy is added to the system.

7.2 Experimental Determination of Energy Change in Reactions

> **Learning Goal 3:** **Describe experiments that yield thermochemical information and calculate fuel values based on experimental data.**

The measurement of the energy demand or energy release in a chemical reaction is **calorimetry**. This technique involves the measurement of the change in the temperature of a quantity of water that is in contact with the reaction of interest and also isolated from its surroundings. A device used in these types of measurements is a calorimeter, which measures heat changes in calories (or energy, since the calorie is a unit of heat energy).

For an exothermic reaction, heat is released by the reaction to the surrounding water, which is isolated from its surroundings. The **specific heat** of water (SH_w) is defined as the number of calories needed to raise the temperature of 1 gram of water 1 degree Celsius. This, coupled with the total number of grams of water (m_w) and the temperature increase (ΔT), allows one to calculate the heat released, Q.

$$Q = m_w \times \Delta T \times SH_w$$

Many chemical reactions that produce heat are combustion reactions. In our bodies, many food substances, principally carbohydrates and fats, are oxidized to produce energy. In these cases, the amount of energy per gram of food is referred to as its fuel value. A special type of calorimeter, a bomb calorimeter, is useful for the measurement of the **fuel value** (calories) of foods.

Example 1

When equal moles of hydrochloric acid (HCl) and sodium hydroxide (NaOH) are mixed in a calorimeter, the temperature of 50.00 grams of solution increased from 25.00°C to 38.00°C. If the specific heat of the solution is the same as the specific heat of water, which is 1.000 calories/(gram)(°C), calculate the quantity of energy in calories involved in this reaction.

Work
1. The change in temperature is $\Delta T = 38.00°C - 25.000°C = 13.00°C$
2. Since heat energy is released from the reaction to the solution, the reaction is exothermic.
3. The quantity of heat absorbed or released by a reaction (Q) is the product of the mass of water or solution in the calorimeter (m_w), the specific heat of the water or solution (SH_w), and the change in temperature (ΔT) of the water or solution as the reaction proceeds from the initial to final states. The above relationship is given by the following equation:

$$Q = (m_w)(SH_w)(\Delta T)$$

4. $Q = 50.00 \text{ g} \times \dfrac{1.000 \text{ calories}}{(1 \text{ gram}[1°C])} \times 13.00 \text{ °C} = 650.0 \text{ calories}$

Example 1 continued

Answer

    650.0 calories are released during the reaction.

Example 2

When 1.000 gram of glucose ($C_6H_{12}O_6$) was burned in a bomb calorimeter, the temperature of 100.0 grams of water in the calorimeter increased from 25.00°C to 63.00°C. Calculate the fuel value of one gram of glucose. Remember that the SH_w for water is 1.000 calorie per gram of water per 1°C temperature change.

Work

1.   Use the equation $Q = (m_w)(SH_w)(\Delta T)$
2. $\Delta T = 63.00°C - 25.00°C = 38.00°C$

3. $Q = 100.0 \text{ g} \times 38.00°C \times \dfrac{1.000 \text{ calories}}{(1 \text{ gram } (1°C))}$

        $Q = 3.800 \times 10^3 \text{ calories}$

4. In food-calorie books, we see the value of 1 gram of carbohydrate (glucose is a pure carbohydrate) to be 4 calories. These calorie values are really kilocalories. Thus, our value of 3800 calories, or 3.8 kilocalories, is close to those given in food-calorie handbooks.

Answer

One gram of glucose (or any carbohydrate) has a fuel value of 3.8×10^3 calories, or 3.8 kilocalories, or just 3.8 nutritional calories.

7.3   Kinetics

    Chemical kinetics is the study of the rate (speed) of a chemical reaction.

Learning Goal 4:   **Describe the concept of reaction rate and the role of kinetics in chemical and physical change.**

The Chemical Reaction

    If the energy made available by the collision of reacting molecules exceeds the bond energies, the bonds will break, and the resulting atoms will recombine in a lower energy configuration. A collision meeting the above conditions and producing one or more product molecules is referred to as an effective collision. Only effective collisions lead to chemical reaction.

Activation Energy and the Activated Complex

Learning Goal 5:	Describe the importance of *activation energy* and the *activated complex* in determining reaction rate.

The minimum amount of energy required to produce a chemical reaction is the **activation energy** for the reaction. As implied above, a large component of the activation energy is the bond energy of the reacting molecules.

The chemical reaction may be represented in terms of the changes in potential energy that occur as a function of the time of the reaction. Several important characteristics of this relationship follow:

1. The reaction proceeds from reactants to products through an extremely unstable intermediate state that we term the **activated complex**.
2. Formation of the activated complex requires energy. The difference between the energy of reactants and activated complex is the activation energy.
3. For an exothermic reaction, the overall energy change must be a net release of energy. The net release of energy is the difference in energy between products and reactants.

Factors That Affect Reaction Rate

Learning Goal 6:	Predict the way reactant structure, concentration, temperature, and catalysis affect the rate of a chemical reaction.

Five major experimental conditions influence the rate of a chemical reaction.

1. Structure of the reacting species
2. The concentration of reactants
3. The temperature of reactants
4. The physical state of reactants
5. The presence of a catalyst

A catalyst is a substance that increases the rate of a reaction. If added to a reaction mixture, the catalytic substance undergoes no net change, nor does it alter the outcome of the reaction.

Mathematical Representation of Reaction Rate

Learning Goal 7:	Write rate equations for elementary processes.

For a generalized reaction:

$$A + B \rightarrow C + D$$

The rate of the reaction is represented by:

$$\text{rate} = k[A]^n[B]^{n'}$$

In this expression, the brackets, [ ], represent molar concentration and

> [A] = molar concentration of species A
> [B] = molar concentration of species B

The exponents, n and n', are experimentally determined. They cannot be deduced from the chemical equation.

The symbol k represents a constant that is unique to the reaction under study. It is termed the **rate constant**.

Knowing the form of the rate expression enables chemists to design chemical reactions that will produce product in the shortest time. In the chemical industry, where the quantity of product produced and sold determines the amount of profit, a thorough understanding of the kinetics of reactions is of paramount importance.

7.4 Equilibrium

Learning Goal 8:	Recognize and describe equilibrium situations.

Many chemical reactions do not proceed to "completion." After a period of time, determined by the kinetics of the reaction, the concentration of reactants no longer decreases and the concentration of products no longer increases. At this point, a mixture of products and reactants exists, and its composition would remain constant unless the experimental conditions were changed. This mixture is in a state of chemical equilibrium.

Rate and Reversibility of Reactions

Many reactions may proceed in either direction, left to right or right to left. The concentration of the various species is fixed at equilibrium because product is being consumed and formed at the same rate. In other words, the reaction continues indefinitely (dynamic) but the concentration of products and reactants is fixed (equilibrium). This is a dynamic equilibrium.

Physical Equilibrium

The most common examples of physical equilibrium involve interconversion between the various states of matter. In the environment, there may be an equilibrium between the water in a lake and water in the surrounding atmosphere (as humidity).

$$H_2O(l) \rightleftharpoons H_2O(g)$$

When the temperature is 32°F (0°C), we may observe both ice and liquid water on the road (a very dangerous situation!).

$$H_2O(s) \rightleftharpoons H_2O(l)$$

Chemical Equilibrium

Chemical change, as well as physical change, can attain equilibrium. For example:

$$H_2(g) + I_2(g) \rightleftharpoons 2HI(g)$$

At equilibrium, the rate of *disappearance* of H_2 and I_2 is equal to the rate of formation of H_2 and I_2. The rates of the forward and reverse reactions are equal, and the concentrations of H_2, I_2, and HI are fixed. The process is dynamic, with a continuous interconversion of products and reactants. We represent this process as:

$$K_{eq} = \frac{[HI]^2}{[H_2][I_2]}$$

The Generalized Equilibrium Constant Expression for a Chemical Reaction

Learning Goal 9: **Write equilibrium-constant expressions and use these expressions to calculate equilibrium constants.**

The most precise description of an equilibrium process is achieved through the use of the equilibrium constant, K_{eq}. We write the general form of an equilibrium chemical reaction as

$$aA + bB \rightleftharpoons cC + dD$$

where A and B represent reactants, C and D represent products, and a, b, c, and d are the coefficients of the balanced equation. The equilibrium constant expression for this general case is

$$K_{eq} = \frac{[C]^c[D]^d}{[A]^a[B]^b}$$

Note:
1. Products of the equilibrium reaction are in the numerator, and reactants are in the denominator.
2. Brackets [ ] represent molar concentration, M (moles/liter).
3. The exponents correspond to the coefficients of the balanced equation.
4. Solids and liquids do not appear in the equilibrium-constant expression.

It does not matter what initial amounts (concentrations) of reactions or products we choose. When the system reaches equilibrium, the calculated value of K_{eq} will not change. The magnitude of K_{eq} can be altered only by changing the temperature; thus, K_{eq} is temperature dependent.

Example 3

Write the equilibrium constant (K_{eq}) for each of the following reactions.

1. $N_2 + 3 H_2 \rightleftharpoons 2 NH_3$
2. $N_2 + O_2 \rightleftharpoons 2 NO$
3. $H_2 + Cl_2 \rightleftharpoons 2 HCl$

Work

The equilibrium constant (K_{eq}) for any equilibrium reaction is equal to the concentration of each product raised to a power equal to its coefficient in the balanced equation divided by all the reactants raised to their coefficients' powers.

Answer

1. $K_{eq} = \dfrac{[NH_3]^2}{[N_2][H_2]^3}$

2. $K_{eq} = \dfrac{[NO]^2}{[N_2][O_2]^3}$

3. $K_{eq} = \dfrac{[HCl]^2}{[H_2][Cl_2]}$

LeChatelier's Principle

Learning Goal 10:   Use LeChatelier's principle to predict changes in equilibrium position.

LeChatelier's principle states that if a stress is placed on an equilibrium system, the system will respond by altering the equilibrium in such a way as to minimize the stress.

Product introduced: equilibrium

$\longleftarrow$

shifted

Reactant introduced: equilibrium

$\longrightarrow$

shifted

Effect of Concentration

Addition of extra product or reactant to a fixed reaction volume is just another way of saying that we have increased the concentration of product or reactant. Removal of material from a fixed volume decreases the concentration. Therefore, changing the concentration of one or more components of a reaction mixture is a way to alter the equilibrium composition of an equilibrium mixture.

Effect of Heat

The change in equilibrium composition caused by the addition or removal of heat from an equilibrium mixture can be explained by treating heat as a product or reactant. Adding heat to an exothermic reaction is similar to increasing the amount of product, shifting the equilibrium to the left. Removing heat from an exothermic reaction shifts the equilibrium to the right.

Heat is a reactant in an endothermic reaction; its removal shifts the equilibrium to the left. Adding heat favors product formation.

Effect of Pressure

Only gases are affected significantly by changes in pressure, because gases are free to expand and compress in accordance with Boyle's law. However, liquids and solids are not compressible, so their volumes are unaffected by pressure.

Therefore, pressure changes will alter equilibrium composition only when they involve a gas or variety of gases as products and/or reactants. If the reaction shifts to conserve volume:

- If the number of moles of gaseous product is <u>greater</u> than the number of moles of gaseous reactant, an increase in pressure shifts the equilibrium to the left.
- If the number of moles of gaseous product is <u>less</u> than the number of moles of gaseous reactant, an increase in pressure shifts the equilibrium to the right.
- If the number of moles of gaseous product and reactant is identical, pressure will have no effect on the equilibrium because there is no volume advantage.

Effect of a Catalyst

A catalyst has no effect on the equilibrium composition. A catalyst increases the rates of both forward and reverse reactions to the same extent. The equilibrium composition and equilibrium concentration do not change when a catalyst is used.

Glossary of Key Terms in Chapter Seven

activated complex (7.3) the arrangement of atoms at the top of the potential energy curve as a reaction proceeds.

activation energy (7.3) the threshold energy that must be overcome to produce a chemical reaction.

calorimetry (7.2) the measurement of changes in heat energy during a chemical reaction.

catalyst (7.3) a substance that speeds up a reaction without undergoing change.

dynamic equilibrium (7.4) the state that exits when the rate of change in the concentration of products and reactants is equal, resulting in no net concentration change.

endothermic reaction (7.1) a chemical or physical change in which energy is absorbed.

enthalpy (7.1) a term that represents heat energy.

entropy (7.1) a term that represents a level of randomness or disorder.

equilibrium constant (7.4) a numerical quantity that summarizes the relationship between the concentration of reactants and products in an equilibrium reaction.

equilibrium reaction (7.4) a reaction that is reversible and the rates of the forward and reverse reactions are equal.

exothermic reaction (7.1) a chemical or physical change that releases energy.

free energy (7.1) the combined contribution of entropy and enthalpy for a chemical reaction.

fuel value (7.2) the amount of energy derived from a given mass of material.

kinetics (7.3) the study of rates of chemical reactions.

LeChatelier's principle (7.4) states that when a system at equilibrium is disturbed, the equilibrium shifts in the direction that minimizes the disturbance.

nutritional Calorie (7.2) equivalent to one kilocalorie (1000 calories); also know as a large Calorie.

order of the reaction (7.3) the exponent of each concentration term in the rate equation.

rate constant (7.3) the proportionality constant that relates the rate of a reaction and the concentration of reactants.

rate equation (7.3) expresses the rate of a reaction in terms of reactant concentration and a rate constant.

rate of chemical reaction (7.3) the change in concentration of a reactant or product per unit time.

reversible reaction (7.4) a reaction that will proceed in either direction, reactants to products or products to reactants.

specific heat (7.2) the quantity of heat (calories) required to raise the temperature of 1 g of a substance one degree Celsius.

surroundings (7.1) the universe outside of the system.

system (7.1) the process under study.

thermodynamics (7.1) the branch of science that deals with the relationship between energies of systems, work, and heat.

Self Test for Chapter Seven

1. When equal moles of HCl and NaOH were mixed in a calorimeter, the temperature of 100.0 grams of water surrounding the reaction increased from 22.5°C to 36.5°C. If the specific heat of water is 1.000 calories/(gram)(°C), calculate the quantity of heat energy in calories that was released.

2. Write the equilibrium constant for each of the following:

   a. $2\,NO_2 \rightleftharpoons N_2O_4$

   b. $PCl_5 \rightleftharpoons PCl_3 + Cl_2$

   c. $2\,NO + 2\,H_2 \rightleftharpoons N_2 + 2\,H_2O$

3. A random, or disordered, system is characterized by what kind of entropy?

4. Are endothermic reactions producing products of lower entropy usually spontaneous or nonspontaneous?

5. Which of the three states of matter has the highest entropy?

6. Are the ΔG values positive or negative for nonspontaneous reactions?

7. What device is used to measure heat changes in calories?

8. What is the amount of energy per gram of food substance?

9. What is the name of an extremely unstable intermediate state of a chemical reaction?

10. Many chemical reactions do not proceed to completion. After a period of time, the concentration of reactants no longer decreases, and the concentration of products ceases to increase. At this point, what state describes the chemical reaction?

11. What constant provides the most precise description of an equilibrium process?

12. What French chemist in the nineteenth century discovered that changes in equilibrium were dependent upon stress that might be applied to the system?

13. Are solid products or reactants a part of the equilibrium-constant expression?

14. In what direction will the addition of heat shift an exothermic reaction?

15. In what direction will the addition of heat shift an endothermic reaction?

Vocabulary Quiz for Chapter Seven

1. _____ is the study of energy, work, and heat.

2. Exothermic reactions (release, absorb) heat.

3. Endothermic reactions (release, absorb) heat.

4. _____ is a measure of the disorder of a system.

5. The amount of energy per gram of food is termed _____.

6. A _____ measures heat changes in chemical reactions.

7. _____ is the study of the rate of chemical reactions.

8. The minimum energy needed for a reaction is the _____.

9. A _____ increases the rate of a reaction.

10. _____ describes a chemical reaction's response to stress.

8 *Acids and Bases and Oxidation-Reduction*

Learning Goals

1 Identify acids and bases and acid-base reactions.
2 Describe the role of the solvent in acid-base reactions.
3 Write equations describing acid-base dissociation and label the conjugate acid-base pairs.
4 Calculate pH from concentration data.
5 Calculate hydronium and/or hydroxide ion concentration from pH data.
6 Provide examples of the importance of pH in chemical and biochemical systems.
7 Describe the meaning and utility of neutralization reactions.
8 State the meaning of the term *buffer* and describe the application of buffers to chemical and biochemical systems, particularly blood chemistry.
9 Describe *oxidation* and *reduction,* and describe some practical examples of redox processes.
10 Diagram a voltaic cell and describe its function.
11 Compare and contrast voltaic and electrolytic cells.

Introduction

A **solution** is a homogeneous (or uniform) mixture of two or more substances. A solution is composed of one or more **solutes**, dissolved in a **solvent**. When the solvent is water, we refer to the homogeneous mixture as an **aqueous solution**.

Two very important classes of solution reactions involve transfer of either positive or negative charge: acid-base reactions (proton transfer) and oxidation-reduction reactions (electron transfer). Each type is described in this chapter and is applied in numerous ways throughout the rest of your study of chemistry.

8.1 Acids and Bases

Learning Goal 1:   Identify acids and bases and acid-base reactions.

Acids have a sour taste, dissolve some metals, and cause vegetable dyes to change color. Release of the hydrogen ion (H^+) is characteristic of acids. Bases have a bitter taste, a slippery feel, and accept hydrogen ions.

Arrhenius Theory of Acids and Bases

An Arrhenius acid dissociates to form hydrogen ions. An Arrhenius base dissociates to form hydroxide ions. The **Arrhenius theory** is less comprehensive than the Brønsted-Lowry theory, discussed below.

Brønsted-Lowry Theory of Acids and Bases

A **Brønsted-Lowry** acid is a proton donor and a Brønsted-Lowry base is a proton acceptor. (Remember that H^+ is simply a proton, so a *proton donor* is giving away an H^+.) Examples include:

$$\text{acid} \quad HCl \ + \ H_2O \ \rightarrow \ H_3O^+ \ + \ Cl^-$$

$$\text{base} \quad NH_3 \ + \ H_2O \ \rightleftharpoons \ NH_4^+ \ + \ OH^-$$

(Note: The double arrow ($\leftrightarrow$) implies that ammonia is a *weak* base. See the following discussion.)

Acid-Base Properties of Water

Learning Goal 2:	Describe the role of the solvent in acid-base reactions, and explain the meaning of the term *pH*.

Water is **amphiprotic**, meaning that it possesses *both* acid and base properties. The fact that water can either accept or donate protons (H^+ ions) makes it an excellent solvent for both acids and bases.

In this reaction, water acts a base, and accepts a proton from hydrochloric acid:

$$HCl \ + \ H_2O \rightarrow \ H_3O^+ \ + \ Cl^-$$

In this reaction, water acts as an acid, and donates a proton to ammonia:

$$NH_3 \ + \ H_2O \ \rightarrow \ NH_4^+ \ + \ OH$$

Acid and Base Strength

The terms acid or base *strength* and acid or base *concentration* are easily confused. Strength is a measure of the degree of dissociation of an acid or base in solution, independent of its concentration. Concentration refers to the amount of acid or base per quantity of solution.

The strength of acids and bases in water is dependent upon the extent to which they react with the solvent, water. Acids and bases are classified as *strong* when the reaction with water is virtually 100% complete and as *weak* when much less than 100% complete. All strong bases are metal hydroxides. Strong bases completely dissociate, or ionize, in aqueous solution to produce hydroxide ions and metal cations. For example, NaOH dissociates to form Na^+ and OH^- ions in water. Weak acids and weak bases dissolve in water principally in the molecular form. Only a small percentage of the molecules dissociate to form the **hydronium** or **hydroxide** ion. For example, the weak base ammonia, NH_3, dissolves in water primarily as NH_3, but a small percentage reacts with water to form NH_4^+ and OH^-.

The most fundamental chemical difference between strong and weak acids and bases is their equilibrium situation. A strong acid, such as HCl, does not exist in equilibrium with its ions, H_3O^+ and Cl^-. A weak acid, such as acetic acid, establishes a dynamic equilibrium with its ions, H_3O^+ and $C_2H_3O_2^-$. (Note that the acetate ion may be written as CH_3COO^- or as the more condensed form $C_2H_3O_2^-$.) This reaction is represented by the equilibrium constant expression:

$$CH_3COOH + H_2O \; \rightleftharpoons \; H_3O^+ + CH_3COO^-$$

$$K_a \; = \; \frac{[H_3O^+][CH_3COO^-]}{[CH_3COOH]}$$

where K_a is the equilibrium constant or acid-dissociation constant for acetic acid. The larger the dissociation constant, the more dissociated (hence, stronger) the acid will be.

The situation for bases is analogous. The equilibrium expression for the weak base ammonia in water is

$$NH_3 + H_2O \; \rightleftharpoons \; NH_4^+ + OH^-$$

$$K_b \; = \; \frac{[NH_4^+][OH^-]}{[NH_3]}$$

Note that in each of the above cases, the solvent, water, is not written as part of the equilibrium constant expression.

Conjugate Acids and Bases

Learning Goal 3: **Write equations describing acid-base dissociation and label the conjugate acid-base pairs.**

Any acid-base reaction can be represented by the general equation

$$HA \; + \; B \; \rightleftharpoons \; BH^+ \; + \; A^-$$

where HA represents an acid and B represents a base. In the forward reaction, the acid donates a proton to the base. The products of this reaction also have acid-base properties. The protonated base can, in fact, donate a proton to the anion, A^-, reforming HA and B. The product acids and bases are termed *conjugate acids and bases*.

A **conjugate acid** is the species formed when a base accepts a proton.

A **conjugate base** is the species formed when an acid donates a proton.

The acid and base on the opposite sides of the equation are collectively termed a **conjugate acid-base pair**. For our general expression, B and BH^+ are a conjugate acid-base pair. Similarly, HA and A^- are a conjugate acid-base pair.

The Dissociation of Water

Although pure water is virtually 100% molecular, a small number of water molecules do ionize. This process occurs by the transfer of a proton from one water molecule to another, producing a hydronium and hydroxide ion. This equilibrium is shown below:

$$H_2O + H_2O \rightleftharpoons H_3O^+ + OH^-$$

This process is the **autoionization**, or self-ionization of water. Water is, therefore, a very weak electrolyte and a poor conductor of electricity. Water has both acid and base properties; the dissociation produces both the hydronium and hydroxide ion.

Water at room temperature has a hydronium ion concentration of 1.0×10^7 M. The hydroxide ion concentration is also 1.0×10^{-7} M.

The product of hydronium and hydroxide ion concentration is the **ion product for water**,

$$\text{ion product} = [H_3O^+][OH^-] = [1.0 \times 10^{-7}][1.0 \times 10^{-7}] = 1.0 \times 10^{-14} = K_w$$

The ion product is a constant because its value is not dependent upon the nature or concentration of the solute, as long as the temperature does not change. This relationship is the basis for the pH scale.

Example 1

If we have a 0.200 molar solution of sodium hydroxide, what are the hydroxide ion (OH^-) and hydronium ion (H_3O^+) concentrations in this solution?

Work
1. If we have a 0.200 molar NaOH solution, it means that we have 0.200 mole of Na^+ ions and 0.200 mole of OH^- ions per liter of this solution.
2. Thus, $[OH^-] = 0.200$ moles/liter
3. Next use the ion product for water (K_w): ion product $= [H_3O^+][OH^-] = 1.0 \times 10^{-14}$

The above equation will be used whenever we know the value of H_3O^+ or OH^- and want to find the other one. The nature and concentration of the solutes added to water do alter the concentrations of H_3O^+ and OH^- present, but the product $[H_3O^+][OH^-]$ always equals 1.0×10^{-14}.
4. Rearrange ion product $= [H_3O^+][OH^-] = 1.0 \times 10^{-14}$ to find $[H_3O^+]$

$$[H_3O^+] = \frac{\text{ion product}}{[OH^-]}$$

Example 1 continued

5. $[H_3O^+]$ = $\dfrac{\text{ion product}}{[OH^-]}$

       $[H_3O^+]$ = $\dfrac{1.0 \times 10^{-14}}{2.00 \times 10^{-1}}$

       $[H_3O^+]$ = 0.50×10^{-13} = 5.0×10^{-14}

Answer

In our solution, the $[OH^-]$ = 0.200 moles/L and the $[H_3O^+]$ = 5 x 10^{-14} moles/L.

8.2 pH: A Measurement Scale for Acids and Bases

Learning Goal 4: Calculate pH from concentration data.

Learning Goal 5: Calculate hydronium and/or hydroxide ion concentration from pH data.

A Definition of pH

The **pH scale** correlates the hydronium ion concentration with a number, the pH, which serves as a useful indicator of the degree of acidity or basicity of a solution. The pH scale specifies "how acidic" or "how basic" a solution is.

1. Addition of an acid (proton donor) to water *increases* $[H_3O^+]$ and decreases $[OH^-]$.

2. Addition of a base (proton acceptor) to water *decreases* $[H_3O^+]$ and increases $[OH^-]$.

3. $[H_3O^+] = [OH^-]$ when *equal* amounts of acid and base are present.

4. In all of the above cases, $[H_3O^+][OH^-] = 1.0 \times 10^{-14}$ = ion product for water at 25°C.

The pH of a solution is defined as the negative logarithm of the molar concentration of the hydronium ion.

$$pH = -\log [H_3O^+]$$

Measuring pH

The pH of a solution may be measured by a pH meter. Also, indicating paper (pH paper) can give an approximate value for the solution pH.

Calculating pH

The pH of a solution can be calculated if the concentration of either H_3O^+ or OH^- is known. Also, $[H_3O^+]$ or $[OH^-]$ can be calculated from the pH.

Example 2
Calculate the pH of a 0.010 molar HCl solution.

Work
1. Since HCl is a strong acid, if 1 mole of HCl dissociates, it produces 1 mole of H_3O^+ ions.

   Therefore a 1.0×10^{-2} M HCl solution has $[H_3O^+] = 1.0 \times 10^{-2}$ M.

2. Use the pH equation:

   pH = −logarithm of the molar concentration of the hydronium ion

   or

   $pH = -\log [H_3O^+]$ (Note: [ ] means molar.)

3. $pH = -\log [H_3O^+]$

   $pH = -\log (1.0 \times 10^{-2})$

   $pH = -(\log 1.0 + \log 10^{-2})$ (Note that log 1.0 = zero; $\log 10^{-2} = -2$)

   $pH = -[0 + (-2)]$

   $pH = -(-2)$

   $pH = 2.0$

Answer
The pH of a 0.010 molar HCl solution is pH = 2.0

Example 3
Calculate the pH of a 1.0×10^{-4} molar solution of NaOH.

Work
1.   NaOH is a strong base; thus, in our problem $[OH^-] = 1.0 \times 10^{-4}$ molar.

2.   Use ion product $= [H_3O^+][OH^-]$ to find $[H_3O^+]$, so we can use the pH equation.

3. $[H_3O^+] = \dfrac{\text{ion product}}{[OH^-]}$

   $[H_3O^+] = \dfrac{1.0 \times 10^{-14}}{1.0 \times 10^{-4}}$ $[H_3O^+] = 1.0 \times 10^{-10}$

4.   Next, use the pH equation:

   $pH = -\log[H_3O^+]$

   $pH = -\log (1 \times 10^{-10})$

   $pH = -(\log 1.0 + \log 10^{-10})$

   $pH = -[0 + (-10)]$

   $pH = 10.0$

Answer
The pH of a 1.0×10^{-4} M NaOH solution is pH = 10.0

The Importance of pH and pH Control

| Learning Goal 6: | Provide examples of the importance of pH in chemical and biochemical systems. |

Solution pH and pH control play a major role in many facets of our everyday lives:

1. Agriculture (soil pH)
2. Physiology (blood pH)
3. Industry
4. Municipal services (purification of drinking water; sewage treatment)
5. Acid rain

8.3 Reactions Between Acids and Bases

Neutralization

| Learning Goal 7: | Describe the meaning and utility of neutralization reactions. |

The reaction of an acid with a base to produce a salt and water is a **neutralization reaction**. In the strictest sense, neutralization requires equal numbers of moles of H_3O^+ and OH^- to produce a neutral solution (no excess acid or base).

A neutralization reaction may be used to determine the concentration of an unknown acid or base solution. The technique of **titration** involves the addition of measured amounts of a standard solution (one whose concentration is known) to neutralize the second, unknown solution. From the volumes of the two solutions and the concentration of the **standard solution**, the concentration of the unknown solution may be determined.

Example 4

A 5.00 g sample of vinegar is titrated with 0.100 M NaOH. If the vinegar requires 40.0 mL of the NaOH for a complete reaction,

    a. Calculate the M of acetic acid in the vinegar
    b. Calculate the % (W / V) of acetic acid in the vinegar.

Assume that the density of the vinegar solution is 1.00 g/mL, and that the only acidic component of the vinegar solution is acetic acid (CH_3COOH). The reaction is:

$$CH_3COOH(aq) + NaOH(aq) \rightarrow CH_3COONa(aq) + H_2O(l)$$

Work, Part A

1. Since the coefficients of both acetic acid and sodium hydroxide (the reactants) in the equation are 1, moles acid = moles base at the equivalence point.

Example 4 continued

2. moles acid $= M_{acid} \times V_{liters\ acid}$

   moles base $= M_{base} \times V_{liters\ base}$

3. Therefore, $M_{acid} \times V_{liters\ acid} = M_{base} \times V_{liters\ base}$
   and,

   $$M_{acid} = M_{base} \times \frac{(V_{liters\ base})}{(V_{liters\ acid})}$$

4. $M_{base} = 0.100M$

   $$V_{liters\ base} = 40.0\ mL \times \frac{1\ L}{10^3\ mL} = 0.0400\ L$$

   $$V_{liters\ acid} = 5.00\ g \times \frac{1\ mL}{1.00\ g} = 5.00\ mL$$

   $$< \text{———— Density conversion}$$

   and

   $$V_{liters\ acid} = 5.00\ mL \times \frac{1\ L}{10^3\ mL} = 0.00500\ L$$

5. Substituting,

   $$M_{acid} = M_{base} \times \frac{(V_{liters\ base})}{(V_{liters\ acid})}$$

   $$M_{acid} = 0.100\ M \times \frac{0.0400\ L}{0.00500\ L}$$

   $$M_{acid} = 0.800\ M\ \text{acetic acid}$$

Work, Part B

1. $\%\ (W/V) = \dfrac{\text{grams acetic acid}}{\text{milliliters of solution}} \times 100\%$

2. grams acetic acid $=$
   $$\frac{0.800\ mL\ \text{acetic acid}}{1\ L} \times \frac{60.0\ g\ \text{acetic acid}}{1\ mol\ \text{acetic acid}} = \frac{48.0\ g\ \text{acetic acid}}{1\ L}$$

Example 4 continued

3. Convert L to mL: $\dfrac{48.0 \text{ g acetic acid}}{1 \text{ L} \times \dfrac{10^3 \text{ mL}}{1 \text{ L}}} = \dfrac{0.0480 \text{ g acetic acid}}{1 \text{ mL}}$

4. Substitute into the initial equation for % (W / V)

$$\% \text{ (W/V)} = \dfrac{0.0480 \text{ g}}{1 \text{ mL}} \times 100\% = 4.80\%$$

Answer
a. The vinegar solution is 0.800 M acetic acid
b. The vinegar solution is 4.80% (W / V)

Polyprotic Substances

Polyprotic substances can donate or accept more that one proton per formula unit. Sulfuric acid, H_2SO_4, is a diprotic acid, and phosphoric acid, H_3PO_4, is a triprotic acid, and barium hydroxide, $Ba(OH)_2$, is a diprotic base. Acid-base reactions may occur in various combining ratios if they involve a polyprotic substance.

One mole of a diprotic acid, such as H_2SO_4 , can react with two moles of NaOH.

$$H_2SO_4 + 2\,NaOH \rightarrow Na_2SO_4 + 2\,H_2O$$

Dissociation of a diprotic acid takes place in two steps.

$$H_2SO_4 + H_2O \rightarrow HSO_4^- + H_3O^+$$

$$HSO_4^- + H_2O \rightleftharpoons HSO_4^{2-} + H_3O^+$$

8.4 Acid-Base Buffers

Learning Goal 8: State the meaning of the term *buffer* and describe the application of buffers to chemical and biochemical systems, particularly blood chemistry.

A buffer solution contains components that enable the solution to resist large changes in pH when acids or bases are added.

The Buffer Process

The basis of buffer action is the establishment of an equilibrium between either a weak acid and its conjugate base, or a weak base and its conjugate acid.

A buffer solution functions in accordance with LeChatelier's Principle (Section 7.4), which states that an equilibrium system, when stressed, will shift its equilibrium to alleviate that stress.

Addition of Base (OH⁻) to a Buffer Solution

For the acetic acid/sodium acetate system:

$$CH_3COOH + H_2O \rightleftharpoons H_3O^+ + CH_3COO^-$$

OH⁻ added, equilibrium shifts to the right

Addition of Acid (H_3O^+) to a Buffer Solution

For the acetic acid/sodium acetate system:

$$CH_3COOH + H_2O \rightleftharpoons H_3O^+ + CH_3COO^-$$

H_3O^+ added, equilibrium shifts to the left

Higher-than-normal CO_2 levels shift the above equilibrium to the right increasing $[H_3O^+]$ and lowering the pH. A situation of high blood CO_2 levels and low pH is **acidosis**.

Lower-than-normal CO_2 levels shift the equilibrium to the left, decreasing $[H_3O^+]$ and making the pH more basic; this condition is termed **alkalosis** (from alkali, implying basic in nature).

Preparation of a Buffer Solution

To calculate the pH of a buffer solution, use either the equilibrium constant expression or the Henderson-Hasselbalch equation (which is derived from the equilibrium constant expression, and is described below). The quantities needed are the equilibrium constant and the buffer solution concentrations of the weak acid and its conjugate base (or, the weak base and its conjugate acid). The equilibrium constant expression is solved for $[H_3O^+]$, which is then used to calculate the pH.

A Practical Note: Consider a buffer made from acetic acid, $HC_2H_3O_2$, and its conjugate base, the acetate anion, $C_2H_3O_2^-$. It's not possible to add simply acetate anions, so the acetate is added as a salt, sodium acetate, which dissolves to form sodium cations and acetate anions. (The sodium cations won't participate in the acid-base reactions, so they can be ignored.) Since the ratio of sodium acetate to acetate anions is equal, we can use the sodium acetate concentration to represent the acetate anion concentration.

$$NaC_2H_3O_2(s) \rightarrow Na^+(aq) + C_2H_3O_2^-(aq)$$

Example 5

Calculate the pH of a buffer solution in which the acetic acid concentration is 0.150 M and the concentration of sodium acetate is 0.090 M. The equilibrium constant, K_a, for acetic acid is 1.75 x 10^{-5}.

Work

1. Write the equilibrium constant expression for acetic acid.

$$CH_3COOH + H_2O \rightleftharpoons H_3O^+ + CH_3COO^-$$

$$K_a = \frac{[H_3O^+][CH_3COO^-]}{[CH_3COOH]}$$

2. Rearrange the equilibrium constant expression to solve for $[H_3O^+]$.

$$[H_3O^+] = \frac{K_a [CH_3COOH]}{[CH_3COO^-]}$$

Substitute the values given in the problem and calculate the concentration of H_3O^+.

$$[H_3O^+] = \frac{(1.75 \text{ x } 10^{-5}) [0.150 \text{ M } CH_3COOH]}{[0.090 \text{ M } CH_3COO^-]}$$

$$[H_3O^+] = 2.9 \text{ x } 10^{-5} \text{ M}$$

3. Calculate the pH using the value of $[H_3O^+]$ determined above.

pH = $-\log[H_3O^+]$

pH = $-\log(2.9 \text{ x } 10^{-5} \text{ M})$

pH = 4.5

Answer

The pH is 4.5 for a buffer solution that is 0.150 M acetic acid and 0.090 M sodium acetate.

The Henderson-Hasselbalch Equation

The solution of the equilibrium-constant expression and the pH may be combined into one expression: the **Henderson-Hasselbalch equation**. The form of this equation is especially amenable to buffer problem calculations.

Henderson-Hasselbalch equation: $\quad pH = pK_a + \log\dfrac{[\text{conjugate base}]}{[\text{weak acid}]}$

Example 6

Solve the previous example using the Henderson-Hasselbalch equation.
Calculate the pH of a buffer solution in which the acetic acid concentration is 0.150 M and the concentration of sodium acetate is 0.090 M. (K_a for acetic acid is 1.75×10^{-5}.)

Work
1. Write the Henderson-Hasselbalch equation for acetic acid.

$$pH = pK_a + \log\frac{[\text{acetate}]}{[\text{acetic acid}]}$$

2. Calculate the pK_a for acetic acid.

$pK_a = -\log K_a$
$pK_a = -\log(1.75 \times 10^{-5}) = 4.76$

3. Use the Henderson-Hasselbalch equation. Substitute the values given in the problem and calculate the pH.

$$pH = 4.76 + \log\frac{[0.090 \text{ M acetate}]}{[0.150 \text{ acetic acid}]}$$

$pH = 4.76 + \log(0.60)$
$pH = 4.76 - 0.22 = 4.54$

Answer
The pH of this buffer solution is 4.54.

8.5 Oxidation - Reduction Processes

Oxidation and Reduction

Learning Goal 9:   **Describe oxidation and reduction, and describe some practical examples of redox processes.**

Oxidation is defined as a loss of electrons. Sodium metal, is, for example, oxidized to a sodium ion, losing one electron:

$$Na \rightarrow Na^+ + e^-$$

Reduction is defined as a gain of electrons. A chlorine atom is reduced to a chloride ion by gaining one electron:

$$Cl + e^- \rightarrow Cl^-$$

Oxidation and reduction are complementary processes. The oxidation half-reaction produces an electron that is the reactant for the reduction half-reaction. The combination of two half-reactions produces the complete reaction:

$$Na + Cl \rightarrow Na^+ + Cl^-$$

Note that the electrons cancel; in the electron-transfer process, no free electrons remain.

In the reaction described above, sodium metal is the reducing agent; it releases electrons for the reduction of chlorine. Chlorine is the oxidizing agent; it accepts electrons from the sodium, which is oxidized.

The characteristics of oxidizing and reducing agents are summarized below:

Oxidizing agent	**Reducing agent**
Is reduced	Is oxidized
Gains electrons	Loses electrons
Causes oxidation	Causes reduction

Applications of Oxidation and Reduction

Corrosion. In one type of corrosion, elemental iron is oxidized to iron(III) oxide:

$$4\ Fe(s)\ +\ 3\ O_2(g)\ \rightarrow\ 2\ Fe_2O_3(s)$$

The oxidation state of iron changes from zero to +3, and oxygen is reduced from zero to −2. Other common examples of oxidation-reduction include **combustion** and **bleaching**.

Biological Processes. There are many examples of biological oxidation-reduction reactions. For example, the electron-transport chain of **respiration** (Chapter 17) involves the reversible oxidation and reduction of iron atoms in cytochrome c:

$$\text{cytochrome c } (Fe^{3+}) + e^- \rightarrow \text{cytochrome c } (Fe^{2+})$$

In many metabolic redox reactions, neither the product nor the reactant carries a charge. To identify whether a particular reactant has been oxidized or reduced, alternative descriptions of these terms are useful:

Oxidation is the *gain* of oxygen or the *loss* of hydrogen.
Reduction is the *loss* of oxygen or the *gain* of hydrogen.

This description makes it much easier to recognize the conversion of ethanol to acetaldehyde as an oxidation, for example.

$$CH_3CH_2OH \longrightarrow \underset{\underset{H}{\displaystyle H_3C}}{\overset{\overset{\displaystyle O}{\parallel}}{C}}$$

Ethanol   Acetaldehyde

(Note that this reaction is not complete, and that only the oxidation half-reaction is shown. Whenever one species is oxidized, something else must be reduced. Metabolic reactions will be discussed in much more detail later in this course.)

Voltaic Cells

Learning Goal 10:   Diagram a voltaic cell and describe its function.

A **voltaic cell** is an electrochemical cell that converts chemical energy into electrical energy. This cell consists of an **anode**, the electrode at which oxidation occurs, and a **cathode**, the site of reduction. Electrons released at the anode travel through an external circuit, where they may do useful work, to the cathode.

The most common application of voltaic cells is in the construction of batteries, some of which are small enough and sufficiently inert to allow implantation in the human body.

Electrolysis

Learning Goal 11:   Compare and contrast voltaic and electrolytic cells.

Electrolysis reactions are the reverse of a voltaic cell; they use electrical energy to cause nonspontaneous redox reactions to occur. The best-known examples of electrolysis include the rechargeable battery and the plating process (silver plating and chrome [chromium] plating).

Glossary of Key Terms in Chapter Eight

acid (8.1) a substance that behaves as a proton donor.
amphiprotic (8.1) a substance that can behave either as a Brønsted acid or a Brønsted base.
anode (8.5) the positively charged electrode in an electrical cell.
Arrhenius theory (8.1) describes an acid as a substance that dissociates to produce H+ and a base as a substance that dissociates to produce OH⁻.
autoionization (8.1) or self-ionization, the reaction of a substance, such as water, with itself to produce a positive and a negative ion.
base (8.1) a substance that behaves as a proton acceptor.
Brønsted-Lowry theory (8.1) describes an acid as a proton donor and a base as a proton acceptor.
buffer capacity (8.4) a measure of the ability of a solution to resist large changes in pH when a strong acid or strong base is added.
buffer solution (8.4) a solution containing a weak acid or base and its salt; it is resistant to large changes in pH upon addition of strong acid or bases.
buret (8.3) a device calibrated to deliver accurately know volumes of liquid, as in a titration.
cathode (8.5) a negatively charged electrode in an electrical cell.
conjugate acid (8.1) substance that has one more proton than the base from which it is derived.

conjugate acid-base pair (8.1) two species related to each other through the gain or loss of a proton.
conjugate base (8.1) substance that has one proton fewer than the acid from which it is derived.
corrosion (8.5) the unwanted oxidation of a metal.
electrolysis (8.5) an electrochemical process that uses electrical energy to cause nonspontaneous oxidation-reduction reactions to occur.
equivalence point (8.3) describes the situation in which reactants have been mixed in the molar ratio corresponding to the balanced equation.
Henderson-Hasselbalch equation (8.4) an equation derived from the equilibrium constant expression that includes pH and is especially amenable to buffer calculations. The Henderson-Hasselbalch equation is:

$$\text{pH} = \text{p}K_a + \log\frac{[\text{conjugate base}]}{[\text{weak acid}]}$$

hydronium ion (8.1) a protonated water molecule.
indicator (8.3) a solute that shows some condition of a solution (such as acidity or basicity) by its color.
ion product for water (8.1) the product of the hydronium and hydroxide ion concentrations in pure water at a specified temperature, 25°C; it has a value of 1.0×10^{-14}.
neutralization (8.3) the reaction between an acid and a base.
oxidation (8.5) the loss of electrons.
oxidizing agent (8.5) a substance that oxidizes, or removes electrons from, another substance.
pH scale (8.2) a numerical representation of acidity or basicity of a solution (pH = -log [H+]).
polyprotic substance (8.3) a substance that can donate or accept more than one proton per molecule.
reducing agent (8.5) a substance that reduces, or donates electrons to, another substance; the reducing agent is itself oxidized in the process.
reduction (8.5) the gain of electrons.
standard solution (8.3) a solution whose concentration is accurately known.
titration (8.3) the process of adding a solution from a buret to a sample until a reaction is completed, at which time the volume is accurately measured and the concentration of the sample is calculated.
voltaic cell (8.5) an electrochemical cell that converts chemical energy into electrical energy.

Self Test for Chapter Eight

1. Calculate the pH of a solution that contains 6.50×10^{-5} moles of hydronium ion per liter.

2. If 14.8 mL of 0.100 M NaOH solution are needed to completely react with 25.0 mL of an unknown HCl solution, what is the molarity of the unknown HCl solution?

3. If we have a 0.200 molar solution of NaOH, what is the molarity of H_3O^+ ions present in the solution?

4. Calculate the pH of a 1.0×10^{-3} molar solution of HCl.

5. What is the term for the reaction of an acid plus a base to yield a salt plus water?

6. Which theory defines an acid as a proton donor and a base as a proton acceptor?

7. What is the measure of the strength of an acid in water?

8. Which one or ones of the following are not strong acids?

   a. HCl

   b. CH_3COOH

   c. H_2SO_4

   d. HNO_3

9. Which one ore ones of the following are not strong bases?

   a. NaOH

   b. NH_4OH

   c. $Ca(OH)_2$

   d. KOH

Vocabulary Quiz for Chapter Eight

1. The _____ theory describes the behavior of bases such as ammonia.

2. _____ is the reaction of a substance, such as water, with itself to produce a positive and negative ion.

3. _____ is a solution that contains an acid-base pair; that acid-base pair is resistant to large changes in pH upon addition of strong acids or bases.

4. The _____ theory describes an acid as a proton donor and a base as a proton acceptor.

5. A _____ substance can donate or accept more than one proton per molecule.

6. _____ is a loss of electrons.

7. _____ is a substance that oxidizes, or removes electrons from, another substance.

8. _____ is a numerical representation of acidity or basicity of a solution.

9. A(n) _____ solution is a solution whose concentration is accurately known.

10. _____ is the process of adding from a buret a solution to a sample until a reaction is completed, at which time the volume is accurately measured.

9 The Nucleus, Radioactivity, and Nuclear Medicine

Learning Goals

1 Enumerate the characteristics of alpha, beta, and gamma radiation.
2 Write balanced equations for common nuclear processes.
3 Calculate the amount of radioactive substance remaining after a specified number of half-lives.
4 Describe the various ways in which nuclear energy may be used to generate electricity: fission, fusion, and the breeder reactor.
5 Explain the process of radiocarbon dating.
6 Cite several examples of the use of radioactive isotopes in medicine.
7 Describe the use of ionizing radiation in cancer therapy.
8 Discuss the preparation of radioisotopes for use in diagnostic imaging studies.
9 Explain the difference between natural and artificial radioactivity.
10 Describe the characteristics of radioactive materials that relate to radiation exposure and safety.
11 Be familiar with common techniques for the detection of radioactivity.
12 Know the common units in which radiation intensity is represented: the curie, roentgen, rad, and rem.

Introduction

This chapter considers the nucleus, nuclear properties, and their applications.

9.1 Natural Radioactivity

Radioactivity is the process by which atoms emit high-energy particles or rays. These particles or rays are termed radiation. Nuclear radiation occurs as a result of an alteration in nuclear composition or structure, which happens because the nucleus is unstable and hence radioactive.

Nuclear symbols are analogous to atomic symbols. The nuclear symbols consist of the elemental symbol, the atomic number (equivalent to the number of protons in the nucleus), and the mass number, which is defined as the sum of the neutrons and protons in the nucleus. Note: Be careful not to confuse the **mass number** (neutrons and protons) with the atomic mass, which includes the contribution of electrons, and is a true mass figure.

Only unstable nuclei undergo change and produce radioactivity. Not all atoms of a particular element undergo radioactive decay. When writing the symbols for a nuclear process, it is important to designate the particular isotope involved.

Three types of natural radiation emitted by unstable nuclei are alpha particles, beta particles, and gamma rays.

Learning Goal 1:   Enumerate the characteristics of alpha, beta, and gamma radiation.

Alpha Particles

Alpha particles (α) contain two protons and two neutrons. An alpha particle is identical to the helium ion (He^{2+}) and is represented as:

$$^4_2 He$$

or

$$\alpha$$

Alpha particles emitted by radioisotopes move relatively slowly (approximately 10% of the speed of light), and they are stopped by very little mass.

Beta Particles

The beta particle (β) is a fast-moving electron traveling at approximately 90% of the speed of light as it leaves the nucleus. The beta particle is represented as

$$^0_{-1} e$$

or

$$\beta$$

the subscript -1 is written in the same position as the atomic number and, like the atomic number (number of protons), indicates the charge of the particle.

Beta particles are smaller, faster, and more energetic and penetrating than alpha particles.

Gamma Rays

Gamma rays (γ) are pure energy. Since energy has no mass or charge, the symbol for a gamma ray is simply:

$$\gamma$$

Gamma radiation is highly energetic and is the most penetrating form of nuclear radiation.

Example 1

List the names, symbols, and characteristics of the three basic types of radiation.

Example 1 continued

Answer

1. The alpha particles have the lowest velocity of the three types of radiation. They have a mass of about 4.0 amu and are composed of two protons and two neutrons. Notice that they have the same structure as the nucleus of a helium atom. Since they have two protons present, they have a charge of +2. The alpha particle may be symbolized as ^4_2He.

2. The beta particle is just a fast-moving electron that has been produced during a nuclear reaction. It has a speed of about 90% the speed of light. Since it is only an electron, it has a charge of −1 and may be symbolized as $^0_{-1}\text{e}$.

3. The gamma rays are pure energy, in contrast to alpha and beta radiation, which are particles. Gamma radiation is highly energetic and is the most penetrating and deadly to living things. It is represented by the gamma symbol, γ.

Properties of Alpha, Beta, and Gamma Radiation

Important properties of α, β, and γ radiation are summarized in Table 9.1 of the textbook.

9.2 Writing a Balanced Nuclear Equation

Learning Goal 2:   Write balanced equations for common nuclear processes.

A **nuclear equation** represents a nuclear process such as radioactive decay. To write a balanced equation, we must remember the following:

1. The total mass on each side of the reaction arrow must be identical.

2. The sum of the charges of the reactant nuclei must be equal to the sum of the charges of the product nuclei.

Alpha Decay

A reaction involving alpha decay is shown below:

$$^{226}_{88}\text{Ra} \quad \rightarrow \quad ^{222}_{86}\text{Rn} \quad + \quad ^4_2\text{He}$$
Radium-226 Radon-222 Helium-4

Beta Decay

Beta decay may be illustrated by the following:

$$^{234}_{90}\text{Th} \quad \rightarrow \quad ^{234}_{91}\text{Pa} \quad + \quad ^0_{-1}\text{e}$$
Thorium-234 Protactinium-234 β particle

Positron Emission

An example of positron emission is the conversion of a proton to a neutron:

$$\ce{^1_1H} \rightarrow \ce{^1_0}n + \ce{^0_1}e$$

proton   neutron   positiron

Gamma Production

If gamma radiation were the only product of nuclear decay, there would be no change in the mass or identity of the radioactive nuclei, since a gamma ray is pure energy, possessing no mass or charge. The gamma emitter has simply gone to a lower energy state. The decay of the **metastable isotope** technetium-99m is shown:

$$\ce{^{99m}_{43}Tc} \rightarrow \ce{^{99}_{43}Tc} + \gamma$$

Predicting Products of Nuclear Decay

It is possible to use a nuclear equation to predict one of the products of a nuclear reaction if the others are known. We know that the mass and charge of the total of all products and reactants must be equal. By difference, we can compute the missing charge and mass that represent the unknown product and deduce its identity.

Example 2

Complete the following nuclear equations:

1. $\ce{^{238}_{92}U} \rightarrow \ce{^{234}_{90}Th}$ + ?

2. $\ce{^{16}_{7}N} \rightarrow \ce{^{16}_{8}O}$ + ?

3. $\ce{^{40}_{19}K} \rightarrow \ce{^{0}_{-1}e}$ + ?

4. $\ce{^{197}_{79}Au}$ + ? $\rightarrow \ce{^{198}_{79}Au}$

Work

Use the following rules:
1.   The total mass on each side of the equation arrow must be identical.
2.   The sum of the atomic numbers on each side of the reaction arrow must be identical.

Answer

The missing parts are given below:

1. $\ce{^4_2He}$ 2. $\ce{^0_{-1}e}$ 3. $\ce{^{40}_{20}Ca}$ 4. $\ce{^1_0}n$

Example 3

Show how molybdenum-98 can be converted by neutron bombardment into technetium-99 plus beta particles. The technetium produced is unstable and can be used as a gamma source for tracer applications.

Work

1. From the periodic table, we find that molybdenum has an atomic number of 42, and technetium's is 43.
2. Next we set up the equation:

$$^{98}_{42}\text{Mo} + ^{1}_{0}\text{n} \rightarrow ^{99m}_{43}\text{Tc} + ^{0}_{-1}\text{e}$$

Notice that the "m" following the mass number means that the isotope is metastable.

Answer

$$^{98}_{42}\text{Mo} + ^{1}_{0}\text{n} \rightarrow ^{99m}_{43}\text{Tc} + ^{0}_{-1}\text{e}$$

9.3 Properties of Radioisotopes

Nuclear Structure and Stability

The energy that holds the protons, neutrons, and other particles together in the nucleus is the **binding energy** of the nucleus. When an isotope decays, some of this binding energy is released.

Factors related to nuclear stability include the following:

1. Nuclear stability correlates with the ratio of neutrons to protons in the isotope.
2. Nuclei with large numbers of protons (84 or more) tend to be unstable.
3. Isotopes containing 2, 8, 20, 50, 82, or 126 protons or neutrons are stable. These "**magic numbers**" seem to indicate the presence of stable energy levels in the nucleus.
4. Isotopes with even numbers of protons or neutrons are generally more stable than those with odd numbers of protons or neutrons.

Example 4

Explain why some isotopes are more stable than others.

Answer

Nuclear stability correlates with the ratio of neutrons to protons in a specific isotope. For elements with atomic numbers less than 84, a neutron/proton ratio of 1 usually means a stable isotope of a specific element. Elements that have an atomic number over 83 usually form only unstable isotopes. Isotopes containing the magic numbers of 2, 8, 20, 50, 82, or 126 protons or neutrons are stable. Isotopes with an even number of protons or neutrons are generally more stable than those with odd numbers of protons or neutrons.

Half-Life

Learning Goal 3:	Calculate the amount of radioactive substance remaining after a specified number of half-lives.

The **half-life,** $t_{1/2}$, is the time required for one-half of a given quantity of a substance to undergo change. Each isotope has its own characteristic half-life.

The degree of stability of an isotope is indicated by the isotope's half-life. Isotopes with short half-lives decay rapidly; they are very unstable.

Decay of a radioisotope that has a reasonably short $t_{1/2}$ is experimentally determined by following its activity as a function of time. Graphing the results produces a radioactive decay curve (see Figure 10.1 in the textbook).

Example 5

Radium-226 ($^{226}_{88}$Ra) has a half-life of 1620 years. If a watch's hands are coated with paint containing Ra-226 in 1920, why is it still radioactive?

Answer

Since only 72 years (a small fraction of one half-life, $\dfrac{72 \text{ years}}{1620 \text{ years}}$) have passed, nearly all the radium-226 used in the paint is still radioactive.

Example 6

The following half-lives are known for different isotopes of radium:

    Ra-223 11   days
    Ra-224 ....3.64   days
    Ra-226 1620   years
    Ra-228 ....6.7   years

Why is there such a difference in their half-lives?

Answer

It is primarily due to the proton/neutron ratio in the isotope, but other as yet undiscovered reasons may play a part as well.

Example 7

How much of a 1.00-gram sample of Ra-226 would remain after three half-lives? The half-life of Ra-226 is 1620 years.

	first		second		third	
1.00g	$\longrightarrow$	0.50 g	$\longrightarrow$	0.25 g	$\longrightarrow$	0.13 g
	half-life		half-life		half-life	

Answer,
0.13 grams of radium-226 remain.

9.4 Nuclear Power

Learning Goal 4:   Describe the various ways in which nuclear energy may be used to generate electricity: fission, fusion, and the breeder reactor.

Energy Production

Einstein's equation relating mass and energy predicts that a small amount of nuclear mass is converted to a very large amount of energy when the nuclear particle breaks apart.
Einstein's equation is as follows:

$$E = mc^2$$

where E = energy
m = mass
c = speed of light

Three major modes of generating electrical power through a nuclear reaction involve the processes of fission, fusion, and the use of what is termed a breeder reactor.

Nuclear Fission

Fission (splitting) occurs when a heavy nuclear particle is split into smaller nuclei and large amounts of energy by a smaller nuclear particle (such as a neutron). The fission process intensifies, producing very large amounts of energy. This process of intensification is referred to as a chain reaction.

Nuclear Fusion

Fusion (meaning to join together) results from the combination of two small nuclei to form a larger nucleus with the concurrent release of large amounts of energy. The best example of a fusion reactor is the sun. Continuous fusion processes furnish our solar system with light and heat.

$$^2_1H \; + \; ^3_1H \; \rightarrow \; ^4_2He \; + \; ^1_0n \; + \; Energy$$

Breeder Reactors

A **breeder reactor** literally manufactures its own fuel. A perceived shortage of fissionable isotopes makes the breeder an attractive alternative to fission reactors. A breeder reactor uses uranium-238, which is abundant but nonfissionable. In a series of steps, the uranium-238 is converted to plutonium-239, which is fissionable and undergoes a fission chain reaction, producing energy.

9.5 Radiocarbon Dating

Learning Goal 5:   Explain the process of radiocarbon dating.

The approximate age of fossils and other objects of archaeological, anthropological, or historical interest may be established through **radiocarbon dating**. Radiocarbon dating is based on the measurement of the relative amounts (or ratio) of carbon-14 and carbon-12 present in an object. The carbon-14, along with the more abundant carbon-12, is converted into living plant material by the process of photosynthesis. Over time, the amount of carbon-14 slowly decreases, because carbon-14 is radioactive ($t_{1/2}$ = 5760 years); the amount of carbon-12, however, remains constant.

When the fossil is found and studied, the relative amounts of carbon-14 and carbon-12 are determined; using suitable equations involving the $t_{1/2}$ of carbon-14, it is possible to calculate (to within ± a few percent) the time that has elapsed since the formation of the object in question.

9.6 Medical Applications of Radioactivity

Learning Goal 6:   Cite several examples of the use of radioactive isotopes in medicine.

The use of radiation in the treatment of various forms of cancer, as well as the newer area of **nuclear medicine**, has become widespread in the past quarter-century.

Cancer Therapy Using Radiation

Learning Goal 7:   Describe the use of ionizing radiation in cancer therapy.

When high-energy radiation, such as gamma radiation, passes through a cell, it may collide with one of the molecules in the cell and cause it to lose one or more electrons, producing an ion pair. For this reason, such radiation is termed **ionizing radiation**. Ions produced in this way may cause subtle changes in cellular biochemical processes, which may result in diminished or altered cell function, or in extreme cases the death of the cell.

An organ that is cancerous is composed of both healthy cells and malignant cells. Cells undergoing division are particularly sensitive to gamma radiation. Therefore, exposing the tumor area to controlled dosages of high-energy gamma radiation from cobalt-60 (a high-energy gamma ray source) will generally kill a higher percentage of abnormal cells than normal cells.

Nuclear Medicine

The diagnosis of a host of biochemical irregularities or diseases of the human body has been made routine through the use of *radioactive tracers*. Tracers are small amounts of radioactive substances used as probes to study internal organs. Because the isotope is radioactive, its path may be followed using suitable detection devices. A "picture" of the organ is obtained, often far more detailed than is possible with conventional X-rays. Such techniques are noninvasive; that is, surgery is not required to investigate the condition of the internal organ, eliminating the risk associated with an operation. These techniques are successful because the radioactive isotope of an element has exactly the same chemical behavior as any other isotope of the same element.

Making Isotopes for Medical Applications

Learning Goal 8:	Discuss the preparation of radioisotopes for use in diagnostic imaging studies.

Learning Goal 9:	Explain the difference between natural and artificial radioactivity.

The radioactivity produced by unstable isotopes is described as **natural radioactivity**. If a normally stable, nonradioactive nucleus is made radioactive through bombardment with protons, neutrons, or alpha particles, the resulting radioactivity is termed **artificial radioactivity**.

The bombardment process is often accomplished in the core of a **nuclear reactor**, where an abundance of small nuclear particles, particularly neutrons, are available. Alternately, extremely high-velocity charged particles (such as a and b) may be produced in a **particle accelerator**, such as a cyclotron.

These product isotopes are often used in hospital laboratories as tracers in nuclear medicine.

9.7 Biological Effects of Radiation

Learning Goal 10:	Describe the characteristics of radioactive materials that relate to radiation exposure and safety.

Radiation Exposure and Safety

Safety considerations are based on the following:

1. The magnitude of the half-life
2. Shielding
3. Distance from the radioactive source
4. Time of exposure
5. Types of radiation emitted

Virtually all applications of nuclear chemistry create radioactive waste and, along with it, the problems of safe handling and disposal. Most disposal sites at present are considered temporary, until a long-term, safe solution can be found.

9.8 Measurement of Radiation

The changes that take place when radiation interacts with matter provide the basis of operation for various radiation detection devices.

Nuclear Imaging

An isotope is administered to a patient, and the isotope begins to concentrate in the organ of interest. Photographs of that region of the body are taken at periodic intervals using a special type of film. Upon development of the series of photographs, a record of the organ's uptake of the isotope as a function of time enables the radiologist to assess the condition of the organ.

Computer Imaging

A specialized television camera, sensitive to emitted radiation from a radioactive substance administered to a patient, develops a continuous and instantaneous record of the voyage of the isotope throughout the body.

The Geiger Counter

A **Geiger counter** is an instrument capable of detecting ionizing radiation.

Film Badges

A **film badge** is merely a piece of photographic film that is sensitive to energies corresponding to radioactive emissions.

Units of Radiation Measurement

The Curie

The **curie** is a measure of the amount of radioactivity in a radioactive source. The curie is independent of the nature of the radiation (α, β, or γ) as well as its effect on biological tissue. A curie is defined as the amount of radioactive material that produces 3.7×10^{10} atomic disintegrations per second.

The Roentgen

The **roentgen** is a measure of ionizing radiation (X ray and gamma ray) only. The roentgen is defined as the amount of radioactive isotope needed to produce 2×10^{9} ion pairs when passing through 1 cubic centimeter of air at $0°C$. The roentgen is a measure of radiation's interaction with air, and gives no information regarding its effect on biological tissue.

The Rad

The **rad**, or radiation absorbed dosage, provides more meaningful information than either of the previous units of measure. It takes into account the nature of the absorbing material. It is defined as the dosage of radiation able to transfer 2.4×10^{-3} calories of energy to 1 kilogram of matter.

The Rem

The **rem**, or roentgen equivalent for man, describes the biological damage caused by the absorption of different kinds of radiation by the body. The rem is obtained by multiplication of the rad by a factor called the relative biological effect, or **RBE**. The RBE is a function of the type of radiation (α, β, or γ). Although a beta particle is more energetic than an alpha particle, an alpha particle is approximately ten times more damaging to biological tissue. As a result, the RBE is 10 for alpha particles and 1 for beta particles.

An estimated lethal dose, symbolized by LD_{50}, is 500 rems. The lethal dose, LD_{50}, is defined as the dose that would be fatal for 50% of the exposed population within 30 days. Some biological effect, however, is detectable at a level as low as 25 rem.

Glossary of Key Terms in Chapter Nine

alpha particle (9.1) consists of two protons and two neutrons (the alpha particle is identical to helium nucleus).
artificial radioactivity (9.6) radiation that results from the conversion of a stable nucleus to another, unstable nucleus.
background radiation (9.7) the radiation that emanates from natural sources.
beta particle (9.1) an electron formed in the nucleus by the conversion of a neutron into a proton.
binding energy (9.3) the energy required to break down the nucleus into its component parts.
breeder reactor (9.4) a nuclear reactor that produces its own fuel in the process of providing electrical energy.
chain reaction (9.4) in a fission reactor, involves neutron production causing subsequent reactions accompanied by the production of more neutrons in a continuing process.
curie (9.8) the quantity of radioactive material that produces 3.7×10^{10} nuclear disintegrations per second.
fission (9.4) involves heavy nuclei split into lighter nuclei accompanied by the release of large quantities of energy.
fusion (9.4) the joining of light nuclei to form heavier nuclei accompanied by the release of large quantities of energy.
gamma ray (9.1) high-energy emission from nuclear processes, traveling at the speed of light.
half-life (9.3) of an isotope, the length of time required for one-half of the initial mass of the isotope to decay to products.
ionizing radiation (9.1) radiation that is sufficiently high in energy to cause ion formation upon impact.
lethal dose (9.8) the dosage of a toxic material, such as radiation, that would be lethal to 50% of a test population.
metastable isotope (9.2) an isotope that will give up some energy to produce a more stable form of the same isotope.
natural radioactivity (9.6) the spontaneous decay of a nucleus to produce high-energy particles or rays.
nuclear equation (9.2) a balanced equation accounting for the products and reactants in a nuclear reaction.
nuclear imaging (9.6) the generation of images of components of the body (organs, tissues) by using techniques based on the measurement of radiation.
nuclear medicine (9.6) a field of medicine that uses radioisotopes for diagnostic and therapeutic purposes.
nuclear reactor (9.6) a device for conversion of nuclear energy into electrical energy.
nuclide (9.1) nuclear species with specified numbers of protons and neutrons.
particle accelerator (9.6) a device for production of high-energy nuclear particles based on the interaction of charged particles with magnetic and electrical fields.
positron (9.2) particle emitted from the nucleus that has the same mass as an electron but has a positive charge.

rad (9.8) or radiation absorbed dosage, the absorption of 2.4 x 10-3 calories of energy per kilogram of absorbing tissue.

radioactivity (9.1) the process by which atoms emit high-energy particles or rays.

radiocarbon dating (9.5) the estimation of the age of an object through measurement of isotopic ratios.

rem (9.8) or Roentgen equivalent for man, the product of rad and RBE.

roentgen (9.8) the dose of radiation producing 2.1 x 109 ions in 1 cm^3 of air at $0°C$ and 1 atmosphere of pressure.

shielding (9.7) material used to provide protection from radiation.

tracer (9.6) a radioisotope that is rapidly and selectively transmitted to the part of the body for which diagnosis is desired.

Self Test for Chapter Nine

1. Complete the following nuclear reactions by supplying the missing part:

   a. $^{27}_{13}\text{Al} + ^{4}_{2}\text{He} \rightarrow ^{30}_{15}\text{P} + ?$

   b. $^{210}_{82}\text{Pb} \rightarrow ^{0}_{-1}\text{e} + ?$

2. How many mg of a 100.0 mg sample of Tc-99m will remain after 30 hours? The half-life of Tc-99m is 6 hours.

3. When using the following nuclear symbols for an isotope of fluorine, which number is equal to the atomic number?

4. Name the three types of natural radiation emitted by unstable nuclei.

5. Which of the types of radiation is really pure energy?

6. Which of the types of radiation is the least-penetrating form of nuclear radiation?

7. What term means that an isotope is unstable?

8. What is defined as the time period required for one-half of a given quantity of a substance to undergo a nuclear change?

9. Nuclear power plants use what nuclear process to produce energy?

10. What term do we use for cells that undergo unnaturally rapid cell division?

11. Where does iodine tend to concentrate in the human body?

12. What is the term that describes the amount of radiation attributable to our surroundings on a day-to-day basis?

13. The intensity of radiation varies in what way with the square of the distance from the source?

14. What do we call the specialized system that is sensitive to emitted radiation from a radioactive substance administered to a patient and produces a continuous picture of the voyage of the isotope throughout the body?

15. What badge is often worn by staff members that shows their exposure to radiation?

16. Which type of radiation travels at the speed of light?

17. What is defined as the spontaneous decay of a nucleus to produce high-energy particles or rays?

18. Radiocarbon dating involves estimation of the age of objects by measuring the amount of carbon-12 plus which radioactive form of carbon?

19. Which type of radiation is similar to a helium atom?

20. Which radioactive element is found in some homes?

Vocabulary Quiz for Chapter Nine

1. The _____ consists of two protons and two neutrons, carries a +2 charge, and results from nuclear decay.

2. The _____ is an electron formed in the nucleus by the conversion of a neutron into a proton.

3. The energy required to break down the nucleus into its component parts is called the _____.

4. _____ is the name of the equation that represents the energy equivalent of mass being equal to the mass times the square of the speed of light.

5. _____ is the process of joining of light nuclei to form heavier nuclei accompanied by the release of large amounts of energy.

6. A form of electromagnetic radiation from nuclear processes is termed _____.

7. The length of time required for one-half of the initial mass of an isotope to decay to products is referred to as the _____.

8. _____ is radiation that is sufficiently high in energy to cause ion formation upon impact.

9. The _____ is the dosage of radiation that would be fatal to 50% of the exposed population.

10. A (n) _____ is a radioisotope that is selectively transmitted to the part of the body for which diagnosis is desired.

10

An Introduction to Organic Chemistry: The Saturated Hydrocarbons

Learning Goals

1. Compare and contrast organic and inorganic compounds.
2. Draw structures that represent each of the families of organic compounds.
3. Write the names and draw the structures of the common functional groups.
4. Write condensed and structural formulas for saturated hydrocarbons.
5. Describe the relationship between structure and physical properties of saturated hydrocarbons.
6. Use the basic rules of the I.U.P.A.C. Nomenclature System to name alkanes and substituted alkanes.
7. Draw constitutional (structural) isomers of simple organic compounds.
8. Write the names and draw the structures of simple cycloalkanes.
9. Draw *cis-* and *trans-* isomers of cycloalkanes
10. Describe conformations of alkanes.
11. Draw the chair and boat conformations of cyclohexane.
12. Write equations for combustion reactions of alkanes.
13. Write equations for halogenation reactions of alkanes.

Introduction

Until 1828 it was thought that all organic compounds were derived from natural sources. All early attempts to synthesize these compounds in the laboratory failed, and it was proposed that a vital force was necessary for their formation. In 1828, Wöhler synthesized urea (an organic compound) from potassium cyanate and ammonium sulfate (two inorganic compounds). Wöhler's experiment is now recognized as the beginning of modern organic chemistry.

Example 1

Draw the dash structures of ammonium cyanate and urea. Why are these two compounds important to the beginning of modern organic chemistry?

Answer

1. Dash structure of ammonium cyanate:

Example 1 continued

Note that the following structure is also acceptable for the isocyanate anion. The two structures differ in their placement of lone pairs of electrons, which are not shown. The structure above has two lone pairs on the N, and two lone pairs on the O, while the structure below has one lone pair on the N, and three lone pairs on the O. Both structures have a full octet of eight electrons.

$$\left[N \equiv C - O \right]^{-}$$

2. Dash structure of urea:

$$
\begin{array}{ccc}
H & O & H \\
| & || & | \\
H-N- & C- & N-H
\end{array}
$$

3. In 1828, the German chemist, Friedrich Wöhler was trying to produce ammonium cyanate, but due to experimental conditions, the compound urea was produced. The two compounds have the molecular formula, CON_2H_4, but they have different structures and properties. The "vital force theory" was laid to rest. Up until his discovery, it was believed that only living things could make compounds like urea, which is found as a waste product in the urine of animals.

10.1 The Chemistry of Carbon

There are several reasons for the existence of hundreds of thousands of organic compounds:

1. Carbon atoms are able to form stable, covalent bonds with other carbon atoms. Consider the family of alkanes as an example. Each molecule in this family contains only carbon and hydrogen, but each is a different chemical substance with unique chemical and physical properties. There are an infinite number of possible alkanes; alkanes with hundreds or even thousands of carbon and hydrogen atoms may be easily synthesized. Even elemental carbon exists in different forms, allotropic forms, which have different properties.

2. Carbon can form stable bonds with other elements, for example nitrogen, oxygen, sulfur, and the halogens.

3. The number of ways that these elements may combine to form unique structures is practically limitless. Compounds having the same molecular formulas but different structures (hence, different properties) are termed **constitutional** or **structural isomers**.

Example 2
List three reasons why there are so many organic compounds and provide examples.

Answer
1. Carbon can form chain, branch, and ring structures with other carbon atoms:
    a. Chain structures with no branches:

    Propane   Butane

    b. Chain structure that also has branch structures:

    Isobutane

    $$H-\overset{\displaystyle H}{\underset{\displaystyle H}{\overset{\displaystyle |}{\underset{\displaystyle |}{C}}}}-\overset{\displaystyle II}{\underset{\displaystyle |}{\overset{\displaystyle |}{C}}}-\overset{\displaystyle H}{\underset{\displaystyle H}{\overset{\displaystyle |}{\underset{\displaystyle |}{C}}}}-H$$

    $$H-\overset{\displaystyle |}{\underset{\displaystyle |}{C}}-H$$

    c. Ring structures:

    Cyclobutane

    $$H-\overset{\displaystyle II}{\underset{\displaystyle |}{\overset{\displaystyle |}{C}}}-\overset{\displaystyle H}{\underset{\displaystyle |}{\overset{\displaystyle |}{C}}}-H$$

    $$H-\overset{\displaystyle |}{\underset{\displaystyle H}{\overset{\displaystyle |}{C}}}-\overset{\displaystyle |}{\underset{\displaystyle H}{\overset{\displaystyle |}{C}}}-H$$

2. Carbon can form compounds that also contain other nonmetals, such as oxygen, sulfur, and nitrogen, in their structure:

    $$H-\overset{\displaystyle H}{\underset{\displaystyle H}{\overset{\displaystyle |}{\underset{\displaystyle |}{C}}}}-O-H$$

    $$H-\overset{\displaystyle H}{\underset{\displaystyle H}{\overset{\displaystyle |}{\underset{\displaystyle |}{C}}}}-\overset{\displaystyle H}{\underset{\displaystyle H}{\overset{\displaystyle |}{\underset{\displaystyle |}{C}}}}-\overset{\displaystyle H}{\underset{}{\overset{\displaystyle |}{N}}}-H$$

    Methyl alcohol   Ethyl amine

3. Carbon can form compounds that have the same molecular formula but a different structure and a different set of properties. Urea and ammonium cyanate are one example of such a set of isomeric compounds.

Important Differences Between Organic and Inorganic Molecules

> **Learning Goal 1:** Compare and contrast organic and inorganic molecules.

The bonds in organic molecules are almost always covalent bonds, while those found in many inorganic substances are ionic bonds. When comparing organic and inorganic compounds, it helps to remember the differences between ionic and covalent bonds.

1. Ionic bonds result from a transfer of one or more electrons. Covalent bonds are formed by a sharing of electrons to form a stable orbital containing two electrons.
2. The ionic bond is electrostatic in nature. It is formed by the attraction of positive and negative ions resulting from the electron transfer process. Charge separation of covalently bonded atoms is much less extreme.
3. Ions are arranged in large, three-dimensional crystals, consisting of many positive and negative ions. Covalently bonded substances exist as discrete units: molecules.
4. Ionic compounds often dissociate in solution (electrolytes), whereas most covalently bonded molecules retain their identity in solution (nonelectrolytes).

Families of Organic Compounds

> **Learning Goal 2:** Draw structures that represent each of the families of organic compounds.

Organic compounds are classified according to groups, or families. The two most general classifications are **hydrocarbons** and **substituted hydrocarbons**. Compounds that contain only carbon and hydrogen are classified as hydrocarbons: hydrogen (hydro) and carbon (carbon). Hydrocarbons are subdivided into two principal classes: **aliphatic** and **aromatic hydrocarbons**. Aliphatic hydrocarbons are further subdivided into three families: alkanes, alkenes, and alkynes.

The **alkanes** (and cycloalkanes) are termed **saturated hydrocarbons**. They are composed solely of C—C and C—H single bonds. The **unsaturated hydrocarbons**, hydrocarbons that contain at least one carbon-carbon double bond or carbon-carbon triple bond, are referred to as **alkenes** and **alkynes**, respectively. Aromatic compounds contain a benzene ring or a derivative of the benzene ring.

Substitution of one or more functional groups for hydrogens in a hydrocarbon brings about major changes in properties. A **functional group** is an atom or group of atoms in a molecule principally responsible for the chemical and physical properties of that molecule. All compounds that contain a particular functional group, for example, the hydroxyl group (—OH), are classified as being in the same family. The hydroxyl group is the functional group of the alcohols.

> **Learning Goal 3:** Write the names and draw the structures of the common functional groups.

Example 3

To which family of organic compounds do each of the following belong? (Refer to Table 10.2 in the text.)

1. CH_3CH_3

2. $CH_3CH=CHCH_3$

3. CH_3OH

4.
$$\begin{array}{c} O \\ \parallel \\ H_3C-C-CH_3 \end{array}$$

5. CH_3Br

Answer
1. alkane
2. alkene
3. alcohol
4. ketone
5. alkyl halide

10.2 Alkanes

Structure and Properties

Learning Goal 4:   **Write condensed and structural formulas for saturated hydrocarbons.**

Alkanes are hydrocarbons that contain only carbon and hydrogen bonded together through carbon-hydrogen and carbon-carbon single bonds. They have the general formula C_nH_{2n+2}.

Several different types of formulas are used to describe organic molecules. The **structural formula** shows all of the atoms in a molecule and shows all bonds as lines. The **molecular formula** provides the atoms and number of each type of atom in a molecule, but gives no information regarding the bonding pattern. The **condensed formula** shows all of the atoms in a molecule and places them in a sequential arrangement that details which atoms are bonded to each other. The simplest is the **line formula.** In this formula, we assume that there is a carbon at any location where two or more lines meet. We also assume that there is a carbon at the end of any line and that each carbon is bonded to the correct number of hydrogen atoms.

A molecular formula tells us the atomic composition of a molecule, but provides no information about the structure of the molecule. For instance, there are three possible structures for a compound with the molecular formula of C_5H_{12}:

Pentane   Methylbutane   2,2-Dimethylpropane

These are the three possible isomers of pentane. These structural formulas provide more information about a molecule.

A suitable compromise between the convenience of the molecular formula and the detail of the structural formula is the condensed formula.

Examples of condensed formulas:

$$CH_3CH_2CH_2CH_2CH_3$$

Pentane

$$CH_3CHCH_2CH_3 \quad (CH_3)$$

Methylbutane

$$H_3C-C-CH_3 \quad (CH_3 \text{ top and bottom})$$

2,2-Dimethylpropane

Example of line formulas:

Pentane   Methylbutane   Dimethylpropane

Example 4

Draw the structural formula of an isomer of ethyl alcohol (CH_3CH_2OH), in which an oxygen atom is placed between two carbon atoms.

Work

Hydrogen always forms one bond; carbon forms four bonds and oxygen forms two bonds.

Example 4 continued
Answer

$$H-\overset{\overset{\displaystyle H}{|}}{\underset{\underset{\displaystyle H}{|}}{C}}-O-\overset{\overset{\displaystyle H}{|}}{\underset{\underset{\displaystyle H}{|}}{C}}-H$$

This compound is dimethyl ether, and it is an isomer of ethyl alcohol.

Example 5
Write the condensed formulas for the following:

1. Propane

$$H-\overset{\overset{\displaystyle H}{|}}{\underset{\underset{\displaystyle H}{|}}{C}}-\overset{\overset{\displaystyle H}{|}}{\underset{\underset{\displaystyle H}{|}}{C}}-\overset{\overset{\displaystyle H}{|}}{\underset{\underset{\displaystyle H}{|}}{C}}-H$$

2. Methylbutane

$$H-\overset{\overset{\displaystyle H}{|}}{\underset{\underset{\displaystyle H}{|}}{C}}-H$$
$$H-\overset{\overset{\displaystyle H}{|}}{\underset{\underset{\displaystyle H}{|}}{C}}-\overset{\overset{\displaystyle H}{|}}{\underset{\underset{\displaystyle H}{|}}{C}}-\overset{\overset{\displaystyle H}{|}}{\underset{\underset{\displaystyle H}{|}}{C}}-H$$

3. Methyl alcohol

$$H-\overset{\overset{\displaystyle H}{|}}{\underset{\underset{\displaystyle H}{|}}{C}}-O-H$$

4. Acetone

$$H-\overset{\overset{\displaystyle H}{|}}{\underset{\underset{\displaystyle H}{|}}{C}}-\overset{\overset{\displaystyle O}{\|}}{C}-\overset{\overset{\displaystyle H}{|}}{\underset{\underset{\displaystyle H}{|}}{C}}-H$$

5. 2-Chloropropane

$$H-\overset{\overset{\displaystyle H}{|}}{\underset{\underset{\displaystyle H}{|}}{C}}-\overset{\overset{\displaystyle H}{|}}{\underset{\underset{\displaystyle Cl}{|}}{C}}-\overset{\overset{\displaystyle H}{|}}{\underset{\underset{\displaystyle H}{|}}{C}}-H$$

Answer

1. $CH_3CH_2CH_3$

2. $CH_3CH(CH_3)CH_2CH_3$

3. CH_3OH

4. CH_3COCH_3

5. $CH_3CH(Cl)CH_3$

Example 6
Write the molecular formulas of the alkanes that contain 4, 8, and 40 carbon atoms.

Work
Use the general formula for finding the molecular formula of any alkane, provided you are given the number of carbon atoms.
C_nH_{2n+2} where n = number of carbon atoms

Answer
4 carbons: C_4H_{10}; 8 carbons: C_8H_{18}; 40 carbons: $C_{40}H_{82}$

Since all the hydrocarbons are composed of nonpolar carbon-carbon and carbon-hydrogen bonds, hydrocarbons are nonpolar molecules. As a result, they are not water soluble, but are readily soluble in nonpolar solvents. Furthermore, virtually all of the hydrocarbons are less dense than water and have relatively low melting points and boiling points.

Alkyl Groups

Alkyl groups result when a hydrogen atom is removed from an alkane. The name is derived from the name of the corresponding alkane by removing the *-ane* ending and replacing it with *-yl*.

Carbon atoms are classified according to the number of carbon atoms to which they are attached. A **primary (1°) carbon** is directly bonded to one other carbon. A **secondary (2°) carbon** is bonded to two other carbon atoms. A **tertiary (3°) carbon** is bonded to three other carbon atoms. Similarly, alkyl groups are classified according to the number of carbon atoms attached to the carbon atom that joins the alkyl group to a molecule.

Nomenclature

The basic rules used for naming compounds in the I.U.P.A.C. Nomenclature System follow:

1. The name of the compound is defined by the longest continuous carbon chain in the compound. This chain is the **parent compound**. (Refer to Table 10.3 in the text for the names of parent compounds.)

2. Each substituent attached to the parent compound is given a name and a number. The number designates the position of the substituent on the main chain, and the name tells what type of substituent is present at that position.

3. The chain must be numbered from one end to the other in order to provide the lowest position number for each substituent. If more than one substituent is present, number from the end that gives the lowest number of the first substituent encountered regardless of the numbers that result for the other substituents.

4. If the same substituent occurs more than once in the compound, a separate position number is supplied for each substituent, and the prefixes *di-, tri-, tetra-, penta-, hexa-, hepta-,* and so forth are used.

5. Place the names of the substituents in alphabetical order before the name of the parent compound. Numbers are separated from each other by commas, and numbers are separated from names by hyphens. By convention halogen substituents are placed before alkyl substituents in this priority sequence regardless of the alphabetization. Names of some common alkyl groups are listed in Tables 10.5 and 10.6.

Example 7

Give the I.U.P.A.C. names of the following branched-chain alkanes:

$$CH_3$$
$$|$$
1. $CH_3CHCH_2CH_3$ \qquad 2. $CH_3CH_2CCH_2CH_2CH_3$
$$|$$ \qquad\qquad\qquad\qquad\qquad $|$
$$CH_3$$ \qquad\qquad\qquad\qquad\qquad\qquad CH_3

Work
1. Determine the longest continuous carbon chain. This is called the parent compound.
2. Number the parent compound chain so that the branches have the lowest possible numbers.

Answer
1. The parent chain is butane. We will number 1, 2, etc. from left to right. The correct name of this branched alkane is 2-methylbutane.
2. The parent chain is hexane. We will number from left to right. The correct name of this branched alkane is 3,3-dimethylhexane.

Example 8

Give the I.U.P.A.C. name of the following:

$$CH_3$$
$$|$$
$$CH_3CHCH_2CCH_2Br$$
$$|\qquad\quad|$$
$$Br\qquad CH_3$$

Work
1. The parent compound is pentane.
2. Number the parent compound from right to left:

$$CH_3$$
$$|$$
$$C-C-C-C-C-Br$$
$$|\qquad\qquad|$$
$$Br\qquad\quad CH_3$$

$$5\quad 4\quad 3\quad 2\quad 1$$

Answer
1,4-dibromo-2,2-dimethylpentane

Constitutional or Structural Isomers

Learning Goal 7:   Draw constitutional (structural) isomers of simple organic compounds.

Two molecules having the same molecular formulas but different structures are called **constitutional** or **structural isomers**. Isomers having the same molecular formula are unique compounds due to their structural differences. They may have similar physical and chemical properties, but in many cases their properties are quite dissimilar.

Example 9

Draw and give the I.U.P.A.C. names of all the alkane isomers of C_5H_{12}.

Work

1. The first isomer you should always draw is the one that has all the carbons in a continuous chain structure:

$$CH_3CH_2CH_2CH_2CH_3 \qquad \text{I.U.P.A.C. name} = \text{pentane}$$

2. Next draw four carbons in a continuous chain, and connect the fifth carbon as a methyl group to this continuous chain:

$$
\begin{array}{cccc}
C- & C- & C- & C \\
 & & | & \\
 & & CH_3 &
\end{array}
$$

Fill in the structure using hydrogen atoms, so that each carbon atom has four bonds.

$$
\begin{array}{ccccc}
 & H & H & H & H \\
 & | & | & | & | \\
H- & C- & C- & C- & C-H \\
 & | & | & | & | \\
 & H & H & | & H \\
 & & & H-C-H & \\
 & & & | & \\
 & & & H &
\end{array}
$$

Then write the condensed structure:

$$
\begin{array}{c}
CH_3 \\
| \\
CH_3CH_2CHCH_3
\end{array}
$$

I.U.P.A.C. name = methylbutane. Note that you can move the methyl group one carbon to the left. However, if you apply the I.U.P.A.C. nomenclature test, the product is still methylbutane. Similarly, you can move the methyl group one carbon to the right. The I.U.P.A.C. nomenclature test reveals that you have drawn pentane, the linear isomer.

Example 9 continued

3. The last isomer of C_5H_{12} is found by drawing three carbons in a continuous chain and connecting the other two carbons as methyl groups to the middle carbon of the continuous chain:

$$CH_3$$
$$|$$
$$C-C-C$$
$$|$$
$$CH_3$$

Fill in the structure using hydrogen atoms as before:

$$
\begin{array}{c}
H \\
| \\
H-C-H \\
H \quad | \quad H \\
| \quad | \quad | \\
H-C-C-C-H \\
| \quad | \quad | \\
H \quad | \quad H \\
H-C-H \\
| \\
H
\end{array}
$$

Then write the condensed structure:

$$CH_3$$
$$|$$
$$CH_3CCH_3$$
$$|$$
$$CH_3$$

Answer

I.U.P.A.C. name = dimethylpropane.

Remember that all isomers must have different I.U.P.A.C. names. To check whether you have drawn duplicate isomers, name them using the I.U.P.A.C. nomenclature system.

10.3 Cycloalkanes

> **Learning Goal 8:**   Write the names and draw the structures of simple cycloalkanes.

The **cycloalkanes** are another family of hydrocarbons closely related to the alkanes. Cycloalkanes have the general molecular formula C_nH_{2n}. Note that they contain two fewer hydrogens than the corresponding alkane.

Cycloalkanes are named by adding the prefix *cyclo-* before the name of the alkane with the same number of carbon atoms. For example, cyclobutane is a cyclic alkane that has four carbon atoms. Substituted cycloalkanes are named by placing the name of the substituent before the name of the cycloalkane. No number is needed if only a single substituent is present. If more than one substituent is present, then the numbers that result in the lowest possible position numbers for the substituents are used.

cis-trans Isomerism in Cycloalkanes

> **Learning Goal 9:**   Draw *cis* and *trans* isomers of cycloalkanes.

Rotation around the bonds in a cyclic structure is limited by the fact that the carbons of the ring are all bonded to other ring carbons. As a result, **geometric isomers**, or *cis-trans* **isomers**, are produced. In the *cis-* isomer, two substituents are on the same side of the ring (either above or below). In the *trans-* isomer, the substituents are on opposite sides of the ring.

10.4 Conformations of Alkanes and Cycloalkanes

> **Learning Goal 10:**   Describe conformations of alkanes.

> **Learning Goal 11:**   Draw the chair and boat conformations of cyclohexane.

There is free rotation around the carbon-carbon single bond, and at room temperature, rotations occur on the order of millions of times per second. Thus, these molecules exist in an infinite variety of arrangements that are rapidly interconverting. These varying arrangements, capable of interconversion by simple rotation about a bond, are called **conformations** or **conformers**.

All cycloalkanes, except cyclopropane, also exist in different conformations. The **chair conformation**, which resembles a lawn chair, is the most energetically favored conformer for six-member ring structures. The **boat conformation** is less stable and resembles a rowboat. The hydrogen atoms that lie above or below the ring are said to be **axial atoms**; those that lie roughly in the plane of the ring are called **equatorial atoms**.

The most stable conformations of alkanes and cycloalkanes are those in which the hydrogen atoms are least crowded.

10.5 Reactions of Alkanes and Cycloalkanes

Combustion

Learning Goal 12: Write equations for combustion reactions of alkanes.

Alkanes and other hydrocarbons may be oxidized (burned in air), producing carbon dioxide and water; this reaction is called **combustion**. During combustion, a large amount of heat energy is released.

Example 10
Show the complete balanced equation for the combustion of octane.

Work
1. Octane: The *-ane* ending tells us that the compound is an alkane. The *oct-* means that it has eight carbon atoms. By using the general molecular formula C_nH_{2n+2}, we can derive the molecular formula of octane. Thus, the molecular formula is C_8H_{18}.
2. The complete combustion of an alkane requires O_2 to react with the alkane to produce $CO_2 + H_2O$.
3. $C_8H_{18} + O_2 \rightarrow CO_2 + H_2O$ This equation is not balanced.

Answer
The balanced equation for the complete combustion of octane is
$$2\,C_8H_{18} + 25\,O_2 \rightarrow 16\,CO_2 + 18\,H_2O$$

Halogenation

Learning Goal 13: Write equations for halogenation reactions of alkanes.

Halogenation of an alkane or cycloalkane is a substitution reaction in which a halogen atom (usually bromine or chlorine) replaces a hydrogen atom.

Typically, when an alkane reacts with certain halogens in the presence of heat and/or light, a substitution reaction results. One of the C—H bonds of the alkane is broken and replaced with a C–X bond in which a halogen atom (X = Br or Cl) has substituted for a hydrogen atom. The products of this reaction are an **alkyl halide** and a hydrogen halide. In more complex alkanes, substitution can occur to some extent at all positions to give a mixture of products.

Example 11
Show the reaction for the stepwise substitution of chlorine atoms for the hydrogen atoms on methane; also, determine the I.U.P.A.C. name for each organic compound produced.

Answer

$$CH_4 + Cl_2 \xrightarrow{\text{UV}} CH_3Cl + HCl$$

$$CH_3Cl + Cl_2 \xrightarrow{\text{UV}} CH_2Cl_2 + HCl$$

$$CH_2Cl_2 + Cl_2 \xrightarrow{\text{UV}} CHCl_3 + HCl$$

$$CHCl_3 + Cl_2 \xrightarrow{\text{UV}} CCl_4 + HCl$$

The products of these reactions are chloromethane, dichloromethane, trichloromethane, and tetrachloromethane, respectively.

Glossary of Key Terms in Chapter Ten

aliphatic hydrocarbon (10.1) any member of the alkanes, alkenes, and alkynes or the substituted alkanes, alkenes, and alkynes.

alkane (10.2) saturated hydrocarbon; a hydrocarbon that contains only carbon and hydrogen and that is bonded together through carbon-hydrogen and carbon-carbon single bonds. Alkanes have the general molecular formula C_nH_{2n+2}.

alkyl group (10.2) a simple hydrocarbon group that results from the removal of one hydrogen from the original hydrocarbon (e.g., methyl, $-CH_3$; ethyl, $-CH_2CH_3$).

alkyl halide (10.6) a substituted hydrocarbon that has the general structure R-X, where R- represents any alkyl group and X is one of the halogens (F-, Cl-, Br-, or I-.)

aromatic hydrocarbon (10.1) an organic compound that contains the benzene ring or a derivative of the benzene ring.

axial atom (10.4) an atom that lies above or below a cycloalkane ring.

boat conformation (10.4) a conformation of the cyclohexane ring that resembles a rowboat.

chair conformation (10.4) the most energetically favorable conformation for a six-carbon cycloalkane; so-called for its resemblance to a lawn chair.

cis-trans isomers (10.3) isomers that differ from one another in the placement of groups on a double bond or a ring.

combustion (10.5) oxidation of hydrocarbons by burning in the presence of air to produce carbon dioxide and water.

condensed formula (10.2) a formula that shows all of the atoms in a molecule and places them in a sequential arrangement that details which atoms are bonded to each other; the bonds themselves are not shown.

conformations (10.4) discrete, distinct isomeric structures that may be converted one to the other by rotation about the bonds in the molecule.

conformers (10.4) discrete, distinct isomeric structures that may be converted one to the other by rotation about the bonds in the molecule.

constitutional isomers (10.2) two molecules having the same molecular formulas but different chemical structures.

cycloalkane (10.3) cyclic alkanes; saturated hydrocarbons, that have the general formula C_nH_{2n}.

equatorial atom (10.4) an atom that lies in the plane of a cycloalkane ring.

functional group (10.1) an atom (or group of atoms and their bonds) that imparts specific chemical and physical properties to a molecule.

geometric isomers (10.3) isomers that differ from one another in the placement of substituents on a double bond or ring.

halogenation (10.5) a reaction in which one of the C-H bonds of a hydrocarbon is replaced with a C-X bond of a halogen atom (X = Br or Cl, generally).

hydrocarbon (10.1) a compound composed solely of the elements carbon and hydrogen.

I.U.P.A.C. Nomenclature System (10.2) The International Union of Pure and Applied Chemistry (I.U.P.A.C.) standard, universal system for the nomenclature of organic compounds.

line formula (10.2) the simplest representation of a molecule in which it is assumed that there is a carbon atom at any location where two or more lines intersect, there is a carbon at the end of any line, and each carbon is bonded to the correct number of hydrogen atoms

molecular formula (10.2) a formula that provides the atoms and number of each type of atom in a molecule but gives no information regarding the bonding pattern involved in the molecule's structure.

parent compound (10.2) In the I.U.P.A.C. Nomenclature System, the parent compound is the longest chain containing the principal functional group (e.g., the hydroxyl group) in the molecule that is being named .

saturated hydrocarbon (10.1) alkane; a hydrocarbon that contains only carbon and hydrogen and that is bonded together through carbon-hydrogen and carbon-carbon single bonds. Alkanes have the general molecular formula C_nH_{2n+2}.

structural formula (10.2) a formula that shows all of the atoms in a molecule and exhibits all bonds as lines.

structural isomers (10.2) molecules having the same molecular formula but different chemical structures.

substituted hydrocarbon (10.1) a hydrocarbon in which one or more hydrogen atoms is replaced by another atom or group of atoms.

substitution reaction (10.5) a reaction that results in the replacement of one group by another.

unsaturated hydrocarbon (10.1) a hydrocarbon containing at least one multiple (double or triple) bond.

Self Test for Chapter Ten

1. Which of the following are isomers of C_5H_{12}? Give letter(s) as your answers.

   a. $CH_3CH_2CH_2CH_3$

   b. $CH_3CH_2CH_2CH_2CH_3$

   c. $CH_3CH(CH_3)CH_2CH_3$

   d.
$$CH_3\underset{\underset{CH_3}{|}}{\overset{\overset{CH_3}{|}}{C}}CH_3$$

   e. $CH_3CH(CH_3)_2$

2. What is the geometry of a carbon surrounded by four single bonds?

3. What is the angle around carbon surrounded by a double bond and two single bonds?

4. What is the hydrogen-carbon-carbon bond angle in ethyne, HCCH?

5. What is the term used to describe the group in an organic compound that is primarily responsible for the chemical and physical properties of that compound?

6. Name the functional group found in all alcohols.

7. Name the functional group found in all aldehydes.

8. Name the family whose simplest member is benzene.

9. Which of the following is an isomer of 1,2-dibromoethane?

    a. Br_2CHCH_3 b. CH_2BrCH_2Br c. $BrCH_2CH_2CH_3$

10. Which of the following is an isomer of C_4H_{10}?

    a. benzene   d. $CH_3CHBrCH_2CH_3$

    b. cyclobutane   e. dimethylpropane

    c.

$$CH_3CHCH_3$$
$$|$$
$$CH_3$$

11. Which of the following is an isomer of CH_3OCH_3?

    a. CH_3CH_2OH d. oxymethylurea

    b. 2-methylbutane   e. dimethylether

    c. $CH_3CH_2OCH_3$

12. Which of the following is the simplest organic compound?

    a. ethyl alcohol   d. urea

    b. methane   e. acetylene

    c. ammonium cyanate

13. Which of the following is not a normal property of an inorganic compound?

    a. flammable   d. reactions are fast

    b. high boiling point   e. high melting point

    c. soluble in water

14. Which of the following formulas have limited usage in organic chemistry?

    a. space filling formula   c. molecular formula

    b. condensed formula   d. structural formula

15. Compounds that contain only carbon and hydrogen atoms are classified in what major subdivision?

    a. alkanes   d. heterocyclic

    b. aliphatics   e. hydrocarbons

    c. aromatics

16. What field of study is defined as the chemistry of all the elements except carbon?

17. What organic compound did Friedrich Wöhler synthesize from inorganic reactants?

18. Alkanes contain only carbon and what other element?

19. Name the functional group found in all alkenes.

20. Give the molecular formula of the alkane that contains nine carbon atoms.

21. Give the I.U.P.A.C. names of the following branched-chain alkanes:

a. $CH_3CHCH_2CH_3$
   |
   CH_3

b. CH_3
   |
   $CH_3CCH_2CH_2CH_3$
   |
   CH_3

c. CH_3 CH_3
   | |
   CH_3 — C — CH — CH_2CH_3
   |
   CH_3

22. Give the I.U.P.A.C. name of the following:

$\quad\quad$ H $\quad$ CH_3
$\quad\quad$ | $\quad\quad$ |
$\quad$ $CH_3CCH_2CCH_2Br$
$\quad\quad$ | $\quad\quad$ |
$\quad\quad$ Cl $\quad$ H

23. Give the I.U.P.A.C. name of the following:

$\quad\quad$ CH_3
$\quad\quad$ |
$\quad$ CH_3CCH_3
$\quad\quad$ |
$\quad\quad$ Br

24. Give the I.U.P.A.C. name of the following:

$\quad\quad\quad$ CH_2CH_3
$\quad\quad\quad$ |
$\quad$ CH_3 — C — CH_3
$\quad\quad\quad$ |
$\quad\quad\quad$ CH_2CHCH_3
$\quad\quad\quad\quad$ |
$\quad\quad\quad\quad$ Br

25. Write an equation representing the complete combustion of propane.

26. Write the names of the monosubstitution products that are produced when Cl_2 and propane react under UV light.

27. Describe the solubility of alkanes in water.

28. Are the melting and boiling points of alkanes generally higher or lower than those of other organic compounds?

29. In the I.U.P.A.C. system of naming compounds, the name of the compound is defined by the longest continuous carbon chain in the compound. What do we call this continuous chain?

30. What is the I.U.P.A.C. name of $CHCl_3$?

31. The general formula C_nH_{2n} can be used to find the molecular formula of any member of what alkane-like class of organic compounds?

32. What does the prefix *cyclo-* mean?

33. What is the term for the oxidation of alkanes and other hydrocarbons to carbon dioxide and water?

34. How many moles of water are produced by the complete combustion of one mole of methane?

35. For methane to react with chlorine, what catalyst is needed?

36. What type of organic reaction involves the replacement of one or more atoms in a molecule by new atoms?

Vocabulary Quiz for Chapter Ten

1. A(n) _____ is a hydrocarbon that contains only carbon and hydrogen bonded together through carbon-hydrogen and carbon-carbon single bonds. They are saturated hydrocarbons that have the general molecular formula C_nH_{2n+2}.

2. Two molecules having the same molecular formulas but different chemical structures are called _____.

3. _____ are isomers (which by definition have the same molecular formula) that differ from one another due to the placement of substituents on a double bond or a ring.

4. A(n) _____ is a simple hydrocarbon group that results from the removal of one hydrogen from the original hydrocarbon (e.g., methyl, $-CH_3$; ethyl, $-CH_2CH_3$).

5. In the I.U.P.A.C. Nomenclature System, the _____ is the longest chain containing the major functional group (e.g., the hydroxyl group) in the molecule.

6. A(n) _____ is any hydrocarbon that contains one or more carbon-carbon multiple bonds.

7. _____ is a reaction in which one of the C-H bonds of a hydrocarbon is replaced with a C-X bond of a halogen atom (X = Br or Cl, generally).

8. A(n) _____ is a compound containing only hydrogen and carbon.

9. An atom or group of atoms that imparts specific chemical and physical properties to a molecule is called a(n) _____.

10. A(n) _____ is a reaction that results in the replacement of one group by another.

11 The Unsaturated Hydrocarbons: Alkenes, Alkynes, and Aromatics

Learning Goals

1 Describe the physical properties of alkenes and alkynes.
2 Draw the structures and write the I.U.P.A.C. names of simple alkenes and alkynes.
3 Write the names and draw the structures of simple geometric isomers of alkenes.
4 Write equations predicting the products of the simple addition reactions of alkenes and alkynes: hydrogenation, halogenation, hydration, and hydro-halogenation.
5 Apply Markovnikov's rule to predict the major and minor products of the hydration and hydrohalogenation reactions of unsymmetrical alkenes.
6 Write equations representing the formation of addition polymers of alkenes.
7 Draw the structures and write the names of common aromatic hydrocarbons.
8 Write equations for substitution reactions involving benzene.
9 Describe heterocyclic aromatic compounds and list several biological molecules in which they are found.

11.1 Alkenes and Alkynes: Structure and Physical Properties

Learning Goal 1: Describe the physical properties of alkenes and alkynes.

Alkenes, alkynes, and **aromatic compounds** contain at least one carbon-carbon double or triple bond. As a result of multiple bonding, these compounds contain fewer hydrogens than alkanes with the same number of carbon atoms. They are referred to as unsaturated because they do not contain as many hydrogens as their carbon skeleton will allow. The structures seen on the next page reveal the structural differences among alkanes, alkenes, and alkynes.

Each of the compounds in the diagram has the same number of carbon atoms, but they differ in the number of hydrogen atoms. Alkynes have the general formula C_nH_{2n-2} and thus contain two fewer hydrogens than the corresponding alkene (general formula C_nH_{2n}). Alkenes have two fewer hydrogens than the corresponding alkane (general formula C_nH_{2n+2}).

Alkanes contain only single bonds; alkenes have at least one carbon-to-carbon double bond; and alkynes contain at least one carbon-to-carbon triple bond. The differences in bond order result in variation in molecular geometry and chemical reactivity among these three families. Alkanes have tetrahedral carbon atoms. When carbon is bonded by one double bond and two single bonds, as in alkenes, the molecule is planar and each bond angle is 120°. When two carbons are bonded by a triple bond, the molecule is linear and the bond angles are 180°.

	Ethane (Ethane)	Ethene (Ethylene)	Ethyne (Acetylene)
Structural Formulas	H H | | H—C—C—H | | H H	H H \ / C = C / \ H H	H—C≡C—H
Molecular Formulas	C_2H_6	C_2H_4	C_2H_2
Condensed Formulas	CH_3CH_3	$CH_2 = CH_2$	$HC≡CH$

Alkenes and alkynes are nonpolar. As a result they are not water-soluble but are readily soluble in nonpolar solvents and in many low-polarity organic solvents such as ether or chloroform. They are also less dense than water and have relatively low boiling points.

11.2 Alkenes and Alkynes: Nomenclature

Learning Goal 2:   **Draw the structures and write the I.U.P.A.C. names of simple alkenes and alkynes.**

The nomenclature of alkenes and alkynes is analogous to that of the alkanes, with the following exceptions. For alkenes, the parent name is derived from the longest continuous carbon chain containing the double bond. Then the *-ane* ending of the alkane is replaced with the *-ene* ending of an alkene. The chain is numbered to give the lowest numbers for the two carbons containing the double bond.

Alkynes are named in the same way as the alkenes, except that the *-ane* ending of the corresponding alkane is replaced with the *-yne* ending of alkynes. The rules used in numbering the alkene chain are also used in alkyne nomenclature.

Example 1
Draw and name the alkane, alkene, and alkyne that contain three carbon atoms.

Work
1. To find the molecular formula of an alkane, use the general formula C_nH_{2n+2}
2. To find the molecular formula of an alkene, use the general formula C_nH_{2n}
3. To find the molecular formula of an alkyne, use the general formula C_nH_{2n-2}

Answer
1. The alkane with three carbon atoms has the molecular formula C_3H_8:

$$CH_3CH_2CH_3 \qquad \text{I.U.P.A.C. name: Propane}$$

2. The alkene with three carbon atoms has the molecular formula C_3H_6:

$$CH_3CH{=}CH_2 \qquad \text{I.U.P.A.C. name: Propene}$$

3. The alkyne with three carbon atoms has the molecular formula C_3H_4:

$$CH_3C{\equiv}CH \qquad \text{I.U.P.A.C. name: Propyne}$$

Example 2
Give the I.U.P.A.C. name of the following molecule:

$$\begin{array}{c} CH_2CH_2CH_2CH_3 \\ | \\ CH_3 - CH{=}C - CH_2 - CH_3 \end{array}$$

Work
1. First determine the parent compound (longest continuous chain) that contains the double bond. The parent compound is heptene. Since the *hept-* means seven carbons and the *-ene* means double bond, we have

$$\begin{array}{c} C - C - C - C \\ | \\ C - C{=}C \end{array}$$

2. Next, label the parent to give the double bond carbons the lowest possible numbers:

$$\begin{array}{cccc} 4 & 5 & 6 & 7 \\ C - C - C - C \\ | \\ C - C{=}C \\ 1 \quad 2 \quad 3 \end{array}$$

2-heptene; the 2 tells us that the double bond is between carbons 2 and 3.

Example 2 continued

3. Finally, list the attached groups, as in naming alkanes.

$$
\begin{array}{cccc}
4 & 5 & 6 & 7 \\
C-C-C-C & & & \\
\end{array}
$$

$$
C-C{=}C-CH_2CH_3
$$

$$
\begin{array}{ccc}
1 & 2 & 3
\end{array}
$$

Answer
3-Ethyl-2-heptene

Example 3
Determine the I.U.P.A.C. name of the following molecule:

$$
CH_3-CH-CH_2-C{\equiv}C-CH_3
$$
$$
\quad\quad\quad |
$$
$$
\quad\quad\quad CH_3
$$

Work
1. First determine the parent compound that contains the triple bond. The parent compound is hexyne. The *hex-* means six carbons, and the *-yne* means a triple bond.
2. Label the compound so that the triple bond carbons have the lowest numbers possible.

$$
C-C-C-C{\equiv}C-C
$$
$$
\begin{array}{cccccc}
6 & 5 & 4 & 3 & 2 & 1
\end{array}
$$

Thus, it is 2-hexyne.
3. Finally, list the attached groups as before:

$$
C-C-C-C{\equiv}C-C
$$
$$
\quad\quad |
$$
$$
\quad\quad CH_3
$$

Answer
5-Methyl-2-hexyne

Example 4
Draw each of the following using structural formulas:

1. 1-Bromo-2-hexyne   2. 2-Butene

Work
1. 1-Bromo-2-hexyne: there are 6 carbons (*hex-*) in the parent compound and a triple bond between carbons 2 and 3 (*2-yne*). Also, there is a bromine (Br–) bonded to carbon 1.

154

Example 4 continued

$$\overset{\displaystyle Br}{\underset{\displaystyle |}{}}$$

C— C≡C— C— C— C

Next, fill in the other bonds with hydrogen atoms.

$$\underset{\displaystyle H}{\overset{\displaystyle Br}{H-\underset{|}{\overset{|}{C}}-}}C\equiv C-\underset{\displaystyle H}{\overset{\displaystyle H}{\underset{|}{\overset{|}{C}}}}-\underset{\displaystyle H}{\overset{\displaystyle H}{\underset{|}{\overset{|}{C}}}}-\underset{\displaystyle H}{\overset{\displaystyle H}{\underset{|}{\overset{|}{C}}}}-H \qquad \text{or} \qquad BrCH_2-C\equiv C-CH_2CH_2CH_3$$

2. **2-Butene:** There are four carbons [*but-*] in the parent compound and a double bond between carbons 2 and 3.

C— C=C— C

Next, fill in the other bonds with hydrogen atoms.

$$H-\underset{\displaystyle H}{\overset{\displaystyle H}{\underset{|}{\overset{|}{C}}}}-\overset{\displaystyle H}{\overset{|}{C}}=\overset{\displaystyle H}{\overset{|}{C}}-\underset{\displaystyle H}{\overset{\displaystyle H}{\underset{|}{\overset{|}{C}}}}-H \qquad \text{or} \qquad CH_3CH=CHCH_3$$

11.3 Geometric Isomers: A Consequence of Unsaturation

Learning Goal 3:   **Write the names and draw the structures of simple geometric isomers of alkenes.**

The carbon-carbon double bond is rigid because there is no free rotation around the double bond. The rigidity is caused by the shapes of the orbitals involved in the double bond. The electrons of one carbon-carbon bond lie in a line between the two nuclei. The electrons of the second bond lie above and below the carbon atoms of the double bond. This bond would have to be broken to allow rotation to occur; thus, the double bond is rigid.

The rigidity of the carbon-carbon double bond in alkenes produces another class of isomers: **geometric isomers**. Geometric isomers are described using the prefixes *cis* and *trans*, which provide an easy method for naming and distinguishing between the two isomeric forms.

Geometric isomers are different molecules with different physical and chemical properties. Although their properties may be similar, they are never identical.

The prefixes *cis* and *trans* refer to the placement of the substituents attached to the carbon-carbon double bond. When identical groups are on the same side of the double bond, the prefix *cis* is used; when identical groups are on opposite sides of the double bond, *trans* is the appropriate prefix.

155

As we saw in the previous chapter, cycloalkanes can also have geometric isomers. When two substituents are on the same side of the plane of the ring, the compound is the *cis* isomer. Alternatively, if the two substituents are on opposite sides of the plane of the ring, the compound is the *trans* isomer.

Example 5

Draw the *cis*- and *trans*- isomers of 2-butene.

Work

1. The carbon-carbon double bond is rigid as a result of the shapes of the orbitals involved in its formation. The specific atoms that are bonded to the two carbon atoms are locked into specific arrangements around the double bond.
2. The *cis* arrangement means that the two reference groups (say A and A) are on the same side of the double bond. The *trans* arrangement means the two A groups are across from each other.

cis-isomer   *trans*-isomer

Answer

1. First we draw 2-butene in a straight form, as before:

$$CH_3CH=CHCH_3$$

2. Next, locate the double bond and place the remaining groups either *cis* or *trans* from one another:

cis-2-Butene   *trans*-2-Butene

11.4 Alkenes in Nature

Ethene, the simplest alkene, is a plant growth substance produced by ripening fruit and in areas of a plant where cell division is occurring.

There are numerous polyenes in nature. These are molecules with multiple double bonds. The basic unit of these molecules, which include steroids, chlorophyll, and the lipid soluble vitamins A, D, E, and K, is the *isoprene* unit:

$$CH_2{=}{=}\overset{\displaystyle \overset{CH_3}{|}}{C}{-}CH{=}{=}CH_2$$

These polyenes are also called *isoprenoids* or *terpenes*. *Geraniol*, the scent of geraniums, is a natural insect repellant. D-*Limonene*, the aroma of oranges, is purified from citrus rinds. It is a useful solvent used in many cleaning products.

Myrcene is found in bayberry and has a spicy aroma used in perfumes and candles. *Farnesol*, found in many flowers, is used in skin care products marketed to increase skin elasticity. *Retinol*, a form of vitamin A, penetrates skin and stimulates collagen and elastin formation, which reduces wrinkles.

11.5 Reactions Involving Alkenes and Alkynes

Learning Goal 4:	Write equations predicting the products of the simple addition reactions of alkenes and alkynes: hydrogenation, halogenation, hydration, and hydrohalogenation.

The major reactions of alkenes and alkynes involve the addition of atoms or molecules to the carbon-carbon double or triple bond. The principal kinds of addition reactions are hydrogenation, halogenation, hydration, and hydrohalogenation.

Hydrogenation: Addition of H_2

Hydrogenation is the addition of a molecule of hydrogen (H_2) to a carbon-carbon double bond to produce an alkane. Two new C—H single bonds are formed as the double bond is broken.

Hydrogenation generally requires heat and/or pressure. The reaction always requires a metal catalyst, such as platinum or nickel, to allow the reaction to occur at a reasonably rapid rate.

The food industry takes advantage of the hydrogenation reaction. Vegetable oils are liquid because the fat molecules have many double bonds. When these oils are hydrogenated, many of the double bonds become carbon-carbon single bonds. This increases the rigidity of the fat molecules and solid fats are formed.

Example 6
Write an equation representing the hydrogenation reaction for 2-butene. Indicate the conditions needed and name all reactants and products.

Example 6 continued

Work

1. The term *hydrogenation* means the reaction of gaseous H_2 with a specific reactant. A catalyst such as Pt, Pd, or Ni is needed to speed up the reaction. Elevated temperature and pressure are also required.

Answer

$$CH_3CH = CHCH_3 \quad + \quad H_2 \quad \xrightarrow[\text{pressure}]{\text{Ni, heat}} \quad CH_3CH_2CH_2CH_3$$

2-Butene   Butane

Hydrogenation of an alkyne requires two moles of H_2.

Halogenation: Addition of X_2

Chlorine (Cl_2) or bromine (Br_2) can be added to a double bond. This reaction, called **halogenation**, proceeds readily and does not require a catalyst.

Example 7

Write an equation representing the halogenation of ethene to produce 1,2-dichloroethane.

Work

1. The term *halogenation* means the addition of a halogen molecule (X_2) across the double bond of an alkene. The double bond is broken and two new C–X bonds are formed.

Answer

Ethene   +   Chlorine   ⟶   1,2-Dichloroethane

Alkynes also react with halogens. Two moles of halogen are required and the product is a tetrahaloalkane.

Halogenation involving bromine (Br_2) can be used to show the presence of double or triple bonds in an organic compound. The reaction mixture is red due to the presence of dissolved bromine. If the red color is lost, it indicates that bromine was consumed, bromination occurred, and the compound was unsaturated. One mole of Br_2 is consumed per mole of alkene, and two moles of Br_2 are consumed per mole of alkyne.

Hydration: Addition of H_2O

Learning Goal 5:   **Apply Markovnikov's rule to predict the major and minor products of the hydration and hydrohalogenation reactions of unsymmetrical alkenes.**

A **hydration** reaction is the addition of a water molecule to a carbon-carbon double bond. When an alkene is reacted with water containing a trace of strong acid, an –OH group bonds to one carbon of the carbon-carbon double bond and an –H atom bonds to the other carbon. The product of the hydration of an alkene is an alcohol.

When an alkene is unsymmetrical (carries two different groups on the double bond carbons) hydration can yield two different products. One is usually favored over the other, as explained by **Markovnikov's rule,** which tells us that the "rich get richer." This means that the carbon of the carbon-carbon double bond that carries the greater number of hydrogen atoms most often will receive the hydrogen atom being added to the double bond. The other carbon becomes bonded to the –OH group.

Example 8
Write an equation representing the hydration of ethene to produce ethanol (ethyl alcohol).
Work
1. The term *hydration* means the addition of a water molecule (HOH) across the double bond of an alkene. The –H bonds to one carbon of the double bond and the –OH attaches to the other carbon of the double bond. The double bond is broken, and in its place are two new single bonds.
Answer

$$Ethene \quad + \quad Water \quad \xrightarrow{\quad acid \quad} \quad Ethanol$$

Example 9

Write an equation representing the hydration of propene, and name and draw the major product.

Work

The hydration of an unsymmetrical alkene (like propene—having different groups on each side of the double bond) favors one product over the other.

$$\begin{array}{c} H \quad \quad H \\ \diagdown \;\; \diagup \\ C \\ \| \\ C \\ \diagup \;\; \diagdown \\ H_3C \quad \quad H \end{array}$$

The carbon atom of the double bond that has the greater number of directly attached hydrogen atoms most often will receive the hydrogen atom being added to the double bond. The remaining carbon atom bonds to the –OH group. This is known as Markovnikov's rule.

Note that propene is unsymmetrical, since one carbon of the double bond has 2 –H atoms bonded to it while the other carbon of the double bond has only one –H atom and one –CH_3 group attached to it.

Answer

Propene   +   Water $\xrightarrow{\text{acid}}$ 2-Propanol (major product)

Hydration of an alkyne produces an *enol* in which the –OH is bonded to one carbon of a carbon-carbon double bond. The enol is quickly isomerized into an aldehyde or ketone.

Hydrohalogenation: Addition of HX

A hydrogen halide (H–Br, H–Cl, or H–I) also can be added to an alkene. This addition reaction is called **hydrohalogenation** and the product is an alkyl halide.

The reaction mechanism of hydrohalogenation is the same as that for hydration, and the nature of the major and minor alkyl halide products can be predicted by Markovnikov's rule. This reaction provides an easy, alternative means of preparing alkyl halides.

Example 10
Write an equation representing the reaction of hydrogen bromide with propene. Include the major and minor products. Also, indicate any other needed conditions.

Work
1. The reaction follows Markovnikov's rule.
2. Since alkenes are insoluble in water, the alkene and the hydrogen bromide (or the HX) are dissolved in CCl_4 as the solvent. Alkenes and CCl_4 are both nonpolar.

Answer

| Propene | Hydrogen bromide | 2-Bromopropane (major product) | 1-Bromopropane (minor product) |

Addition Polymers of Alkenes

Learning Goal 6:   Write equations representing the formation of addition polymers of alkenes.

Addition polymers are produced by the sequential addition of an alkene monomer to produce the polymer, which is a macromolecule composed of many repeating structural units (monomers). Many useful plastics are addition polymers produced from alkene monomers. Table 11.2 of the text presents a number of common addition polymers.

Example 11
Write an equation representing the reaction that produces the addition polymer polyvinyl chloride (PVC) from vinyl chloride monomers. What is the I.U.P.A.C. name for vinyl chloride?

Example 11 continued

Work

1. Begin by determining the structure of vinyl chloride (Table 11.2):

$$\begin{array}{c} H \\ \diagdown \\ H \diagup \end{array} C = C \begin{array}{c} Cl \\ \diagup \\ \diagdown H \end{array}$$

2. Now determine the structure of the addition polymer:

$$H - \overset{\displaystyle H}{\underset{\displaystyle H}{C}} - \overset{\displaystyle Cl}{\underset{\displaystyle H}{C}} - \overset{\displaystyle H}{\underset{\displaystyle H}{C}} - \overset{\displaystyle Cl}{\underset{\displaystyle H}{C}} - \overset{\displaystyle H}{\underset{\displaystyle H}{C}} - \overset{\displaystyle Cl}{\underset{\displaystyle H}{C}} - \text{etc.}$$

Using the simplified general form for the polymer, write the equation representing the reaction.

Answer

$$n \left[\begin{array}{c} H \\ \diagdown \\ H \diagup \end{array} C = C \begin{array}{c} Cl \\ \diagup \\ \diagdown H \end{array} \right] \xrightarrow[\text{pressure}]{\text{heat}} \left[\begin{array}{c} H \quad Cl \\ | \quad | \\ \sim\sim C - C \sim\sim \\ | \quad | \\ H \quad H \end{array} \right]_n$$

The I.U.P.A.C. name for vinyl chloride is chloropropene.

11.6 Aromatic Hydrocarbons

Structure and Physical Properties

Learning Goal 7:	Draw the structures and write the names of common aromatic hydrocarbons.

    Aromatic compounds are all characterized by the presence of an aromatic ring within the structure. The simplest, and most common, aromatic compound is benzene, which consists of a six carbon ring in a planar hexagonal arrangement. Each carbon is bonded to two other carbon atoms and to a single hydrogen atom.

    In 1865, Friedrich Kekulé proposed that single and double bonds alternated around the hexagonal ring. Since benzene did not decolorize bromine, he further proposed that the double and single bonds shifted position rapidly. Today we show this structure as a resonance model.

The current model of benzene structure proposes that each carbon atom is bonded to two other carbon atoms and to a hydrogen atom. The remaining six electrons are located in p orbitals that overlap to produce a cloud of electrons above and below the ring. Because the electrons are delocalized within the cloud, benzene is unusually stable and resists addition reactions typical of alkenes.

Nomenclature

Most simple aromatic compounds are named as derivatives of benzene. Others, for example phenol and toluene, have common historical names that must simply be memorized.

When there are two groups on the ring, there are three possible arrangements. Each arrangement is indicated by a prefix, as follows:

ortho, o: substituents are on adjacent carbon atoms
meta, m: substituents are on carbon atoms separated by one carbon atom
para, p: substituents are on carbon atoms separated by two carbon atoms

When more than two substituents are attached to the ring, a numerical system must be used in naming the compound.

Example 12
Name the following aromatic compounds:

1.　　　　　　2.　　　　　　3.　　　　　　4.

Work
The term *aromatic* means that there is a benzene ring present.
If only one group is bonded to the benzene structure, no number is needed in its name.
If two groups are bonded to the benzene ring, the following naming method is used:
   • if the groups are on adjacent carbons, use *ortho-*
   • if groups are separated by one carbon, use *meta-*
   • if two carbons are between groups, use *para-*
For three groups, it is best to use numbers.

Answer
1. Chlorobenzene
2. 1,4-Dichlorobenzene or *para*-dichlorobenzene
3. 1,3,5-Tribromobenzene
4. 1,2-Dibromobenzene or *ortho*-dibromobenzene

In I.U.P.A.C. nomenclature, the group that results from the removal of a single hydrogen atom from the benzene ring is called the **phenyl group**. Aromatic hydrocarbons having long aliphatic side chains are often named as phenyl-substituted hydrocarbons.

Reactions Involving Benzene

Learning Goal 8: Write equations for substitution reactions involving benzene.

The typical reactions of benzene are **substitution reactions**. In these reactions, a hydrogen atom is replaced by another atom or group of atoms. Benzene can react with chlorine or bromine.

Benzene   Bromine   Bromobenzene

Benzene also reacts with sulfur trioxide to undergo substitution, as seen in the following equation:

Benzene   Sulfur trioxide   Benzenesulfonic acid

Benzene can also undergo nitration in a reaction with concentrated nitric acid dissolved in concentrated sulfuric acid. This reaction requires a temperature of 50-55°C.

Benzene   Nitric acid   Nitrobenzene

164

Polynuclear Aromatic Hydrocarbons

Polynuclear aromatic hydrocarbons are composed of two or more aromatic rings joined to one another. Naphthalene has a distinctive aroma and is used as mothballs. Benzopyrene, found in cigarette smoke, is on of the most potent carcinogens known.

Section 11.7 Heterocyclic Aromatic Compounds

Learning Goal 9: **Describe heterocyclic aromatic compounds and list several biological molecules in which they are found.**

Heterocyclic aromatic compounds are those that have at least one atom other than carbon as part of the structure of the aromatic ring. Molecules such as pyridine, pyrimidines, purines, and pyrrole are heterocyclic aromatic compounds that are found in nicotine, DNA, RNA, and the porphyrin ring found in chlorophyll and hemoglobin.

Glossary of Key Terms in Chapter Eleven

addition polymer (11.5) a polymer prepared by sequential addition of monomers.

addition reaction (11.5) a reaction in which two molecules add together to form a new molecule; often involves the addition of one molecule to a double or triple bond in an unsaturated molecule.

alkene (11.1) a hydrocarbon that contains one or more carbon-carbon double bonds; an unsaturated hydrocarbon with the general formula C_nH_{2n}.

alkyne (11.1) a hydrocarbon that contains one or more carbon-carbon triple bonds; an unsaturated hydrocarbon with the general formula C_nH_{2n-2}.

aromatic compound (11.6) a hydrocarbon that contains a benzene ring or that has properties that are similar to those exhibited by benzene.

geometric isomers (11.3) isomers that differ from one another in the placement of substituents on a double bond or a ring.

halogenation (11.5) a reaction in which one of the C-H bonds of a hydrocarbon is replaced with a C-X bond of a halogen atom.

heterocyclic aromatic compound (11.7) a compound having at least one atom other than carbon in the structure of the aromatic ring.

hydration (11.5) a reaction in which water is added to a molecule.

hydrogenation (11.5) a reaction in which hydrogen (H_2) is added to a double or a triple bond.

hydrohalogenation (11.5) the addition of a hydrohalogen (HCl, HBr) to an unsaturated bond.

Markovnikov's rule (11.5) the rule that states that a hydrogen atom, adding to a carbon-carbon double bond, is more likely to add to the carbon having the larger number of hydrogens attached to it already.

monomer (11.5) the individual molecules from which a polymer is formed.

phenyl group (11.6) a benzene ring that has had a hydrogen atom removed.

polymer (11.5) a very large molecule formed by the combination of many small molecules (called monomers).

substitution reaction (11.6) a reaction in which a hydrogen atom attached to a carbon atom is replaced by another atom or group of atoms.

unsaturated compound (Intro) any hydrocarbon that contains one or more carbon-carbon double or triple bonds.

Self Test for Chapter Eleven

1. Write the molecular formula for and name the straight chain alkane, alkene, and alkyne that contain four carbon atoms.

2. Write the I.U.P.A.C. name of the following compound:

$$CH_3 - CH = C - CH_2CH_3$$
$$|$$
$$CH_2CH_3$$

3. Write the I.U.P.A.C. name of the following compound:

$$Br$$
$$|$$
$$CH_3CHCHC \equiv CCH_3$$
$$|$$
$$Br$$

4. Draw the structure of 2-methyl-2-pentene.

5. Draw the structure of *cis*-3-methyl-2-pentene.

6. Name the following compound using the prefixes *cis*- or *trans*-:

7. Write an equation for the hydrogenation reaction for 1-butene.

8. Write an equation for the hydration of propene. Indicate both the major and minor products.

9. Write an equation for the reaction of hydrogen chloride with 2-methyl-2-pentene.

10. Name the following aromatic compounds:

a.   b.   c.   d.   e.

11. What is the common name of ethyne?

12. What is the I.U.P.A.C. name of the following molecule?

13. What is the I.U.P.A.C. name of the following molecule?

$$CH_3 \diagdown \quad \diagup Br$$
$$C=C$$
$$Br \diagup \quad \diagdown H$$

14. What reaction is used to convert alkenes into alkanes?

15. What two catalysts are often used in the hydrogenation of alkenes?

16. Provide the product of the reaction represented by the following equation:

$$CH_2 = CH_2 \ + \ H_2O \ \xrightarrow{\text{acid}} \ ?$$

17. What compound makes up the basic unit of aromatic compounds?

18. What is the common name of trichloromethane?

19. In ethane, the two carbon atoms are joined by what type of bond?

20. Alkenes are characterized by what type of reaction?

21. What does the ending -*ane* mean?

22. What does the ending -*yne* mean?

23. Write an equation showing the reaction of propyne with 2 moles of HBr.

24. Write an equation showing the reaction of 1-butyne with 2 moles of hydrogen gas.

25. Provide the product of the reaction represented by the following equation:

$$CH_3CH_2CH = CH_2 \ + \ H_2 \ \xrightarrow{\text{Pd, heat}} \ ?$$

26. What materials are hydrogenated to produce a product such as Crisco?

Vocabulary Quiz for Chapter Eleven

1. A(n)_____ is a hydrocarbon that contains at least one carbon-carbon double bond. They are also described as unsaturated hydrocarbons with the general formula C_nH_{2n}.

2. A large molecule formed by the sequential addition of small units is called a (an)

   _____.

3. _____ are isomers (which by definition have the same molecular formula) that differ from one another due to the placement of substituents on a double bond or a ring.

4. A(n)_____ is a benzene ring that has had a hydrogen atom removed.

5. In the I.U.P.A.C. Nomenclature System, the _____ is the longest chain containing the major functional group in the molecule that is being named.

6. A(n)_____ is any hydrocarbon that contains one or more carbon-carbon triple bonds.

7. _____ is the addition of a hydrohalogen (HCl, HBr) to an unsaturated bond.

8. A(n) _____ is a reaction in which two molecules add together to form a new molecule. This often involves the addition of one molecule to a double or triple bond in an unsaturated molecule.

9. A(n) _____ is a compound having at least one atom other than carbon in the structure of the aromatic ring.

10. A(n) _____ is a reaction that results in the replacement of one group by another.

12 *Alcohols, Phenols, Thiols, and Ethers*

Learning Goals

1 Rank selected alcohols by relative water solubility, boiling points, or melting points.
2 Write the names and draw the structures of common alcohols.
3 Discuss the biological, medical, or environmental significance of several alcohols.
4 Classify alcohols as primary, secondary, or tertiary.
5 Write equations representing the preparation of alcohols by the hydration of an alkene.
6 Write equations representing the preparation of alcohols by the hydrogenation (reduction) of aldehydes and ketones.
7 Write equations showing the dehydration of an alcohol.
8 Write equations representing the oxidation of alcohols.
9 Discuss the role of oxidation and reduction reactions in the chemistry of living systems.
10 Discuss the use of phenols as germicides.
11 Write names and draw structures for common ethers and discuss their use in medicine.
12 Write equations representing the dehydration reaction between two alcohol molecules.
13 Write names and draw structures for simple thiols and discuss their biological significance.

Introduction

The functional group of the **alcohols** and **phenols** is the **hydroxyl group** (–OH). Alcohols have the general formula R–OH, in which R represents an alkyl group. Thus, the simplest alcohol is methyl alcohol: a methyl group bonded to a hydroxyl group, CH_3OH. Phenols have an aryl group in place of the alkyl group of the alcohols, and thus have the general formula Ar–OH. Phenol, the simplest member of this family, consists of a phenyl group (a benzene ring missing one hydrogen atom) bonded to a hydroxyl group.

Ethers contain two alkyl or aryl groups attached to an oxygen atom. Thus, the functional group of the ethers is R–O–R or Ar–O–Ar. **Thiols** are a family of compounds in which the sulfur atom has been substituted for the oxygen atom of an alcohol.

12.1 Alcohols: Structure and Physical Properties

Learning Goal 1:	Rank selected alcohols by relative water solubility, boiling points, or melting points.

We can think of alcohols (and phenols) as substituted water molecules: an H–O–H molecule in which one hydrogen atom has been replaced by an alkyl or aryl group. The R–O–H portion of an alcohol is planar, as is H–O–H. In addition, the bond angles of both R–O–H and H–O–H are 104.5°.

Alcohols are polar molecules owing to the hydroxyl group. The oxygen and hydrogen atoms have very different electronegativities (oxygen is 3.5 and hydrogen is 2.1). Consequently, the oxygen atom displays a partial negative charge and the hydrogen atom bears a partial positive charge. Alcohol molecules can form hydrogen bonds with other alcohol molecules and with the molecules of a polar solvent.

Both the polarity of alcohol molecules and the ability to form intermolecular hydrogen bonds exert a strong influence on the physical properties of alcohols. They have higher boiling points than hydrocarbons or ethers of similar molecular weight and show greater solubility in water. In fact, the smallest alcohols (1-4 carbons) are highly soluble in water; those having 5 or 6 carbons have decreased solubility; and those with more than 6 carbons are insoluble in water.

Example 1
For the following pairs of compounds, indicate the one that has the higher boiling point:

1. CH_4 (methane) or CH_3OH (methanol)
2. $CH_3CH_2CH_2CH_3$ (butane) or $CH_3OCH_2CH_3$ (ethyl methyl ether)
3. ethyl methyl ether or 1-propanol

Work
When comparing the boiling points of alkanes, ethers, and alcohols of similar molecular weight, remember that the alcohols have the highest boiling points. The ethers are found to have higher boiling points than the alkanes:

$$alcohol > ethers > alkanes$$

Answer
1. Methanol
2. Ethyl methyl ether
3. 1-Propanol

Example 2
Predict the solubility of each of the following alcohols in water solutions:

1. $CH_3CH_2CH_2CH_2CH_2CH_2OH$ 2. CH_3CH_2OH 3. $CH_3(CH_2)_8CH_2OH$

Work
1. Alcohols of 1-4 carbons are very soluble in water. As the carbon chain length increases, the alcohol becomes less soluble. Those with more than 6 carbons are insoluble in water.

Answer
1. Not very soluble 2. Very soluble 3. Insoluble in water

12.2 Alcohols: Nomenclature

> **Learning Goal 2:** Write the names and draw the structures for common alcohols.

I.U.P.A.C. Nomenclature

Determine the *parent compound*, in this case the longest continuous carbon chain containing the –OH group. Drop the *-e* ending of the alkane chain and replace it with *-ol*. Number the chain so that the hydroxyl group has the lowest possible number. Add substituents, named and numbered appropriately, as prefixes to the alcohol name.

If two hydroxyl groups are present, the suffix *-diol* is used; if three hydroxyl groups are present, the suffix used is *-triol*. The positions of hydroxyl groups are numbered as usual.

Example 3

Name the following compounds using the I.U.P.A.C. method:

a.

CH_3CHCH_3
  |
  OH

b.

$CH_3CHCH_2CHCH_3$
  | |
  OH CH_3

c.

CH_2CH_2
  | |
  OH OH

d.

$CH_3CHCHCH_3$
  | |
  Br OH

Work

1. Determine the parent compound that contains the –OH bonded to it.
2. Change the *-e* ending of the parent alkane's name to *-ol*.
3. Next, number the parent carbon chain so that the –OH has the lowest possible number.
4. If the alcohol has two or more –OH groups, add *-diol* or *-triol* to the parent's name.
5. Then use all the other rules we have used previously to finish naming the other groups that may be present.

Answer

a. 2-Propanol
b. 4-Methyl-2-pentanol
c. 1,2-Ethanediol
d. 3-Bromo-2-butanol

Common Names

The common names for the alcohols are derived from the name of the corresponding alkyl group. The name of the alkyl group is followed by the word *alcohol*.

Example 4
Give the common names of each of the following alcohols:

1. CH_3OH
2. CH_3CH_2OH
3.

    CH_3CHCH_3
          |
        OH

4. $CH_3CH_2CH_2CH_2OH$
5.

          CH_3
          |
   $H_3C - C - CH_3$
          |
         OH

Work
To determine the common names for the alcohols, first find which alkyl group is bonded to the –OH group. Next, write down the alkyl group name and add the word alcohol to complete the common name of that alcohol.

Answer

1. Methyl alcohol	3. Isopropyl alcohol	5. *t*-Butyl alcohol
2. Ethyl alcohol	4. Butyl alcohol	

12.3 Medically Important Alcohols

Learning Goal 3:   **Discuss the biological, medical, or environmental significance of several alcohols.**

Several of the small alcohols are important in medicine and industry. The smallest, *methanol* (methyl alcohol, wood alcohol) is a common solvent. It is extremely toxic if ingested.

Ethanol (ethyl alcohol, grain alcohol) is a solvent and disinfectant. Ethanol for human consumption is produced by fermentation of sugars by yeast. Denatured alcohol is ethanol to which a denaturing agent has been added to make it undrinkable.

2-Propanol (isopropyl alcohol, rubbing alcohol) is commonly used as a disinfectant and solvent. Like methanol, it is very toxic if ingested.

1,2-Ethanediol (ethylene glycol) is a common antifreeze for cars.

1,2,3-Propanetriol (glycerol) is a component of stored fats in the body. It is used in cosmetics and pharmaceuticals.

12.4 Classification of Alcohols

Learning Goal 4:	Classify alcohols as primary, secondary, or tertiary.

Alcohols are classified as methyl, **primary (1°), secondary (2°),** or **tertiary (3°)** depending on the number of alkyl groups attached to the **carbinol carbon**, the carbon that bears the hydroxyl group. If there is one alkyl group attached to the carbinol carbon, the alcohol is a primary alcohol. An alcohol with two alkyl groups bonded to the carbinol carbon is a secondary alcohol. A tertiary alcohol has three alkyl groups bonded to the carbinol carbon.

Example 5

Classify the following alcohols as either primary, secondary, or tertiary:

1. CH_3CH_2OH

2. CH_3CHCH_3
    |
    OH

3. $CH_3CH_2CHCH_3$
    |
    OH

4. $CH_3CH_2CH_2CCH_3$ with CH_3 above the fourth carbon and OH below it

Answer
1. Primary
2. Secondary
3. Secondary
4. Tertiary

12.5 Reactions Involving Alcohols

Preparation of Alcohols

Learning Goal 5:	Write equations representing the preparation of alcohols by the hydration of an alkene.

Learning Goal 6:	Write equations representing the preparation of alcohols by the hydrogenation (reduction) of aldehydes and ketones.

In the laboratory, alcohols are prepared by the hydration of alkenes or the reduction of aldehydes and ketones.

Example 6

Show the reduction reactions for the conversions of formaldehyde and acetaldehyde into their corresponding alcohols.

Answer

1.

Formaldehyde
(methanal)

Methyl alcohol
(methanol)

2.

Acetaldehyde
(ethanal)

Ethyl alcohol
(ethanol)

For an example of the hydration of an alkene, see Example 8 in Chapter 11.

Dehydration of Alcohols

Learning Goal 7:   Write equations showing the dehydration of an alcohol.

Alcohols undergo **dehydration** (loss of water) when heated in the presence of concentrated sulfuric or phosphoric acids. The products of dehydration are an alkene and a water molecule. Quite simply, a dehydration reaction is the reverse of the hydration reactions that produce an alcohol from an alkene and water.

In some cases, dehydration of an alcohol may produce a mixture of products. **Zaitsev's rule** tells us that the more highly substituted alkene will be the major product.

Example 7

Write an equation representing the dehydration reactions for the following alcohols:

1. 1-Propanol
2. Ethanol
3. 2-Butanol

Example 7 continued

Work

The dehydration of an alcohol produces an alkene plus water. Remember that in some cases the dehydration of an alcohol yields a mixture of products. In those instances, it is the more highly branched alkene that is the major product.

Answer

1. $CH_3CH_2CH_2OH$ $\xrightarrow[\text{heat}]{\text{acid}}$ $CH_3CH = CH_2$ $+$ H_2O

       1-Propanol   Propene

2. CH_3CH_2OH $\xrightarrow[\text{heat}]{\text{acid}}$ $CH_2 = CH_2$ $+$ H_2O

       Ethanol   Ethene

3.
$$CH_3CH_2 - \overset{\overset{\displaystyle OH}{|}}{\underset{\underset{\displaystyle H}{|}}{C}} - CH_3 \xrightarrow[\text{heat}]{\text{acid}} CH_3CH_2CH = CH_2 + H_2O$$

                                                         1-Butene

      2-Butanol

$$CH_3CH = CHCH_3 + H_2O$$

                                         2-Butene

Two products are formed. The major product is 2-butene.

Oxidation Reactions

Learning Goal 8:   **Write equations representing the oxidation of alcohols.**

Some alcohols can be **oxidized** to produce aldehydes, ketones or carboxylic acids. The most commonly used oxidizing agents are basic potassium permanganate ($KMnO_4/OH^-$) and chromic acid (H_2CrO_4).

Methanol and all primary alcohols produce aldehydes, while secondary alcohols form ketones. Tertiary alcohols cannot be oxidized. This is because the carbon bonded to the hydroxyl group must contain at least one C–H bond in order for oxidation to occur. Since tertiary alcohols contain three C–C bonds to the carbon bonded to the hydroxyl group they cannot be oxidized. Aldehydes can undergo further oxidation to produce carboxylic acids.

Example 8

Write an equation representing the oxidation of methanol and ethanol by the liver. Name all products.

Work

The oxidation of an alcohol first produces an aldehyde or ketone. If an aldehyde is formed, it can be further oxidized to produce a carboxylic acid.

Answer

1. CH_3OH → (liver enzymes) → Formaldehyde (very toxic) → (liver enzymes) → Formic acid (very toxic)

Methanol

2. CH_3CH_2OH → (liver enzymes) → Acetaldehyde → (liver enzymes) → Acetic Acid

Ethanol

Example 9

Provide the oxidation products for the following compounds.

1. primary alcohol
2. secondary alcohol
3. tertiary alcohol
4. aldehyde
5. 2-propanol
6. 2-methyl-2-propanol

Answer

1. aldehyde, then to a carboxylic acid
2. ketone
3. no reaction
4. carboxylic acid
5. ketone—called acetone
6. no reaction—a tertiary alcohol

176

12.6 Oxidation and Reduction in Living Systems

Learning Goal 9: **Discuss the role of oxidation and reduction reactions in the chemistry of living systems**

In organic and biological chemistry, **oxidation** involves a gain of oxygen or a loss of hydrogen from a compound. A **reduction** reaction involves a loss of oxygen or gain of hydrogen. To recognize an oxidation or reduction reaction, simply count the hydrogen and oxygen atoms in the products and reactants and apply the definitions presented above.

Oxidation and reduction reactions are extremely important in the metabolic pathways that harvest energy for use by our bodies. A class of enzymes called **oxidoreductases** catalyze these reactions. **Coenzymes** serve as acceptors, donors, and carriers of hydrogen in these cellular oxidation-reduction reactions. Nicotinamide adenine dinucleotide, NAD^+, is one such coenzyme.

12.7 Phenols

Learning Goal 10: **Discuss the use of phenols as germicides.**

Phenols are compounds in which the hydroxyl group is attached to an aryl group. Owing to the polar hydroxyl group, the phenols are also polar compounds.

The simplest member of this family is known by the common name phenol. This compound is of interest in the history of medicine. Joseph Lister, a British physician, observed that the incidence of post-surgical infections could be radically decreased if the surgical instruments and the incision were treated with an antimicrobial chemical. The agent he used was carbolic acid, a dilute solution of phenol. As a result of his observations, the use of antiseptics and disinfectants has become routine medical practice.

12.8 Ethers

Learning Goal 11: **Write names and draw structures for common ethers and discuss their use in medicine.**

Ethers are structurally related to alcohols. However, a quick look at the geometry of the functional group characteristic of the ethers (R–O–R) reveals that these compounds are much less polar than alcohols. Indeed, they are much less water-soluble and have much lower boiling points than the comparable alcohols.

Ethers are chemically inert. Under normal conditions, they will not react with oxidizing agents, reducing agents, or bases.

Common names for ethers are derived from the names of the alkyl groups attached to the ether oxygen. The alkyl group names are listed as prefixes before the word ether and may be ordered by size (small to large) or alphabetically. For instance, the compound

CH_3–O–CH_2CH_3 can be called methyl ethyl ether (prefixes arranged by size of the alkyl group) or ethyl methyl ether (prefixes ordered alphabetically).

This same compound would be named methoxyethane using the I.U.P.A.C. Nomenclature System. With this system, an ether is named as a substituted hydrocarbon, in this case an alkane. In this system, the –OR group is named as an alkoxy group.

Example 10

Name the following ethers using the I.U.P.A.C. Nomenclature System:

$$\overset{\displaystyle OCH_3}{\underset{\displaystyle}{|}}$$

a. $CH_3CH_2\overset{|}{C}HCH_2CH_2CH_2CH_3$

$$\overset{\displaystyle OCH_2CH_3}{\underset{\displaystyle}{|}}$$

b. $CH_3\overset{|}{C}HCH_2CH_2CH_3$

Work

Parent compounds:	a. Heptane	b. Pentane
Position of alkoxy group:	a. Carbon-3	b. Carbon-2
Substituents:	a. Methoxy	b. Ethoxy

Answer

a. 3-Methoxyheptane   b. 2-Ethoxypentane

Learning Goal 12:	Write equations representing the dehydration reaction between two alcohol molecules.

Ethers may be prepared by a dehydration reaction between two alcohol molecules.

Example 11

Write an equation showing the synthesis of diethyl ether.

Work

Diethyl ether would be prepared by a dehydration reaction between two molecules of ethanol.

Answer

$$CH_3CH_2OH + CH_3CH_2OH \xrightarrow{H^+} CH_3CH_2-O-CH_2CH_3 + H_2O$$

Ethanol   Ethanol   Diethyl ether

Diethyl ether was the first general anesthetic used in medical practice. However, since ethers are highly flammable and can form explosive peroxides upon storage, diethyl ether has largely been replaced by halogenated ethers, such as penthrane and ethrane.

12.9 Thiols

> **Learning Goal 13:** **Write names and draw structures for simple thiols and discuss their biological significance.**

Compounds that contain the –SH group are known as **thiols**. Although similar to alcohols in structure, thiols generally have lower boiling points than corresponding alcohols, though they are higher in molecular weight. Furthermore, thiols, and many other sulfur compounds, have nauseating aromas.

Thiols are also involved in protein structure and conformation. It is the ability of two thiol groups to easily undergo oxidation to a –S–S–(disulfide) bond that helps maintain the correct shape of a protein. The amino acid cysteine has a thiol group and can participate in disulfide bond formation in proteins.

Coenzyme A is a thiol that serves as a "carrier" of acetyl groups and fatty acids in cellular metabolic reactions.

Glossary of Key Terms in Chapter Twelve

alcohol (12.1) an organic compound that contains a hydroxyl group (-OH) attached to an alkyl group.

carbinol carbon (12.4) in an alcohol, the carbon to which the hydroxyl group is attached.

dehydration (of alcohols) (12.5) a reaction that involves the loss of a water molecule; e.g., the loss of water from an alcohol and the concomitant formation of an alkene.

disulfide (12.9) an organic compound that contains the disulfide group (-S-S-).

elimination reaction (12.5) a reaction in which a molecule loses atoms or ions from its structure.

ether (12.8) an organic compound that contains two alkyl and/or aryl groups attached to an oxygen atom; R-O-R, Ar-O-R, and Ar-O-Ar.

fermentation (12.3) the anaerobic (in the absence of oxygen) metabolism or degradation of glucose by microorganisms.

hydration (of alkenes) (12.5) a reaction in which water is added to a molecule; e.g., the addition of water to an alkene to form an alcohol.

hydroxyl group (12.1) the -OH functional group.

oxidation (of alcohols) (12.6) the loss of electrons by a molecule, atom, or ion; in organic compounds, the gain of oxygen or loss of hydrogen; e.g., the conversion of an alcohol to an aldehyde or ketone via the use of an oxidizing agent.

phenol (12.7) an organic compound that contains a hydroxyl group (-OH) attached to a benzene ring.

primary (1°) alcohol (12.4) an alcohol with the general formula RCH_2OH.

reduction (of carbonyl compounds) (12.6) a gain of electrons by a molecule, ion, or atom; in organic compounds, the loss of oxygen or gain of hydrogen; e.g., the conversion of a carbonyl compound to an alcohol via the use of a reducing agent.

secondary (2°) alcohol (12.4) an alcohol with the general formula R_2CHOH.

tertiary (3°) alcohol (12.4) an alcohol with the general formula R_2COH.

thiol (12.9) an organic compound that contains a thiol group (-SH).

Zaitsev's rule (12.5) a rule that states that in an elimination reaction, the alkene with the greatest number of alkyl groups on the double bonded carbon (the more highly substituted alkene) is the major product of the reaction.

Self Test for Chapter Twelve

1. From each of the following pairs of compounds, pick the one that has the higher boiling point:

   a. 1-nonanol or 1-hexanol

   b. ethane or methanol

   c. methoxyethane or 1-propanol

   d. methane or dimethyl ether

2. Which of the following alcohols has the greatest solubility in water?

   a. 1-propanol

   b. 1-butanol

   c. 1-heptanol

   d. 1-decanol

3. Provide the I.U.P.A.C. name of the following compound:

$$\underset{\quad\quad OH\quad CH_3}{CH_3\overset{\displaystyle Br}{\overset{|}{C}}HCHCHCH_3}$$

4. Provide the I.U.P.A.C. name of the following compound:

$$\underset{OH}{CH_3\overset{|}{C}HCH_2CH_2CH_2Cl}$$

5. Determine the I.U.P.A.C. name of the trialcohol called glycerol:
   $HOCH_2CHOHCH_2OH$

6. Provide the common names of the following:

   a. $CH_3CH_2CH_2OH$

   b. $CH_3CHOHCH_3$

   c. $C(CH_3)_3OH$

7. Complete the following chemical reactions:

   a.

   b. $CH_2{=}CH_2$ + H_2O $\xrightarrow[\text{catalyst}]{H^+}$?

8. Complete the oxidation reactions of each of the following alcohols to form aldehydes:

   a. CH_3OH $\xrightarrow{\text{oxidation}}$

   b. CH_3CH_2OH $\xrightarrow{\text{oxidation}}$

9. What are the products of the dehydration of each of the following alcohols?

    a. ethanol

    b. 1-propanol

    c. 2-butanol

10. Determine the oxidation products of the following alcohols:

    a. CH_3OH c. 2-propanol

    b. ethanol   d. 2-methyl-2-propanol

11. Provide the common names of each of the following ethers:

    a. CH_3OCH_3 c. $CH_3OCH_2CH_3$

    b. $CH_3CH_2OCH_2CH_3$

12. Which of the following compounds is the most oxidized?

    a. propane   d. acetic acid

    b. 1-propanol   e. propanone

    c. 2-propanol

13. Which of the following compounds is the most reduced?

    a. methane   c. dichloromethane

    b. chloromethane   d. trichloromethane

14. What class of compounds is represented by the general formula ROH?

15. What class of compounds is represented by the general formula RSH?

16. Why do alcohols boil at higher temperatures than pure hydrocarbons or ethers of similar molecular weights?

17. What alcohol is produced by the fermentation of sugars and starches? Use the I.U.P.A.C. name.

18. If ethanol has some agent added to it to make it unfit to drink, what do we call that solution?

19. Write the common name of the alcohol that is a viscous, sweet-tasting, nontoxic liquid. (Refer to the text.)

20. In a secondary alcohol, how many hydrogen atoms are directly attached to the carbinol carbon atom?

21. Which kind of alcohol will not undergo oxidation under normal conditions?

22. Potassium permanganate and potassium dichromate are used with alcohols in what kind of chemical reaction?

23. Which product of the oxidation of ethanol causes many of the adverse effects of the "morning-after hangover"?

24. Which class of compounds with a strong odor is found in the warning scent of skunks?

25. Which of the following is the least polar? Give the letter as the answer.
    a. alcohols
    b. ethers
    c. water

26. Which of the following cannot hydrogen bond to other molecules?
    a. alcohols   b. ethers   c. water

Vocabulary Quiz for Chapter Twelve

1.   Organic compounds that contain hydroxyl groups attached to alkyl groups are
      _____.

2.   A family of organic compounds that contain a sulfhydryl group is the _____.

3.   The carbon atom bonded to the hydroxyl group is called the _____.

4.   _____ is the anaerobic (in the absence of oxygen) metabolism or
      degradation of glucose by microorganisms.

5.   A(n) _____ is an organic compound that contains two alkyl and/or aryl
      groups attached to an oxygen atom: R-O-R, Ar-O-R, and Ar-O-Ar.

6.   A(n) _____ is an organic compound that contains a hydroxyl group (-OH)
      attached to a benzene ring.

7.   The -OH function group is called the _____.

8.   _____ states that in an elimination reaction, the alkene with the greatest
      number of alkyl groups on the double bonded carbon (the more highly substituted
      alkene) is the major product of the reaction.

9.   In inorganic chemistry, _____ is the loss of electrons accompanied by a
      change in charge. In organic chemistry, this is more generally seen as a gain of
      oxygen or loss of hydrogen.

10.   An alcohol with a carbinol carbon bonded to three alkyl groups is a _____
      alcohol.

13 Aldehydes and Ketones

Learning Goals

1 Draw the structures and discuss the physical properties of aldehydes and ketones.
2 From the structures, write the common and I.U.P.A.C. names of aldehydes and ketones.
3 List several aldehydes and ketones that are of natural, commercial, health, and environmental interest and describe their significance.
4 Write equations for the preparation of aldehydes and ketones by the oxidation of alcohols.
5 Write equations representing the oxidation of carbonyl compounds.
6 Write equations representing the reduction of carbonyl compounds.
7 Write equations for the preparation of hemiacetals, hemiketals, acetals, and ketals.
8 Draw the keto and enol forms of aldehydes and ketones.
9 Write equations showing the aldol condenstation.

Introduction

The **carbonyl group** (–C = O) is characteristic of many groups of organic compounds, including the **aldehydes** and **ketones**, carboxylic acids and esters. The carbonyl group and the two atoms attached to it are coplanar.

Aldehydes and ketones are carbonyl-containing compounds that differ from one another in the type of atom or atoms attached to the carbonyl carbon. In ketones, the carbonyl carbon is attached to two carbon atoms, whereas in aldehydes the carbonyl carbon is attached to at least one hydrogen atom; the second atom attached to the carbonyl carbon in aldehydes may be another hydrogen or a carbon atom.

13.1 Structure and Physical Properties

Learning Goal 1:	Draw the structures and discuss the physical properties of aldehydes and ketones.

Owing to the polar carbonyl group, aldehydes and ketones are moderately polar compounds. As a result, they boil at higher temperatures than hydrocarbons or ethers having the same number of carbon atoms, but at temperatures lower than comparable alcohols.

Aldehydes and ketones composed of five or fewer carbon atoms are reasonably soluble in water because of the hydrogen bonding between the carbonyl group and water molecules. Larger members of these carbonyl-containing compounds are less polar, more hydrocarbon-like, and thus more soluble in nonpolar organic solvents.

Example 1
Which member of each of the following pairs of compounds that have similar molecular weights has the higher boiling point?

1. butane or methoxyethane
2. methoxyethane or propanol
3. propanone or methoxyethane

Work
Aldehydes and ketones are moderately polar compounds and boil at higher temperatures than hydrocarbons or ethers of similar molecular weights. The aldehydes and ketones, however, boil at lower temperatures than similar alcohols.

Answer
1. methoxyethane
2. propanol
3. propanone

Example 2
Show the polar attraction that is present between two acetone molecules.

Answer

13.2 I.U.P.A.C. Nomenclature and Common Names

Learning Goal 2:	From the structures, write the common and I.U.P.A.C. names of aldehydes and ketones.

Naming Aldehydes

Determine the parent compound, that is, the longest continuous carbon chain containing the carbonyl group. Drop the final -e of the parent alkane and replace it with -al. The parent chain is always numbered beginning with the carbonyl carbon as carbon-1. All other substituents are named and numbered as usual.

Common names for aldehydes are derived from the same Latin root as the corresponding carboxylic acids (see Table 13.1). Substituted aldehydes are named as derivatives of the straight-chain parent compound, using Greek letters to indicate the positions of substituents. The carbon atom bonded to the carbonyl carbon is referred to as the α-carbon.

Naming Ketones

Determine the parent compound. Replace the *–e* ending of the parent alkane with the *-one* suffix of the ketone family. The longest carbon chain is numbered to give the carbonyl carbon the lowest possible number. All other substituents are named and numbered as usual.

The common names for ketones are derived by naming the alkyl groups that are bonded to the carbonyl carbon. These are used as prefixes followed by the word *ketone*. Alkyl groups may be arranged alphabetically or by size.

Example 3
Determine the I.U.P.A.C. and common names of the following aldehydes and ketones:

$$
\begin{array}{c}
\qquad\qquad\quad O \\
\qquad\qquad\quad \| \\
1.\quad H-C-H
\end{array}
\qquad\qquad
\begin{array}{c}
\qquad\qquad\qquad O \\
\qquad\qquad\qquad \| \\
3.\quad CH_3CH_2-C-CH_3
\end{array}
$$

$$
\begin{array}{c}
\qquad\qquad\qquad O \\
\qquad\qquad\qquad \| \\
2.\quad CH_3CH_2-C-H
\end{array}
\qquad\qquad
\begin{array}{c}
\qquad\qquad\qquad\qquad O \\
\qquad\qquad\qquad\qquad \| \\
4.\quad CH_3CH_2CH_2-C-CH_3
\end{array}
$$

Work
Aldehydes are named according to the longest continuous carbon chain (parent) that contains the carbonyl group. The final *-e* of the parent alkane is dropped and replaced by *-al* for aldehyde. Numbers are not needed for unsubstituted aldehydes, since the carbonyl group is always at the end of the molecule. For ketones, the final *-e* of the alkane name is replaced with *-one*, and numbers may be needed to show the position of the carbonyl group in the compound.

Answer
1. methanal (common name: formaldehyde)
2. propanal (common name: propionaldehyde)
3. butanone (common name: ethyl methyl ketone)
4. 2-pentanone (common name: methyl propyl ketone)

Example 4
Write structural formulas for each of the following:

1. 5-bromohexanal
2. 3-methylheptanal
3. 3-bromo-2-pentanone
4. cyclohexanone

Example 4 continued

Answer

1. 5-bromohexanal: $CH_3CHCH_2CH_2CH_2 - \overset{\displaystyle O}{\overset{\displaystyle \|}{C}} - H$

   $\underset{\displaystyle Br}{\vert}$

2. 3-methylheptanal: $CH_3CH_2CH_2CH_2CHCH_2 - \overset{\displaystyle O}{\overset{\displaystyle \|}{C}} - H$

   $\underset{\displaystyle CH_3}{\backslash}$

3. 3-bromo-2-pentanone: $CH_3CH_2CH - \overset{\displaystyle O}{\overset{\displaystyle \|}{C}} - CH_3$

   $\overset{\displaystyle Br}{\overset{\displaystyle \vert}{}}$

4. cyclohexanone: $= O$

13.3 Important Aldehydes and Ketones

Learning Goal 3:	List several aldehydes and ketones that are of natural, commercial, health, and environmental interest and describe their significance.

Methanal (formaldehyde) is a gas. It is available commercially as an aqueous solution (formalin) that is used to preserve tissue samples.

Ethanal (acetaldehyde) is produced from ethanol in the liver and is responsible for the symptoms of a hangover.

Propanone (acetone) is the simplest ketone. It is an important solvent because it can dissolve organic compounds and is also miscible (mixes) with water. It is found as a solvent in adhesives, paints, and nail polish remover. Many complex members of the ketone family are important in the food industry as food additives. Others are useful as medicinals and agricultural chemicals.

13.4 Reactions Involving Aldehydes and Ketones

Preparation of Aldehydes and Ketones

> **Learning Goal 4:** Write equations for the preparation of aldehydes and ketones by the oxidation of alcohols.

In the laboratory, aldehydes and ketones are often prepared by the oxidation of the corresponding alcohol. Any aldehyde or ketone can be prepared if the correct alcohol is available.

Oxidation of methyl alcohol gives methanal. Oxidation of a primary alcohol produces an aldehyde; oxidation of a secondary alcohol yields a ketone. Tertiary alcohols do not undergo oxidation. The conclusion we can draw from this information is that the carbonyl carbon must have at least one hydrogen substituent for oxidation to occur.

Example 5

Provide the I.U.P.A.C. names of the oxidation products for the following alcohols:
1. methanol
2. 2-propanol

Answer
1. methanal (or methanoic acid, if methanal is further oxidized)
2. propanone

Oxidation Reactions

> **Learning Goal 5:** Write equations representing the oxidation of carbonyl compounds.

Aldehydes are very easily oxidized further to carboxylic acids. In fact, they are so easily oxidized that it is often very difficult to prepare or store them. Oxidation of an aldehyde yields a carboxylic acid:

Ketones do not undergo further oxidation reactions because a carbon-hydrogen bond to the carbonyl carbon is necessary for the reaction to occur.

Aldehydes and ketones can be distinguished from one another based on their ability to undergo oxidation reactions. The **Tollens' test**, or Tollens' silver mirror test, is the most common such test. Tollens' reagent consists of a basic solution of $Ag(NH_3)_2{}^+$. An aldehyde will undergo an oxidation-reduction reaction in which the silver ion (Ag^+) is reduced to silver metal (Ag^0) as the aldehyde is oxidized to a carboxylic acid. The silver metal precipitates from solution and coats the vessel, giving a smooth silver mirror. Because ketones cannot undergo further oxidation, they do not react with the Tollens' reagent.

Example 6

Write the major products for each of the following reactions, using Tollens' reagent as the oxidizing agent:

1.
$$CH_3 - \overset{\displaystyle O}{\overset{\displaystyle \|}{C}} - H \;\; + \;\; Ag(NH_3)_2^+ \longrightarrow \quad ?$$

2.
$$CH_3CH_2 - \overset{\displaystyle O}{\overset{\displaystyle \|}{C}} - H \;\; + \;\; Ag(NH_3)_2^+ \longrightarrow \quad ?$$

3.
$$CH_3 - \overset{\displaystyle O}{\overset{\displaystyle \|}{C}} - CH_3 \;\; + \;\; Ag(NH_3)_2^+ \longrightarrow \quad ?$$

Work

Treatment of an aldehyde with Tollens' reagent gives an oxidation-reduction reaction. The aldehyde is oxidized to the carboxylic acid, and the silver ion (Ag^+) is reduced to silver metal (Ag^0).

Answer
1. $CH_3COOH + Ag^0$
2. $CH_3CH_2COOH + Ag^0$
3. No reaction—Tollens' reagent can not oxidize ketones.

Another test used to distinguish between aldehydes and ketones is **Benedict's test.** In this test, Cu^{2+} is reduced to Cu^+. Cu^{2+} is soluble and gives a blue solution, while the Cu^+ precipitates as the red solid, Cu_2O. Benedict's test has been used to determine the concentrations of glucose in urine.

Reduction Reactions

Learning Goal 6:	Write equations representing the reduction of carbonyl compounds.

Aldehydes and ketones are easily reduced to the corresponding alcohol by a large number of different reducing agents, designated [H]. The general reduction reaction is shown:

$$R-\overset{\overset{\displaystyle O}{\|}}{C}-R' \quad \xrightarrow{[H]} \quad R-\overset{\overset{\displaystyle OH}{|}}{\underset{\underset{\displaystyle H}{|}}{C}}-R'$$

Aldehydeor ketone   Alcohol

The classical reaction for aldehyde and ketone reduction is **hydrogenation**, in which the carbonyl compound is reacted with hydrogen gas. This reaction requires a metal catalyst, pressure and/or heat. The carbon-oxygen double bond (the carbonyl group) is reduced to a carbon-oxygen single bond.

$$R-\overset{\overset{\displaystyle O}{\|}}{C}-R' \quad + \quad H_2 \quad \xrightarrow{Pt} \quad R-\overset{\overset{\displaystyle OH}{|}}{\underset{\underset{\displaystyle H}{|}}{C}}-R'$$

Aldehydeor ketone   Alcohol

Example 7

Label the following as oxidation or reduction reactions:

1. methanol to methanoic acid
2. methanol to formaldehyde
3. benzoic acid to benzaldehyde
4. 2-pentanol to 2-pentanone
5. acetone to 2-propanol

Answer

1. oxidation
2. oxidation
3. reduction
4. oxidation
5. reduction

Addition Reactions

Learning Goal 7: Write equations for the preparation of hemiacetals, hemiketals, acetals, and ketals.

The most common reaction of the carbonyl group is **addition** across the carbon-oxygen double bond. The reaction requires catalytic amounts of acid, represented as H^+ over the reaction arrow.

The addition of one alcohol molecule to an aldehyde produces a **hemiacetal**. Addition of a second alcohol molecule yields an **acetal**.

$$
\underset{\text{Aldehyde}}{\overset{\displaystyle O}{\underset{R}{\overset{\|}{C}}_{\diagup}}\diagdown_H} \; + \; R'OH \;\underset{\xrightarrow{H^+}}{\rightleftharpoons}\; \underset{\text{Hemiacetal}}{R-\underset{H}{\overset{OH}{\underset{|}{\overset{|}{C}}}}-O-R'} \; + \; R''OH \;\underset{\xrightarrow{H^+}}{\rightleftharpoons}\; \underset{\text{Acetal}}{R-\underset{H}{\overset{OR''}{\underset{|}{\overset{|}{C}}}}-O-R'} \; + H_2O
$$

Similarly, the reaction of a ketone with one alcohol molecule produces a **hemiketal**. Addition of a second alcohol molecule yields a **ketal**.

$$
\underset{\text{Ketone}}{\overset{\displaystyle O}{\underset{R}{\overset{\|}{C}}_{\diagup}}\diagdown_R} \; + \; R'OH \;\underset{\xrightarrow{H^+}}{\rightleftharpoons}\; \underset{\text{Hemiketal}}{R-\underset{R}{\overset{OH}{\underset{|}{\overset{|}{C}}}}-O-R'} \; + \; R''OH \;\underset{\xrightarrow{H^+}}{\rightleftharpoons}\; \underset{\text{Ketal}}{R-\underset{R}{\overset{OR''}{\underset{|}{\overset{|}{C}}}}-O-R'} \; + H_2O
$$

Hemiacetals and hemiketals are commonly found in the structures of carbohydrates. Linear sugar molecules have many hydroxyl groups and at least one carbonyl group that can undergo an intramolecular reaction in solution to give cyclic hemiacetals or hemiketals.

Example 8

Show the reaction of acetone with methyl alcohol to form a hemiketal compound.

Answer

$$
\underset{\text{Acetone}}{\overset{\displaystyle O}{H_3C\overset{\|}{\underset{}{C}}CH_3}} \; + \; CH_3OH \;\underset{\xrightarrow{H^+}}{\rightleftharpoons}\; \underset{\text{Hemiketal}}{CH_3-\underset{CH_3}{\overset{OH}{\underset{|}{\overset{|}{C}}}}-O-CH_3}
$$

Keto-Enol Tautomerism

An aldehyde or a ketone may exist in an equilibrium mixture of two isomers called **tautomers**. One isomer is the **keto** form; the other is the **enol** form. The two forms are shown below:

$$
\begin{array}{ccc}
R-\overset{\overset{\displaystyle H}{|}}{\underset{\underset{\displaystyle R'}{|}}{C}}-\overset{\overset{\displaystyle O}{\|}}{C}-R'' & \rightleftharpoons & \overset{\displaystyle R}{\underset{\displaystyle R'}{}}C=C\overset{\displaystyle OH}{\underset{\displaystyle R''}{}}
\end{array}
$$

$$\text{Keto form} \qquad\qquad\qquad \text{Enol form}$$

Typically, the keto form is more stable than the enol form.

Aldol Condensation

The **aldol condensation** is a reaction in which aldehydes or ketones react to form larger molecules. The α-carbon of one aldehyde forms a bond with the carbonyl carbon of a second aldehyde. One biological example of the aldol condensation occurs in the gluconeogenesis pathway. This pathway accomplishes the synthesis of the sugar glucose. The enzyme *aldolase* catalyzes the condensation of the aldehyde glyceraldehyde-3-phosphate and the ketone dihydroxyacetone phosphate.

Glossary of Key Terms in Chapter Thirteen

acetal (13.4) the family of organic compounds formed via the reaction of two molecules of alcohol with an aldehyde in the presence of an acid catalyst.

addition reaction (13.4) a reaction in which two molecules add together to form a new molecule; often involves the addition of one molecule to a double or triple bond in an unsaturated molecule; e.g., the addition of alcohol to an aldehyde or ketone to form a hemiacetal or hemiketal.

aldehyde (13.1) a class of organic molecules characterized by a carbonyl group; the carbonyl carbon is bonded to a hydrogen atom and to another hydrogen or an alkyl or aryl group.

aldol condensation (13.4) a reaction in which aldehydes or ketones react to form a larger molecule.

Benedict's Test (13.4) a test reagent used to distinguish aldehydes and ketones; a test for reducing sugars.

carbonyl group (Intro) the functional group that contains a carbon-oxygen double bond; -C=O; the functional group found in aldehydes and ketones.

hemiacetal (13.4) the family of organic compounds formed via the reaction of one molecule of an alcohol with an aldehyde in the presence of an acid catalyst.

hemiketal (13.4) the family of organic compounds formed via the reaction of one molecule of alcohol with a ketone in the presence of an acid catalyst.

hydrogenation (13.4) a reaction in which hydrogen (H_2) is added to a double or a triple bond.

ketal (13.4) the family of organic compounds formed via the reaction of two molecules of alcohol with a ketone in the presence of an acid catalyst.

ketone (13.1) a family of organic molecules characterized by a carbonyl group; the carbonyl carbon is bonded to two alkyl groups, two aryl groups, or one alkyl and one aryl group.

oxidation (of carbonyl compounds) (13.4) the loss of electrons by a molecule, ion, or atom; in organic compounds, it can be recognized by the gain of oxygen or the loss of hydrogen; e.g., the conversion of an alcohol to an aldehyde or ketone via the use of an oxidizing agent.

Tollens' Test (13.4) a test reagent (silver nitrate in ammonium hydroxide) used to distinguish between aldehydes and ketones; also called the Tollens' silver mirror test.

Self Test for Chapter Thirteen

1. Which of the following compounds of each set has the highest boiling point?

   a. 1-propanol or propane or propanone

   b. pentane or methoxyethane or propanal

2. Draw the polar attraction between two butanone molecules.

3. Give the I.U.P.A.C. names of the following:

   a. $CH_3 - \overset{\overset{\displaystyle O}{\|}}{C} - H$

   d. $CH_3CH_2 - \overset{\overset{\displaystyle O}{\|}}{C} - CH_2CH_3$

   b. $CH_3 - \overset{\overset{\displaystyle O}{\|}}{C} - CH_3$

   e. $CH_3CH_2\underset{\underset{\displaystyle Br}{|}}{CH}CH_2CH_2 - \overset{\overset{\displaystyle O}{\|}}{C} - H$

   c. $CH_3CH_2 - \overset{\overset{\displaystyle O}{\|}}{C} - H$

4. Why are 2-propanone and 2-butanone not correct I.U.P.A.C. names?

5. Write structural formulas for the following:

   a. 2-bromobutanal

   b. 2-methyl-3-pentanone

6. Show the reactions for the oxidation of methanol and ethanol by the liver.

7. Complete the following oxidation reactions:

   a. $H - \overset{\overset{\displaystyle O}{\|}}{C} - H \xrightarrow{[O]}$?

   b. $CH_3 - \overset{\overset{\displaystyle O}{\|}}{C} - H \xrightarrow{[O]}$?

   c. $CH_3 - \overset{\overset{\displaystyle O}{\|}}{C} - CH_3 \xrightarrow{[O]}$?

d. $CH_3CH_2 - \overset{\overset{\displaystyle O}{\|}}{C} - H \quad \xrightarrow{[O]} \quad ?$

e. $CH_3 - \overset{\overset{\displaystyle OH}{|}}{\underset{\underset{\displaystyle CH_3}{|}}{C}} - CH_3 \quad \xrightarrow{[O]} \quad ?$

8. Give the I.U.P.A.C. names of the oxidation products for the following alcohols:

   a. CH_3CH_2OH

   b. 1-propanol

   c. 2-propanol

   d. 2-methyl-2-propanol

9. Show the major products for each of the following oxidation reactions:

   a. $CH_3 - \overset{\overset{\displaystyle O}{\|}}{C} - H \quad + \quad Ag(NH_3)_2{}^+ \quad \longrightarrow \quad ?$

   b. $CH_3 - \overset{\overset{\displaystyle O}{\|}}{C} - CH_3 \quad + \quad Ag(NH_3)_2{}^+ \quad \longrightarrow \quad ?$

   c. $CH_3CH_2 - \overset{\overset{\displaystyle O}{\|}}{C} - CH_3 \quad + \quad Ag(NH_3)_2{}^+ \quad \longrightarrow \quad ?$

10. Label the following as oxidation or reduction reactions.

    a. conversion of methyl alcohol by the liver into formaldehyde

    b. conversion of acetaldehyde by the liver into acetic acid

    c. conversion of 3-pentanone into 3-pentanol

    d. conversion of ethylene glycol (1,2-ethanediol) to oxalic acid (ethanedioic acid)

11. Complete the following reactions:

    a. $RCHO + ROH \rightleftharpoons ?$

    b. $NAD^+ + CH_3OH \rightarrow ?$

    c. $NAD^+ + CH_3CH_2OH \rightarrow ?$

12. Which class(es) of organic compounds contain a carbonyl functional group?

13. The general formula RCOR represents what class of compounds?

14. Give the common name of the two-carbon aldehyde.

15. Formaldehyde is available commercially as an aqueous solution known as what?

16. Aldehydes are easily further oxidized to what compounds?

17. What is the common name of ethanoic acid?

18. The oxidation of benzaldehyde produces what compound?

19. List two common laboratory tests used to distinguish between aldehydes and ketones.

20. Name the product of the following reaction:

$$R-\underset{\underset{H}{|}}{\overset{\overset{OH}{|}}{C}}-O-R \quad + \quad ROH \quad \xrightarrow{\text{acid}} \quad ?$$

21. When β-carotene is cleaved, what vitamin is produced? (Hint: Review *A Human Perspective: The Chemistry of Vision.*)

22. Why are ketones polar compounds?

23. RCHO is in what class of compounds?

24. What do we call a benzene group that has one hydrogen atom removed from it?

25. The ending *-al* on the name of a compound means that it is what kind of compound?

26. The ending *-one* on the name of a compound means that it is what kind of compound?

Vocabulary Quiz for Chapter Thirteen

1. The functional group that consists of a carbon double bonded to an oxygen is the _____.

2. A test for the presence of aldehydes that uses a buffered solution of Cu(II) and that can be used to test for the presence of glucose in the urine is _____.

3. The product of the reaction between an aldehyde and an alcohol is called a(n) _____.

4. A(n) _____ is the product of the reaction between a hemiketal and an alcohol.

5. _____ is the reaction in which H_2 is added to a double or triple bond.

6. A(n) _____ is formed via the reaction of two molecules of alcohol with an aldehyde in the presence of an acid catalyst.

7. The conversion of an alcohol to an aldehyde or ketone is a(n) _____ reaction.

8. A(n) _____ is a reaction in which two molecules add together to form a new molecule; it often involves the addition of one molecule to a double or triple bond in an unsaturated molecule.

9. _____ is a test reagent (silver nitrate in ammonium hydroxide) used to test for the carbonyl functional group.

10. _____ is a reaction in which aldehydes or ketones react to form a larger molecule.

14

Carboxylic Acids and Carboxylic Acid Derivatives

Learning Goals

1 Write structures and describe the physical properties of carboxylic acids.
2 Determine the common and I.U.P.A.C. names of carboxylic acids.
3 Describe the biological, medical, or environmental significance of several carboxylic acids.
4 Write equations that show the synthesis of a carboxylic acid.
5 Write equations representing acid-base reactions of carboxylic acids.
6 Write equations representing the preparation of an ester.
7 Write structures and describe the physical properties of esters.
8 Determine the common and I.U.P.A.C. names of esters.
9 Write equations representing the hydrolysis of an ester.
10 Define the term *saponification* and describe how soap works in the emulsification of grease and oil.
11 Determine the common and I.U.P.A.C. names of acid chlorides.
12 Write equations representing the synthesis of acid chlorides.
13 Determine the common and I.U.P.A.C. names of acid anhydrides.
14 Write equations representing the synthesis of acid anhydrides.
15 Discuss the significance of thioesters and phosphoesters in biological systems.

Introduction

Carboxylic acids are characterized by the **carboxyl group**, and have the following general structure:

$$\underset{\text{Aromatic carboxylic acid}}{Ar - \overset{\displaystyle O}{\overset{\displaystyle \|}{C}} - OH} \qquad \underset{\text{Aliphatic carboxylic acid}}{R - \overset{\displaystyle O}{\overset{\displaystyle \|}{C}} - OH}$$

The term *carboxylic acid* tells us that the carboxyl group is derived from a carbonyl group and hydroxyl group. It further tells us that these molecules are acids.

Esters are produced in the reaction between a carboxylic acid and an alcohol and have the following general structure:

$$R - \overset{\displaystyle O}{\overset{\displaystyle \|}{C}} - O - R \qquad Ar - \overset{\displaystyle O}{\overset{\displaystyle \|}{C}} - O - Ar \qquad Ar - \overset{\displaystyle O}{\overset{\displaystyle \|}{C}} - O - R$$

Examples of aliphatic and aromatic esters

The following structure is called the **acyl group**:

$$\left[\text{(Ar) or} \quad R - \overset{\displaystyle O}{\overset{\displaystyle \|}{C}} - \right]$$

The acyl group is the functional group of the carboxylic acid derivatives, including the esters and amides.

Example 1
The following general formulas are used to represent different classes of organic compounds. Name each of the appropriate families.

1. ROH   4. ROR
2. RCHO   5. RCOOH
3. RCOR   6. RCOOR

Answer
1. alcohol   4. ether
2. aldehyde   5. carboxylic acid
3. ketone   6. ester

14.1 Carboxylic Acids

Structure and Physical Properties

Learning Goal 1:	Write structures and describe the physical properties of carboxylic acids.

The carboxyl group consists of two very polar functional groups, the carbonyl group and the hydroxyl group. Thus, carboxylic acids are very polar compounds. In addition, carboxylic acids can hydrogen bond to one another. As a result, they boil at higher temperatures than aldehydes, ketones, or even alcohols of comparable molecular weight.

Carboxylic acids can form intermolecular hydrogen bonds with water molecules. Thus, small carboxylic acids are water-soluble. However, solubility falls off dramatically as the carbon chain length increases.

The lower molecular weight carboxylic acids have sharp, sour tastes and unpleasant aromas. The longer chain carboxylic acids are called fatty acids and are components of many biologically important lipids.

Example 2

Which member of each of the following sets of compounds has the highest boiling point?

1. ethane or ethanol or ethanoic acid
2. propanal or propanone or 1-propanol
3. methanol or methanal or methanoic acid

Work

The boiling points of most alkanes, alcohols, aldehydes, and carboxylic acids that have similar molecular weights obey the following relationship:

> alkane < aldehyde or ketone < alcohol < carboxylic acid
> lowest   highest

Answer

1. ethanoic acid   2. 1-propanol   3. methanoic acid

Nomenclature

Learning Goal 2:	Determine the common and I.U.P.A.C. names of carboxylic acids.

In the I.U.P.A.C. System, carboxylic acids are named by replacing the -*e* ending of the parent alkane with the suffix -*oic acid*. The suffix –*dioic* acid is used if there are two carboxyl groups. The parent chain is numbered so that the carboxyl carbon is carbon-1. Other groups are named and numbered in the usual way.

To name the acyl group of a carboxylic acid, replace the -*oic* acid suffix with –*yl*. For instance, the acyl group of acetic acid is the acetyl group.

The carboxylic acid derivatives of cycloalkanes are named by adding the suffix *carboxylic acid* to the name of the cycloalkane or substituted cycloalkane. The carboxyl group is defined to be on carbon-1.

Common names of carboxylic acids are frequently used. Table 14.1 in the text shows the I.U.P.A.C. and common names of several carboxylic acids, as well as their sources and the Latin or Greek words that gave rise to the common names.

Aromatic carboxylic acids are usually named as derivatives of benzoic acid. The -*oic acid* or -*ic acid* suffix is attached to the appropriate prefix. However, common names of substituted benzoic acids are frequently used.

Example 3

Give the I.U.P.A.C. names for the following carboxylic acids:

1. CH_3COOH 2. CH_3CH_2COOH 3. $CH_3CHBrCH_2COOH$

Example 3 continued

Work
1. Determine the longest continuous chain (parent) of carbon atoms that includes the carboxyl group. The carbon atom of the carboxyl group is defined as 1 and no number is included in the name.
2. Then drop the *-e* of the parent alkane name and add *-oic acid*.

Answer
1. ethanoic acid   2. propanoic acid   3. 3-bromobutanoic acid

---

Example 4
Give the common name of each of the following carboxylic acids:

1. CH_3COOH 3. $HCOOH$
2. $CH_3CH_2CH_2COOH$ 4. $C_{17}H_{35}COOH$

Answer
1. acetic acid   3. formic acid
2. butyric acid   4. stearic acid

Some Important Carboxylic Acids

Learning Goal 3:	Describe the biological, medical, or environmental significance of several carboxylic acids.

Many carboxylic acids found in nature are listed in Table 14.1 of the text. Fatty acids can be isolated from a variety of fats and oils. More complex carboxylic acids are also found in a variety of foodstuffs. Citric acid is found in citrus fruits and is added to foods to give a sharp taste (sour candies) or as a preservative and antioxidant. Adipic acid adds tartness to soft drinks and helps to retard spoilage. Bacteria in milk produce lactic acid as a product of fermentation of sugars. It contributes a tangy flavor to yogurt and buttermilk and acts as a food preservative. Lactic acid is also produced by muscles during strenuous exercise.

Reactions Involving Carboxylic Acids

Preparation of Carboxylic Acids

Learning Goal 4:	Write equations that show the synthesis of a carboxylic acid.

Many of the small carboxylic acids are prepared by the oxidation of the corresponding primary alcohol or aldehyde. A variety of oxidizing agents can be used in these reactions and a catalyst is often required. The general reaction is:

$$RCH_2OH \xrightarrow{[O]} R\overset{\displaystyle O}{\overset{\|}{-C}}-H \xrightarrow{[O]} R\overset{\displaystyle O}{\overset{\|}{-C}}-OH$$

Primary alcohol   Aldehyde   Carboxylic acid

Acid-Base Reactions

Learning Goal 5:	Write equations representing acid-base reactions of carboxylic acids.

The carboxylic acids behave as weak acids because they are proton donors. They are weak acids (typically less than 5% dissociation) that produce a carboxylate ion and a hydrogen ion in water, as seen in the following example:

$$R\overset{\displaystyle O}{\overset{\|}{-C}}-OH \ + \ H_2O \ \rightleftharpoons \ R\overset{\displaystyle O}{\overset{\|}{-C}}-O^- \ + \ H_3O^+$$

Carboxylic acid   Water   Carboxylate anion   Hydronium ion

When strong bases are added to a carboxylic acid, neutralization occurs:

$$R\overset{\displaystyle O}{\overset{\|}{-C}}-OH \ + \ NaOH \ \longrightarrow \ R\ \overset{\displaystyle O}{\overset{\|}{-C}}-O \ Na^! \ + \ H_2O$$

Carboxylic acid   Strong base   Carboxylic acid salt   Water

The salt of a carboxylic acid is named by replacing the *-ic acid* suffix with *-ate*. This name is preceded by the name of the appropriate cation, for instance, sodium.

The salts are ionic substances and, hence, are quite soluble in water. The long-chain carboxylic acid salts (fatty acid salts) are good **soaps**. Soaps are made from water, strong base, and natural fats and oils obtained from animals or plants. Soaps form micelles around grease and oil. The hydrophobic alkane tail of the soap dissolves the oils and grease, and the hydrophilic carboxylate end of the molecule remains associated with water. This produces a micelle, a tiny sphere with the alkyl group tails of the soap molecules, grease, and oils in the center and the carboxylate on the outside of the sphere, dissolved in the water.

Example 5
Show the reactions of the following carboxylic acids with water:

1. CH_3COOH
2. $HCOOH$
3. $CH_3CH_2CH_2COOH$

Answer
1. Acetic acid dissociates to produce the acetate ion and hydronium ion.
    $$CH_3COOH + H_2O \rightleftharpoons CH_3COO^- + H_3O^+$$

2. Formic acid dissociates to produce the formate ion and hydronium ion.
    $$HCOOH + H_2O \rightleftharpoons HCOO^- + H_3O^+$$

3. Butyric acid dissociates to produce the butyrate ion and hydronium ion.
    $$CH_3CH_2CH_2COOH + H_2O \rightleftharpoons CH_3CH_2CH_2COO^- + H_3O^+$$

Esterification

Learning Goal 6:	Write equations representing the preparation of an ester.

Carboxylic acids react with alcohols to form esters and water according to the following general reaction:

14.2 Esters

Structure and Physical Properties

Learning Goal 7:	Write structures and describe the physical properties of esters.

Esters are mildly polar and have pleasant aromas. Many esters are found in natural foodstuffs. Esters boil at approximately the same temperature as aldehydes or ketones of comparable molecular weight. The simpler ones are reasonably soluble and nonreactive in water.

Nomenclature

Esters are formed in the reaction of a carboxylic acid with an alcohol and both of these families are reflected in the naming of the ester. The *alkyl* or *aryl* portion of the alcohol name is used as the prefix and the *-ic acid* ending of the name of the carboxylic acid is replaced with *-ate*.

Example 6

Show the general reaction of an alcohol (R' OH) with a carboxylic acid (RCOOH) to produce an ester.

Answer

$$R-\underset{\substack{\|\\O}}{C}-OH \quad + \quad R'-OH \quad \underset{}{\overset{H^+, \text{ heat}}{\rightleftharpoons}} \quad R-\underset{\substack{\|\\O}}{C}-OR' \quad + \quad H_2O$$

   Carboxylic acid   Alcohol   Ester   Water

Example 7

Give the common names of the following esters:

1. $CH_3-O-\underset{\substack{\|\\O}}{C}-H$

2. $CH_3CH_2-O-\underset{\substack{\|\\O}}{C}-CH_3$

3. $\langle\!\!\!\bigcirc\!\!\!\rangle-\underset{\substack{\|\\O}}{C}-O-CH_3$

4. $CH_3CH_2-O-\underset{\substack{\|\\O}}{C}-CH_2CH_2CH_3$

Work

Esters are formed from the reaction of a carboxylic acid with an alcohol. The alkyl or aryl portion of the alcohol name is used as a prefix and is followed by the name of the carboxylic acid in which the *-ic acid* (common or I.U.P.A.C.) is replaced by *-ate*.

Answer
1. methyl formate
2. ethyl acetate
3. methyl benzoate
4. ethyl butyrate

Reactions Involving Esters

Preparation of Esters

The conversion of a carboxylic acid to an ester requires heat and is catalyzed by a trace of acid (H^+). The general equation representing this reaction was shown above.

Hydrolysis of Esters

Learning Goal 9:   Write equations representing the hydrolysis of an ester.

Esters undergo **hydrolysis** reactions in water. This reaction requires heat and a small amount of acid (H^+) or base (OH^-) to catalyze the reaction, as seen in the following general equations:

| Ester | | Water | | Carboxylic acid | Alcohol |

| Ester | Strong base | | Carboxylic acid salt | Alcohol |

Learning Goal 10:   Define the term *saponification* and describe how soap works in the emulsification of grease and oil.

The base-catalyzed hydrolysis of an ester is called **saponification**, which produces a carboxylic acid salt. The carboxylic acid is formed when the reaction mixture is neutralized with an acid. Saponification is used to hydrolyze fats and oils, which are esters, to the salts of long chain fatty acids – *soaps*.

Condensation Polymers

Condensation polymers are formed by the polymerization of monomers in a reaction that forms a small molecule, such as water or an alcohol. Polyesters are formed by reacting a dicarboxylic acid and a dialcohol.

Because each of the reactants has two reacting groups, the new molecule formed still has two reactive groups, —OH and —COOH. So the polymerization reaction continues, forming very long polyester molecules.

When formed as fibers, polyesters can be used to make fabrics. They can also be formed into films, such as mylar, which is used as the base for recording tapes and photographic film.

14.3 Acid Chlorides and Acid Anhydrides

Acid Chlorides

Learning Goal 11:   Determine the common and I.U.P.A.C. names of acid chlorides.

Acid chlorides are named by dropping the *-ic acid* ending of the common or the *-oic* acid ending of the I.U.P.A.C. name of the carboxylic acid and replacing it with *-oyl chloride*.

Example 8
Name the following acid chlorides:

$$
\begin{array}{cc}
\quad\quad\quad \overset{\textstyle O}{\underset{\textstyle \|}{}} & \quad\quad\quad \overset{\textstyle O}{\underset{\textstyle \|}{}} \\
1.\quad CH_3CH_2 - C - Cl & 2.\quad BrCH_2 - C - Cl
\end{array}
$$

Work
Acid chlorides are named by dropping the *-ic acid* ending of the common name or the *-oic acid* ending of the I.U.P.A.C. name and replacing it with *-oyl chloride*.

Answer
1. propanoyl chloride   2. 2-bromoethanoyl chloride

Acid chlorides are noxious, irritating chemicals. They are slightly polar and boil at about the same temperature as the corresponding aldehyde or ketone. They cannot be dissolved in water because they react violently with it. The reaction is a hydrolysis reaction that releases HCl and the carboxylic acid.

Learning Goal 12:   Write equations representing the synthesis of acid chlorides.

Acid chlorides are prepared from the corresponding carboxylic acid by reaction with one of the following reagents: PCl_3, PCl_5, or $SOCl_2$.

$$
\overset{\textstyle O}{\underset{\textstyle \|}{R - C - OH}} \xrightarrow[\text{chloride}]{\substack{\text{inorganic} \\ \text{acid}}} \overset{\textstyle O}{\underset{\textstyle \|}{R - C - Cl}} + \text{inorganic products}
$$

Acid Anhydrides

Learning Goal 13:   Determine the common and I.U.P.A.C. names of acid anhydrides.

Acid anhydrides are named by replacing the *acid* ending of the carboxylic acid name with the term *anhydride*.

Example 9

Draw the structures of the following acid anhydrides:

1. acetic anhydride   2. propanoic anhydride   3. 3-bromobutanoic anhydride

Work

Anhydrides are formed in a reaction between an acid chloride and a carboxylate ion. (The structure appears to be two carboxylic acids that combine with the loss of a water molecule.)

Answer

1. $$CH_3 - \overset{\overset{\displaystyle O}{\|}}{C} - O - \overset{\overset{\displaystyle O}{\|}}{C} - CH_3$$ 2. $$CH_3CH_2 - \overset{\overset{\displaystyle O}{\|}}{C} - O - \overset{\overset{\displaystyle O}{\|}}{C} - CH_2CH_3$$

3. $$CH_3CHBrCH_2 - \overset{\overset{\displaystyle O}{\|}}{C} - O - \overset{\overset{\displaystyle O}{\|}}{C} - CH_2CHBrCH_3$$

Learning Goal 14:   Write equations representing the synthesis of acid anhydrides.

Anhydrides are formed in a reaction between an acid chloride and a carboxylate ion. They boil at much lower temperatures than the corresponding carboxylic acids. Generally they are hydrolyzed by water. This reaction is speeded up by the addition of acid.

14.4 Nature's High Energy Compounds: Phosphoesters and Thioesters

Learning Goal 15: Discuss the significance of thioesters and phosphoesters in biological systems.

An alcohol can react with phosphoric acid to produce a phosphate ester, or **phosphoester**, as seen here:

$$R-OH \quad + \quad HO-\overset{\overset{\displaystyle O}{\|}}{\underset{\underset{\displaystyle OH}{|}}{P}}-OH \quad \longrightarrow \quad R-O-\overset{\overset{\displaystyle O}{\|}}{\underset{\underset{\displaystyle OH}{|}}{P}}-OH \quad + \quad H_2O$$

| Alcohol | Phosphoric acid | Phosphate ester | Water |

Phosphoesters of monosaccharides are very important in energy harvesting reactions in the cell.

Adenosine triphosphate (ATP) is the universal energy currency for all living organisms. It consists of a nitrogenous base (adenine) and a phosphoester of the sugar ribose. There are three phosphoryl groups attached to C-5 of ribose (a triphosphate group). The combination of two phosphate groups results in the loss of a water molecule. Because water is lost, the resulting bond is called a **phosphoanhydride bond**.

$$R-O-\overset{\overset{\displaystyle O}{\|}}{\underset{\underset{\displaystyle OH}{|}}{P}}-OH \quad + \quad HO-\overset{\overset{\displaystyle O}{\|}}{\underset{\underset{\displaystyle OH}{|}}{P}}-OH \quad \longrightarrow \quad R-O-\overset{\overset{\displaystyle O}{\|}}{\underset{\underset{\displaystyle OH}{|}}{P}}-O-\overset{\overset{\displaystyle O}{\|}}{\underset{\underset{\displaystyle OH}{|}}{P}}-OH \quad + \quad H_2O$$

| Phosphate ester | Phosphoric acid | Phosphoanhydride bond |

The energy of ATP is made available through hydrolysis of either of the two phosphoanhydride bonds. Energy is released in this exothermic reaction. When the phosphoryl group is transferred to a sugar molecule, some of that energy resides in the phosphorylated sugar, thereby "energizing" it.

Cellular enzymes can carry out a reaction between a carboxylic acid and a thiol to produce a **thioester**:

$$R-S-\overset{\overset{\displaystyle O}{\|}}{C}-R'$$

Thioester

This reaction is essential in energy-generating pathways as a means of "activating" acyl groups for subsequent breakdown reactions. The complex thiol coenzyme A is the most important acyl group activator in the cell. The most common thioester is the acetyl ester, called **acetyl coenzyme A** (acetyl CoA). Acetyl CoA transfers the acetyl group from glycolysis to an intermediate of the citric acid cycle. Coenzyme A also serves to activate the acyl group of fatty acids during β-oxidation, the pathway by which fatty acids are oxidized to produce ATP.

Glossary of Key Terms in Chapter Fourteen

acetyl coenzyme A (14.4) a molecule composed of coenzyme A and an acetyl group; the intermediate that provides acetyl groups for complete oxidation by aerobic respiration.

acid anhydride (14.3) the product formed in a reaction between an acid chloride and a carboxylate ion.

acid chloride (14.3) a member of the family of organic compounds with the following general formula:

$$\underset{\text{R}-\text{C}-\text{Cl}}{\overset{\overset{\textstyle O}{\textstyle \|}}{}} \qquad \underset{\text{Ar}-\text{C}-\text{Cl}}{\overset{\overset{\textstyle O}{\textstyle \|}}{}}$$

acyl group (Intro) the functional group that contains the carbonyl group attached to one alkyl or aryl group.

adenosine triphosphate (14.4) ATP; the triphosphate ester of the nucleoside adenosine; the primary energy transport molecule used by the cells in cellular metabolism.

carboxyl group (14.1) the -COOH functional group; the functional group found in carboxylic acids.

carboxylic acid (14.1) the family of organic compounds that contains the -COOH functional group.

carboxylic acid derivative (14.2) any of several families of organic compounds that are derived from carboxylic acids; includes the acid chlorides, esters, amides, and acid anhydrides.

condensation polymer (14.2) a polymer, which is a large molecule formed by combination of many small molecules (monomers) in a repeating pattern, that resulted from joining of monomers in a reaction that forms a small molecule, such as water or an alcohol.

ester (14.2) a carboxylic acid derivative formed by the reaction of a carboxylic acid and an alcohol.

fatty acid (14.1) any member of the family of continuous-chain carboxylic acids that generally contain 4 to 20 carbon atoms; the naturally occurring members of this class contain an even number of carbon atoms.

hydrolysis (14.2) a chemical reaction that involves the reaction of a molecule with water; results in the splitting off of the water molecule as new bonds are formed.

oxidation (of alcohols) (14.1) the loss of electrons by a molecule, ion, or atom; in organic molecules, it can be recognized as a gain of oxygen or a loss of hydrogen; e.g., the conversion of an alcohol to a carboxylic acid via the use of a oxidizing agent.

phosphoanhydride (14.4) the bond formed when two phosphate groups react with one another and a water molecule is lost.

phosphoester (14.4) the product of the reaction between phosphoric acid and an alcohol.

saponification (14.2) a reaction in which a soap is produced; generally, the base-catalyzed hydrolysis of an ester.

soap (14.2) any of a variety of the alkali metal salts of fatty acids.

thioester (14.4) the product of a reaction between a thiol and a carboxylic acid.

Self Test for Chapter Fourteen

1. Give the general formula for each of the following classes of organic compounds:

    a. alcohol   d. aldehyde

    b. ether   e. carboxylic acid

    c. ester   f. ketone

2. Draw the general structure of an acyl group.

3. Give the I.U.P.A.C. names of the following:

    a. HCOOH

    b. $CH_3(CH_2)_4COOH$

c. CH_3COOH

d. $CH_3CH(CH_3)CH_2COOH$

4. Draw the following carboxylic acids:

    a. acetic acid

    b. formic acid

    c. benzoic acid

    d. capric acid

5. Provide the common names of each of the following carboxylic acids:

    a. $CH_3(CH_2)_2\overset{\displaystyle O}{\overset{\displaystyle \|}{C}}-OH$

    b. CH_3CH_2COOH

    c. $CH_3\underset{\displaystyle CH_3}{\overset{\displaystyle O}{\overset{\displaystyle \|}{CH}-C}}-OH$

6. Give the common name of the following esters:

    a. $H-\overset{\displaystyle O}{\overset{\displaystyle \|}{C}}-O-CH_2CH_3$

    b. $CH_3CH_2-\overset{\displaystyle O}{\overset{\displaystyle \|}{C}}-O-CH_3$

    c. $CH_3CH_2CH_2CH_2CH_2-O-\overset{\displaystyle O}{\overset{\displaystyle \|}{C}}-CH_3$

7. Give the I.U.P.A.C. names of the following esters:

    a. methyl acetate

    b. $CH_3CH_2-\overset{\displaystyle O}{\overset{\displaystyle \|}{C}}-O-CH_2CH_3$

8. Which of the following compounds that have nearly the same molecular weight has the highest boiling point?

    a. acetic acid

    b. propanal

    c. propanone

    d. 1-propanol

9. Complete the following reactions:

    a. $CH_3COOH + H_2O \rightleftharpoons$?

    b. $CH_3COOH + NaOH$?

    c. $CH_3COOH + CH_3OH$ (acid as a catalyst) ?

10. What are the hydrolysis products of an ester?

11. What two ions are formed when a soluble carboxylic acid reacts with a molecule of water?

12. For the following reactions, name the products that are formed:

    a. butyric acid + NaOH

    b. acetic acid + $Ca(OH)_2$

13. The long-chain carboxylic acid salts (fatty acid salts) are good _____.

14. What is the term for the large, nonpolar hydrocarbon end of a soap?

15. What two compounds must be reacted to form methyl propanoate?

16. What molecule is the universal energy currency for all cells?

17. What is the name of the base-catalyzed hydrolysis of an ester?

18. Name the functional group found in all carboxylic acids.

19. Which carboxylic acid is found in fire-ants? Give the common name.

20. What is the term for the product of a reaction between a thiol and a carboxylic acid?

21. Provide the I.U.P.A.C. name for the product of the reaction between ethanoic acid and methanol in the presence of heat and a trace of acid.

22. A primary alcohol is mixed with potassium permanganate solution. What reaction occurs and what is the function of the $KMnO_4$?

23. Which class of organic compounds is a weak acid in a water solution?

24. What is the name of the molecule that carries 2-carbon acetyl groups to the citric acid cycle?

25. Aqueous bases, such as a mixture of potassium carbonate and potassium hydroxide, are obtained by leaching wood ashes with what substance?

26. Is the carboxylate part of soap hydrophobic or hydrophilic?

27. What is the term for the "particles" formed when molelcules of oil or grease are surrounded by soap molecules?

28. Provide the missing reactant in the following equation:

$$CH_3CH_2CH_2COOH \ + \ ? \ \longrightarrow \ CH_3CH_2CH_2 - \overset{\overset{\displaystyle O}{\|}}{C} - O - CH_3$$

29. What is the product of the reaction between an alcohol and phosphoric acid?

30. What is the term for the bond formed between two phosphate groups?

Vocabulary Quiz for Chapter Fourteen

1. The universal energy currency for all cells is _____.

2. A chemical reaction in which a bond is broken and a water molecule is released is called _____.

3. _____ is the hydrolysis of an ester by an aqueous base.

4. The long chain carboxylic acids are called _____.

5. The conversion of an alcohol to a carboxylic acid is an example of a(n) _____ reaction.

6. _____ is the product of the reaction between a thiol and a carboxylic acid.

7. The characteristic functional group of the carboxylic acid derivatives is the _____.

8. The characteristic functional group of the carboxylic acids is the _____.

9. The product of the reaction between phosphoric acid and an alcohol is a(n) _____.

10. The alkali metal salt of a fatty acid is a(n) _____.

15 Amines and Amides

Learning Goals

1. Classify amines as primary, secondary, or tertiary.
2. Describe the physical properties of amines.
3. Draw and name simple amines using the systematic and common nomenclature systems.
4. Write equations representing the synthesis of amines.
5. Write equations showing the basicity and neutralization of amines.
6. Describe the structure of quaternary ammonium salts and discuss their use as antiseptics and disinfectants.
7. Discuss the biological significance of heterocyclic amines.
8. Describe the physical properties of amides.
9. Draw the structure and write the common and I.U.P.A.C. names of amides.
10. Write equations representing the preparation of amides.
11. Write equations showing the hydrolysis of amides.
12. Draw the general structure of an amino acid.
13. Draw and discuss the structure of a peptide bond.
14. Describe the function of neurotransmitters.

Introduction

Amines are organic molecules that contain the amino group,–NH_2, or substituted amino group. They may be aromatic or aliphatic and have the general formula (Ar–) R–NH_2. You can think of the amines as substituted ammonia molecules in which one or more of the hydrogens has been substituted by an organic group. Amides are carboxylic acid derivatives. The amide group is made up of two portions: one portion from a carboxyl group and the other from an amine. The amino group is found in many important biological molecules, particularly the proteins and nucleic acids (DNA and RNA), and is important in understanding the properties of these large, complex molecules.

15.1 Amines

Structure and Physical Properties

Learning Goal 1: Classify amines as primary, secondary, or tertiary.

Like ammonia, amines are pyramidal molecules. The nitrogen atom is attached to three groups and has a nonbonding pair of electrons. They are classified by the number of hydrocarbon groups attached to the nitrogen. **Primary (1°) amines** have one R group; **secondary (2°) amines** have two R groups; and **tertiary (3°) amines** have three R groups.

Example 1

Classify the following amines as either primary, secondary, or tertiary:

1. CH_3NH_2
2. CH_3NHCH_3
3. $CH_3N(CH_3)_2$

Work

Amines are classified according to the number of alkyl or aryl groups that are directly attached to the nitrogen atom.

Answer

1.

$$\overset{\displaystyle H}{\underset{\displaystyle |}{H_3C-N-H}}$$

primary: one directly attached carbon atom

2.

$$\overset{\displaystyle H}{\underset{\displaystyle |}{H_3C-N-CH_3}}$$

secondary: two attached carbon atoms

3.

$$\overset{\displaystyle CH_3}{\underset{\displaystyle |}{H_3C-N-CH_3}}$$

tertiary: three attached carbon atoms

Example 2

Write the general formulas for the following:
1. primary amine
2. secondary amine
3. tertiary amine

Example 2 continued

Answer
1. RNH_2
2. R_2NH
3. R_3N

Learning Goal 2:   Describe the physical properties of amines.

The N–H bond is polar, and therefore hydrogen bonding occurs between amine molecules. This feature determines the physical properties of the amines, such as boiling point and water solubility. The –NH group is less polar than the –OH group. As a result, the boiling points of amines are lower than comparable alcohols, but higher than comparable ethers or alkanes. The smaller amines are readily soluble in water, but as the size of the hydrocarbon groups increases, their solubility in water decreases.

Example 3
Which member of each of the following sets of compounds with similar molecular weights has the highest boiling point?
1. propane or dimethyl ether or ethanamine
2. ethanol or dimetyl ether or ethanamine

Work
The normal boiling points of alkanes, ethers, primary amines, and primary alcohols obey the following relationship:

$$alkane < ether < primary\ amine < primary\ alcohol$$

Answer
1. ethanamine
2. ethanol

Example 4
Show the hydrogen bonding of methylamine with water.

Work
Since nitrogen is more electronegative than hydrogen, the N–H group is polar.

Example 4 continued

Answer

$$H_3C-N-H$$

(Structure showing hydrogen bonding between methylamine and three water molecules, with N-H and N lone pair interactions labeled with δ^- and δ^+ partial charges)

Example 5

The boiling points of methylamine, dimethylamine, and trimethylamine are –6.3°C, 7°C, and 3.5°C, respectively. What causes these differences?

Answer

There is a molecular weight difference between each of these molecules (15 grams/mole) due to differences in the number of $-CH_3$ units. The boiling point of trimethylamine is lower than that of dimethylamine because trimethylamine is less polar owing to its symmetrical structure.

Nomenclature

Learning Goal 3:   Draw and name simple amines using the systematic and common nomenclature systems.

In the systematic nomenclature system, the final *-e* of the name of the parent compound is dropped and the suffix *-amine* is added. For secondary or tertiary amines, the prefix *N*-alkyl is added to the name of the parent compound. Many aromatic amines have special names. An example of this is aniline, a benzene molecule with a substituent amino group.

Common names are also used, especially for the simple amines. The common names of the alkyl groups bonded to the amine nitrogen are followed by the suffix *-amine*. Each alkyl group is listed alphabetically as one continuous word followed by the ending *-amine*.

Example 6
Determine the systematic names of the following amine compounds:

1. CH_3NH_2

2. $CH_3 - \overset{\overset{\displaystyle H}{|}}{\underset{\underset{\displaystyle NH_2}{|}}{C}} - CH_3$

3. $CH_3CH_2CH_2\overset{\overset{}{}}{\underset{\underset{\displaystyle NH_2}{|}}{C}}HCH_3$

4. $CH_3CH_2CH_2NHCH_3$

Work
1. The parent compound of an amine is the longest continuous chain of carbon atoms to which the amino group is bonded.
2. Drop the final -*e* of the parent alkane and add the suffix -*amine*.
3. If a substituent is present on the nitrogen (other carbon atoms are directly bonded to the nitrogen), it is designated by the prefix *N*-.

Answer
1. methanamine
2. 2- propanamine
3. 2- pentanamine
4. *N*-methyl-1-propanamine

Example 7
Determine the common names for each of the following amines.

1. CH_3NH_2

2. $(CH_3)_2NH$

3. $(CH_3CH_2)_2NH$

4. $CH_3 - \overset{\overset{\displaystyle H}{|}}{N} - \overset{\overset{\displaystyle CH_3}{|}}{\underset{\underset{\displaystyle H}{|}}{C}} - CH_2CH_3$

Work
The common names of the amines are derived from the various alkyl groups bonded to the amine nitrogen atom.

Answer
1. methylamine
2. dimethylamine
3. diethylamine
4. methyl-*sec*-butylamine

Medically important amines include decongestants, antibiotics, and anesthetics.

Reactions Involving Amines

Preparation of Amines

Learning Goal 4:   Write equations representing the synthesis of amines.

Amines are prepared by the reduction of amides and nitro compounds. If two of the R groups of an amide are hydrogen atoms, the product will be a *primary amine*. If only one R group is a hydrogen atom and the other is an organic group, the product will be a *secondary amine*. If both R groups are organic groups, the product will be a *tertiary amine*.

Amines may also be formed by the reduction of an amide, as shown in the following equation:

$$
\underset{\text{Amide}}{R-\overset{\overset{\displaystyle O}{\|}}{C}-\overset{\overset{\displaystyle R}{|}}{N}-R} \quad\xrightarrow{\ [H]\ }\quad \underset{\text{Amine}}{R-CH_2-\overset{\overset{\displaystyle R}{|}}{N}-R}
$$

(R may represent a hydrogen atom or an organic group.)

Basicity

Learning Goal 5:   Write equations showing the basicity and neutralization of amines.

Amines have a nonbonding pair of electrons that can be shared with an electron-deficient group to form a new bond. For instance, an amine may react with a proton (H^+), producing a new N–H bond. The original unshared pair of electrons on the nitrogen atom has been shared with the electron-deficient proton. The product is an **alkylammonium ion**.

Neutralization

Since amines are moderately strong bases, they react with most acids to form alkylammonium salts. The generalized reaction is represented here:

$$
R-\overset{\overset{\displaystyle H}{|}}{\underset{\underset{\displaystyle H}{|}}{N}}\!: \ + \ HCl \ \longrightarrow \ R-\overset{\overset{\displaystyle H}{|}}{\underset{\underset{\displaystyle H}{|}}{\overset{+}{N}}}-H \ \ Cl^-
$$

The product, an *alkylammonium salt*, is named by replacing the term *amine* with the term *ammonium* followed by the name of the anion. The salts are ionic species and hence are quite soluble in water.

Example 8

Complete the following reactions and name the products:

1. $CH_3CH_2NH_2 + HCl \rightarrow$?
2. $(CH_3)_2NH + HI \rightarrow$?
3. $CH_3NH_2 + H_2O \rightleftharpoons$

Work

Amines are weak to moderately strong bases and will react with most acids to form salts.

Answer

1. $CH_3CH_2NH_3^+ Cl^-$ ethylammonium chloride
2. $(CH_3)_2NH_2^+ I^-$ dimethylammonium iodide
3. $CH_3NH_3^+ OH^-$ methylammonium hydroxide

Because amines act as bases (proton acceptors) and amine salts act as acids (proton donors), they are useful as *buffers*. Many naturally occurring amines serve as biological buffers. For instance, the protein hemoglobin helps maintain the acid/base (pH) balance of the blood.

Several important drugs are amines. They are generally administered as alkylammonium salts because the salts are more soluble in water and body fluids.

Quaternary Ammonium Salts

Learning Goal 6:   **Describe the structure of quaternary ammonium salts and discuss their use as antiseptics and disinfectants.**

Quaternary ammonium salts are ammonium salts having four organic groups bonded to the nitrogen. They have the general structure $R_4N^+ A^-$.

Quaternary ammonium salts with very long carbon chains are often used as disinfectants and antiseptics because they have detergent activity.

15.2 Heterocyclic Amines

Learning Goal 7:   **Discuss the biological significance of heterocyclic amines.**

Heterocyclic amines are cyclic amines in which one of the atoms of the backbone of the ring is nitrogen. They are found in many important cellular macromolecules, including DNA, RNA, and coenzymes. Many drugs, including cocaine, nicotine, quinine, morphine, heroin, and LSD (lysergic acid diethylamide), are heterocyclic amines, as is the artificial sweetener saccharin. Several examples are provided in Figure 15.4 in the text.

15.3 Amides

Amides are an important class of nitrogen-containing organic compounds with the functional group shown here:

$$\text{(Ar)} \quad R - \overset{\displaystyle O}{\overset{\displaystyle \|}{C}} - NH_2$$

As careful inspection of this structure shows, amides are carboxylic acid derivatives.

The amide bond is the central feature in the structure of proteins. The amide bond between two amino acids is called a peptide bond.

Structure and Physical Properties

Learning Goal 8:   Describe the physical properties of amides.

Most amides are solids at room temperature, with boiling points even higher than the corresponding carboxylic acid. The simpler amides are quite soluble in water. Both of these properties reflect the strong intermolecular hydrogen bonding. The water solubility decreases as the molecular weight increases.

Nomenclature

Learning Goal 9:   Draw the structure and write the common and I.U.P.A.C. names of amides.

Amides are named by removing the *-ic acid* ending of the common name or the *-oic acid* ending of the I.U.P.A.C. name of the carboxylic acid and replacing it with *-amide*. If there are substituents on the nitrogen, they are placed as prefixes and are indicated by *N-*, followed by the name of the substituent. There are no spaces between the prefix and the amide name.

Example 9

Determine the I.U.P.A.C. names of the following amides:

1. $CH_3 - \overset{\displaystyle O}{\overset{\displaystyle \|}{C}} - NH_2$

2. $CH_3CH_2 - \overset{\displaystyle O}{\overset{\displaystyle \|}{C}} - \overset{\displaystyle H}{\overset{\displaystyle |}{N}} - CH_3$

3. $CH_3CH_2CH_2CH_2 - \overset{\displaystyle O}{\overset{\displaystyle \|}{C}} - \overset{\displaystyle H}{\overset{\displaystyle |}{N}} - CH_2CH_3$

4. $CH_3CH_2CONH_2$

Example 9 continued

Work

1. The common and I.U.P.A.C. names of the amides are derived from the common and I.U.P.A.C. names of the corresponding carboxylic acids.
2. Amides are named by removing the *-ic acid* ending of the common name or the *-oic acid* ending of the I.U.P.A.C. name of the carboxylic acid and replacing it with *-amide*.
3. Substituents on the nitrogen atom are placed as prefixes and are indicated by *N-* followed by the name of the substituent.

Answer

1. ethanamide
2. *N*-methylpropanamide
3. *N*-ethylpentanamide
4. propanamide

Medically important amides include barbiturates and acetaminophen.

Reactions Involving Amides

Preparation of Amides

| Learning Goal 10: Write equations representing the preparation of amides. |

Amides can be prepared from carboxylic acid derivatives. The product of the reaction between an acid chloride and either ammonia or an amine produces an amide and either ammonium chloride or an alkylammonium chloride. Two molar equivalents of ammonia or amine are required for this reaction, which is termed an **acyl group transfer** because the **acyl group** of the acid chloride is transferred from the Cl atom of the acid chloride to the N atom of the amine.

$$
\underset{\substack{\text{Acid} \\ \text{chloride}}}{H_3C-\overset{\overset{\textstyle O}{\|}}{C}-Cl} \quad + \quad \underset{\text{Ammonia}}{2\,NH_3} \quad \xrightarrow{\text{heat}} \quad \underset{\text{Amide}}{R-\overset{\overset{\textstyle O}{\|}}{C}-NH_2} \quad + \quad \underset{\substack{\text{Ammonium} \\ \text{chloride}}}{NH_4{}^+Cl^-}
$$

Amides may also be prepared by the reaction between an acid anhydride and ammonia or an amine.

$$\underset{\substack{\text{Acid anhydride}}}{R-\overset{\overset{\displaystyle O}{\|}}{C}-O-\overset{\overset{\displaystyle O}{\|}}{C}-R} \quad + \quad \underset{\substack{\text{Ammonia}}}{2\,NH_3} \quad \longrightarrow \quad \underset{\substack{\text{Amide}}}{R-\overset{\overset{\displaystyle O}{\|}}{C}-NH_2} \quad + \quad \underset{\substack{\text{Ammonium}\\\text{salt}}}{R-\overset{\overset{\displaystyle O}{\|}}{C}-O^-NH_4^+}$$

When subjected to heat, the ammonium salt loses a water molecule and produces a second amide molecule.

Hydrolysis of Amides

Learning Goal 11:   Write equations showing the hydrolysis of amides.

Hydrolysis of an amide results in breaking the amide bond to produce the carboxylic acid and ammonia or an amine. This reaction requires heat and the presence of a strong acid or base.

$$\underset{\substack{\text{Amide}}}{R-\overset{\overset{\displaystyle O}{\|}}{C}-NH-R'} \quad + \quad \underset{\substack{\text{Strong acid}}}{H_3O^+} \quad \longrightarrow \quad \underset{\substack{\text{Carboxylic}\\\text{acid}}}{R-\overset{\overset{\displaystyle O}{\|}}{C}-OH} \quad + \quad \underset{\substack{\text{Amine}}}{R'-NH_3^+}$$

Example 10

Complete the following reactions:

1. $\quad H_3C-\overset{\overset{\displaystyle O}{\|}}{C}-Cl \quad + \quad 2\,NH_3 \quad \xrightarrow{\text{heat}} \quad ?$

2. $\quad CH_3CH_2-\overset{\overset{\displaystyle O}{\|}}{C}-NH-CH_3 \quad + \quad H_3O^+ \quad \longrightarrow \quad ?$

Answer

1. $\quad H_3C-\overset{\overset{\displaystyle O}{\|}}{C}-NH_2 \quad + \quad NH_4^+Cl^-$

2. $\quad CH_3CH_2-\overset{\overset{\displaystyle O}{\|}}{C}-OH \quad + \quad CH_3N^+H_3$

15.4 A Preview of Amino Acids, Proteins, and Protein Synthesis

Learning Goal 12:	Draw the general structure of an amino acid.

Learning Goal 13:	Draw and discuss the structure of a peptide bond.

Proteins are polymers of amino acids. Amino acids have an amino group and a carboxyl group and typically have the following general structure:

$$H_2N - \overset{\overset{\displaystyle H}{|}}{\underset{\underset{\displaystyle R}{|}}{C}} - \overset{\overset{\displaystyle O}{||}}{C} - OH$$

The R group may be a hydrogen atom or an organic group. The primary structure of a protein consists of many amino acids bonded to one another by amide bonds called **peptide bonds**. Each peptide bond is formed by a reaction joining the amino group of one amino acid and the carboxyl group of another.

15.5 Neurotransmitters

Neurotransmitters are chemicals that carry messages, or signals, from a nerve cell to a target cell, which may be another nerve cell or a muscle cell. They may be *inhibitory* or *excitatory* and all are nitrogen-containing compounds.

Catecholamines
The catecholamines include dopamine, norepinephrine, and epinephrine (adrenaline). Too little dopamine results in Parkinson's disease. Too much is associated with schizophrenia. Dopamine is also associated with addictive behavior. Norepinephrine and epinephrine are involved in the "fight or flight" response. All catecholamines are synthesized from the amino acid tyrosine.

Serotonin
A deficiency of serotonin is associated with depression and eating disorders. Serotonin is involved in pain perception, regulation of body temperature, and sleep. Serotonin is synthesized from the amino acid tryptophan.

Histamine
Histamine contributes to allergy symptoms. Antihistamines block histamines and provide relief from allergies. Histamine is synthesized from the amino acid histidine.

γ-Aminobutyric Acid and Glycine
γ-Aminobutyric acid (GABA) and glycine are inhibitory neurotransmitters. It is believed that GABA is involved in control of aggressive behavior. GABA is synthesized from the amino acid glutamate; glycine is an amino acid.

Acetylcholine

Acetylcholine is a neurotransmitter that functions at the neuromuscular junction, carrying signals from the nerve to the muscle. Acetylcholine is synthesized in a reaction between choline and acetyl coenzyme A.

Nitric Oxide (NO) and Glutamate

Nitric oxide and glutamate function in a positive feedback loop that is thought to be involved in learning and the formation of memories. NO is synthesized from the amino acid arginine; glutamate is an amino acid.

Glossary of Key Terms in Chapter Fifteen

acyl group (15.3) the functional group found in the carboxylic acid derivatives that contains the carbonyl group attached to one alkyl or aryl group.

alkaloid (15.2) a class of naturally occurring compounds that contain one or more nitrogen heterocyclic rings; many of the alkaloids have medicinal and other physiological effects.

alkylammonium ion (15.1) the ion formed when the lone pair of electrons of the nitrogen atom of an amine is shared with a proton from a water molecule.

amide (15.3) the family of organic compounds formed by the reaction between a carboxylic acid derivative and an amine.

amide bond (15.3) the bond between the carbonyl carbon of a carboxylic acid and the amino nitrogen of an amine.

amine (15.1) the family of organic molecules with the general formula $R-NH_2$, R_2NH, or R_3N (R- can equal R- or Ar-); they may be viewed as substituted ammonia molecules in which one or more of the ammonia hydrogens has been substituted by a more complex organic moiety.

aminoacyl group (15.4) the functional group that is characteristic of an amino acid:

$$^+H_3N - CH - C - \atop {\displaystyle |} \atop {\displaystyle R}$$

anesthetic (15.2) any drug that causes lack of sensation in any part of the body (local) or causes unconsciousness (general); there is generally an analgesic effect also associated with these compounds.

analgesic (15.2) any drug that acts as a painkiller.

heterocyclic amine (15.2) a heterocyclic compound that contains nitrogen in at least one position in the ring skeleton.

neurotransmitter (15.5) a chemical that carries a message, or signal, from a nerve cell to a target cell.

peptide bond (15.4) the amide bond between two amino acids in a peptide chain.

primary (1°) amine (15.1) an amine with the general formula RNH_2.

quaternary ammonium salt (15.1) an amine salt with the general formula $R_4N^+ A^-$ (in which R– can be an alkyl or aryl group or a hydrogen atom and A^- can be any anion.

secondary (2°) amine (15.1) an amine with the general formula R_2-NH.

tertiary (3°) amine (15.1) an amine with the general formula R_3-N.

transfer RNA (tRNA) (15.4) a small RNA that binds to a specific amino acid at the 3' end and mediates its addition at the appropriate site in a growing peptide chain. This is accomplished by recognition of the correct codon on the mRNA by the complementary anticodon on the tRNA.

Self Test for Chapter Fifteen

1. Identify the following as either a primary, secondary, or tertiary amine or as a quaternary ammonium salt.

   a. $(CH_3)_3N$

   b. $(CH_3)_3N^+CH_2CH_3$ Cl^-

   c. $CH_3CH_2NH_2$

   d. $CH_3CH_2\text{-}NH\text{-}CH_2CH_3$

2. What functional groups of amino acids react to form the peptide bond?

3. Provide the systematic names of the following amine compounds:

   a. $CH_3CH_2NH_2$

   b. $(CH_3)_2CHNH_2$

   c. $(CH_3)_3N$

4. What is the name of the amine of benzene?

5. Complete each of the following reactions:

   a. $CH_3CONH_2 + H_3O^+$

   b. $CH_3CH_2\text{-}CONH\text{-}CH_3 + H_3O^+$

6. Complete the following reactions:

   a. $(CH_3)_2NH + HI \rightarrow$?

   b. $(CH_3)_3N + HCl \rightarrow$?

7. Write the general formulas for the following:

   a. primary amine

   b. tertiary amine

8. Determine the I.U.P.A.C. names of the following amides:

   a. acetamide

   b. $CH_3(CH_2)_4\text{-}CONH\text{-}CH_2CH_2CH_3$

   c. $CH_3CH_2CH_2\text{-}CONH\text{-}CH_3$

9. Which neurotransmitter is involved in allergic reactions?

10. The central nervous system refers to what parts of the body?

11. Name two important neurotransmitters.

12. Why are most drugs that are amines delivered as alkylammonium salts?

13. What disease is caused by a deficiency of dopamine in the human body?

14. Which functional group is found in amines?

15. Amines may be viewed as compounds that are substituted products of which inorganic compound?

16. Amines are classified according to the number of alkyl or aryl groups directly attached to what atom?

17. In what biological molecules are the heterocyclic compounds, pyrimidines and purines, found?

18. What is the function of a neurotransmitter?

19. What product is formed in the reaction of an amine with an acid?

20. Name the product that results when methylamine reacts with HCl.

21. The ability of the amines to act as bases and of amine salts to act as acids makes a mixture of these useful as what in solution?

22. Which functional group is composed of one portion from a carboxylic acid and a second portion from an amine?

23. What is the physical state of most amides at room temperature?

24. What two products are released by the hydrolysis of an amide?

25. Why are quaternary ammonium salts effective as disinfectants and antiseptics?

26. What is the term for an ammonium salt that has four organic groups bonded to the nitrogen?

27. The amide bond is the central feature in the structure of what class of biological molecules?

28. Nicotine is an agonist of which neurotransmitter?

29. What is the general term for a drug, such a morphine, that acts as a painkiller?

Vocabulary Quiz for Chapter Fifteen

1. The amide bond joining two amino acids is a(n) _____ __.

2. An amine having the general formula RNH_2 is a(n) _____ _____.

3. The functional group that is produced by a reaction between a carboxylic acid and an amine is the _____.

4. A(n) _____ is an amine with three alkyl groups attached to the nitrogen atom.

5. A(n) _____ is a ring structure that has a nitrogen atom as one of the ring atoms.

6. A drug that can act as a pain killer is a(n) _____.

7. The _____ are naturally occurring compounds that contain one or more nitrogen heterocyclic rings. These compounds often have powerful effects on the human body and mind.

8. A(n) _____ is the cation formed when the nonbonding electron pair of an amine nitrogen is shared with a proton.

9. A(n) _____ has the general formula $R_4N^+A^-$.

10. The collection of organic molecules with the general formulas RNH_2, R_2NH, or R_3N is the _____ family.

16 *Carbohydrates*

Learning Goals

1 Explain the difference between complex and simple carbohydrates and know the amounts of each recommended in the daily diet.
2 Apply the systems of classifying and naming monosaccharides according to the functional group and number of carbons in the chain.
3 Determine whether a molecule has a chiral center.
4 Explain stereoisomerism.
5 Identify monosaccharides as either D- or L-.
6 Draw and name the common monosaccharides using structural formulas.
7 Given the linear structure of a monosaccharide, draw the Haworth projection of its α- and β-cyclic forms and vice versa.
8 By inspection of the structure, predict whether a sugar is a reducing or a non-reducing sugar.
9 Discuss the use of Benedict's reagent to measure the level of glucose in urine.
10 Draw and name the common disaccharides and discuss their significance in biological systems.
11 Describe the difference between galactosemia and lactose intolerance.
12 Discuss the structural, chemical, and biochemical properties of starch, glycogen, and cellulose.

Introduction

Learning Goal 1:	Explain the difference between complex and simple carbohydrates and know the amounts of each recommended in the daily diet.

Carbohydrates are produced in plants by photosynthesis. They are the main source of energy for both plants and animals. They are found in many natural sources, such as grains and cereals, breads, fruits, sugar cane, and sugar beets. A healthy diet should include both complex and simple carbohydrates. It is recommended that 58% of the diet should be carbohydrate and that no more than 10% of the daily caloric intake should be sucrose.

16.1 Types of Carbohydrates

Carbohydrates may be categorized by size. **Monosaccharides** are composed of a single (mono-) sugar (saccharide) unit. A **disaccharide** consists of two monosaccharides. Intermediate in size are the **oligosaccharides**, polymers consisting of three to ten monosaccharide units. The largest and most complex carbohydrates are the **polysaccharides**; polymers consisting of greater than ten monosaccharide units. The largest and most complex carbohydrates are the polysaccharides. Oligosaccharides and polysaccharides are chains of monosaccharides held together by **glycosidic bonds** through "bridging" oxygen atoms.

Example 1

List each of the following as either a monosaccharide, disaccharide, or polysaccharide:

1. starch   5. glycogen
2. glucose   6. sucrose
3. ribose   7. cellulose
4. maltose   8. glyceraldehyde

Work

Monosaccharides are those carbohydrates that cannot be broken down into any simpler substance by hydrolysis. *Disaccharides* are composed of two monosaccharides joined together by a glycosidic bond. *Polysaccharides* are composed of many monosaccharides joined together by glycosidic bonds between each pair.

Answer
1. polysaccharide   5. polysaccharide
2. monosaccharide   6. disaccharide
3. monosaccharide   7. polysaccharide
4. disaccharide   8. monosaccharide

16.2 Monosaccharides

Nomenclature

Learning Goal 2:   **Apply the systems of classifying and naming monosaccharides according to the functional group and number of carbons in the chain.**

If a monosaccharide is a ketone, it is called a **ketose**. If it is an aldehyde, it is called an **aldose**. All carbohydrates contain a large number of hydroxyl groups and are, therefore, *polyhydroxyaldehydes* or *polyhydroxyketones*.

A second system of nomenclature is based on the number of carbon atoms in the main skeleton. A **triose** has three carbons, a **tetrose** has four carbons, and so on. By combining the two systems, a name is derived that provides information about both the structure and composition of a sugar. For instance, a sugar may be an aldotriose, aldohexose, ketotriose, ketotetrose, etc.

In addition, each carbohydrate also has a specific unique name, such as glucose and fructose. Finally, it is important to indicate the isomeric form of the sugar; thus, a D- or L- designation is placed in front of the name.

$$\begin{array}{cc}
\begin{array}{c}
\quad\;\; O \\
\quad\;\; \| \\
\quad\;\; C-H \\
\quad\;\; | \\
H-C-OH \\
\quad\;\; | \\
HO-C-H \\
\quad\;\; | \\
H-C-OH \\
\quad\;\; | \\
H-C-OH \\
\quad\;\; | \\
\quad\;\; CH_2OH
\end{array}
&
\begin{array}{c}
CH_2OH \\
\quad\;\; | \\
\quad\;\; C=O \\
\quad\;\; | \\
HO-C-H \\
\quad\;\; | \\
H-C-OH \\
\quad\;\; | \\
H-C-OH \\
\quad\;\; | \\
\quad\;\; CH_2OH
\end{array}
\end{array}$$

Aldose	Ketose
Hexose	Hexose
Aldohexose	Ketohexose
D-Glucose	**D-Fructose**

Example 2

Classify each of the following monosaccharides as an aldose or a ketose, and also give the number of carbon atoms found in each:

1. ribose   3. galactose
2. glyceraldehyde   4. fructose

Work

A more detailed system of nomenclature may be used for monosaccharides. They may be classified as an aldose (which means they have an aldehyde structure) or as a ketose (ketone structure). Also, by using suffixes, we can indicate the number of carbon atoms found in each. For example, glucose is classified as an aldohexose, since it has six carbon atoms, and it also contains an aldehyde structure.

Answer

1. aldopentose   3. aldohexose
2. aldotriose   4. ketohexose

16.3 Stereoisomers and Stereochemistry

Learning Goal 3:	**Determine whether a molecule has a chiral center.**

Learning Goal 4:	**Explain stereoisomerism.**

Learning Goal 5:	**Identify monosaccharides as either D- or L-.**

Stereochemistry is the study of the spatial arrangement of atoms in a molecule. The prefixes D- and L- are used to distinguish **stereoisomers**. In each member of a pair of stereoisomers, all of the atoms are bonded together using the same bonding pattern; they differ only in the arrangements of their atoms in space. D- and L-stereoisomers are **enantiomers**, nonsuperimposable mirror images of one another.

A carbon atom that has *four different* groups bonded to it is called an asymmetric or **chiral carbon**. A molecule that has a chiral carbon (a chiral molecule) can exist as a pair of enantiomers.

Rotation of Plane-Polarized Light and the Relationship Between Molecular Structure and Optical Activity

Each member of a pair of stereoisomers will rotate a plane of polarized light in different directions. A polarimeter is used to measure the direction of rotation of plane-polarized light. Compounds that rotate light in a clockwise direction are termed **dextrorotatory** and are designated by a plus sign (+). Compounds that rotate light in a counterclockwise direction are called **levorotatory** and are indicated by a minus sign (–). So, (+)-glyceraldehyde will rotate plane-polarized light to the right, and (–)-glyceraldehyde will rotate plane-polarized light by the same amount to the left.

Fischer Projection Formulas

The **Fischer Projection** is a two-dimensional drawing of a molecule that shows a chiral carbon at the intersection of two lines. Horizontal lines represent bonds projecting out of the page and vertical lines represent bonds that project into the page. The most oxidized carbon is always represented at the "top" of the structure.

```
                          O
                          ||
                          C — H
    Fischer Projection
           of        H ——|—— OH
      D-Glucose
                     HO ——|—— H

                      H ——|—— OH

                      H ——|—— OH

                         CH₂OH
```

The D- and L- System of Nomenclature

In Figure 16.4 of the text, the stereoisomers of glyceraldehyde are presented. When the hydroxyl group is drawn to the right of the chiral carbon, the isomer is the D-enantiomer. When the hydroxyl group is drawn to the left of the chiral carbon, the molecule is the L-enantiomer.

Example 3
Draw and label the two different stereoisomers of glyceraldehyde.

Answer

$$
\begin{array}{cc}
\begin{array}{c}
\quad O \\
\quad \| \\
\quad C-H \\
\quad | \\
H-C-OH \\
\quad | \\
\quad CH_2OH
\end{array}
&
\begin{array}{c}
\quad O \\
\quad \| \\
\quad C-H \\
\quad | \\
HO-C-H \\
\quad | \\
\quad CH_2OH
\end{array}
\end{array}
$$

D-Glyceraldehyde   L-Glyceraldehyde

For sugars that have more than three carbons, the D- or L- configuration is determined by the chiral carbon farthest from the aldehyde or ketone group. If the –OH on this carbon is to the right, the stereoisomer is of the D- family. If the –OH is on to the left, the stereoisomer is of the L- family.

Example 4
Determine the configuration (D- or L-) for each of the following:

1.
$$
\begin{array}{c}
\quad O \\
\quad \| \\
\quad C-H \\
\quad | \\
HO-C-H \\
\quad | \\
\quad CH_2OH
\end{array}
$$

2.
$$
\begin{array}{c}
\quad O \\
\quad \| \\
\quad C-H \\
\quad | \\
H-C-OH \\
\quad | \\
HO-C-H \\
\quad | \\
HO-C-H \\
\quad | \\
H-C-OH \\
\quad | \\
\quad CH_2OH
\end{array}
$$

3.
$$
\begin{array}{c}
\quad O \\
\quad \| \\
\quad C-H \\
\quad | \\
H-C-OH \\
\quad | \\
H-C-OH \\
\quad | \\
H-C-OH \\
\quad | \\
\quad CH_2OH
\end{array}
$$

4.
$$
\begin{array}{c}
\quad O \\
\quad \| \\
\quad C-H \\
\quad | \\
H-C-H \\
\quad | \\
H-C-OH \\
\quad | \\
H-C-OH \\
\quad | \\
\quad CH_2OH
\end{array}
$$

Work

By convention, the D- isomer is the one in which the –OH on the chiral carbon farthest from the carbonyl group is on the right-hand side of the structure. The L- form is just the opposite.

Answer
1. L-glyceraldehyde
2. D-galactose
3. D-ribose
4. D-2-deoxyribose

Note that the (+) and (−) designations are a result of a physical measurement with a polarimeter, but the D- and L- designations are based solely on the configuration of the chiral carbon furthest from the aldehyde or ketone group. To determine whether a specific D-sugar is (+) or (−), it must be analyzed using a polarimeter. To determine whether a specific (+) sugar is D- or L- (if the chemical structure is unknown), further experimentation will need to be done.

As an example, given a sample of D-galactose, it is impossible to predict whether it will rotate plane-polarized light to the right or to the left until the sample is actually measured in a polarimeter. It is certain, however, that L-galactose will rotate plane-polarized light by the same amount as D-galactose, but *in the opposite direction.*

16.4 Biologically Important Monosaccharides

Learning Goal 6:	Draw and name the common monosaccharides using structural formulas.

Glucose

Glucose (dextrose), found in fruits and honey, is the preferred energy source of many tissues, especially the brain. Thus, the blood glucose concentration must be carefully controlled to allow the body to function optimally. The enantiomer of glucose found in nature is D-glucose.

Learning Goal 7:	Given the linear structure of a monosaccharide, draw the Haworth projection of its α- and β- cyclic forms and vice versa.

Generally sugars of five or more carbons exist in cyclic form under physiological conditions. This results from the reaction of the carbonyl group at C-1 of glucose with the hydroxyl group at C-5 to produce a six-membered ring. Recall that the reaction between an aldehyde and an alcohol yields a **hemiacetal**. In the case of the cyclic form of glucose, the product is a cyclic *intramolecular hemiacetal.*

Two isomers are formed in this reaction because cyclization creates a new asymmetric carbon, in this case C-1. These isomers, called **anomers**, are designated as either α- or β- forms. In the α-isomer, the C-1 hydroxyl group is below the ring. In the β-isomer, it is above the ring.

Haworth projections are used to depict the three-dimensional configuration of cyclic monosaccharide molecules. To draw a Haworth projection, begin with the structural formula of the linear form of the sugar (refer to Example 16.3 in the text). Chemical groups to the left of the carbon chain are placed above the ring in the Haworth projection. Chemical groups to the right of the carbon chain are drawn beneath the carbon ring in the Haworth projection.

D-Glucose

The oxygen atom attached to C-5 attacks the aldehyde carbon. This O can attack from the front or the back because the aldehyde group is planar. That means that the cyclic form can have the aldehyde oxygen either above (β) or below (α) the plane of the ring.

Haworth projection of α-D-Glucose

This is from the -OH attached to C-5.

This was the aldehyde oxygen.

Haworth projection of β-D-Glucose

Hint: To remember the structure of glucose, look closely at the OH groups of β-D-Glucose. The OH on C-1 is above the plane of the ring, or "up", the OH on C-2 is below the plane of the ring, or "down." Following around the ring gives the pattern "up-down-up-down-up."

Fructose

Fructose, the sweetest of all sugars, is found in honey and fruits. It is also called levulose and fruit sugar. Fructose is a ketohexose.

Cyclization of D-fructose produces α- and β-D-fructose. In this case the product is an intramolecular **hemiketal**. The most common ring structure formed by D-fructose consists of a five-carbon skeleton.

Galactose

Galactose is a hexose (six-carbon sugar) found most commonly as a component of the disaccharide lactose. Lactose, or milk sugar, is the most abundant sugar found in milk. β-D-galactose and a modified form, *N*-acetyl-β-D-galactosamine, are found in the oligosaccharides on the surface of red blood cells. These oligosaccharides are referred to as the *blood group antigens*. They determine whether an individual has Type A, B, AB, or O blood.

Ribose and Deoxyribose, Five-Carbon Sugars

Ribose, a **pentose**, is a component of ribonucleic acids (RNA) and of several coenzymes, compounds required for the action of some enzymes. The molecule that carries the genetic information in the cell is deoxyribonucleic acid (DNA). DNA contains a modified form of D-ribose in which the –OH group at C-2 has been replaced by a hydrogen. This sugar is called *2-deoxyribose*.

Reducing Sugars

Learning Goal 8:	By inspection of the structure, predict whether a sugar is a reducing or a nonreducing sugar.

Learning Goal 9:	Discuss the use of Benedict's reagent to measure the level of glucose in urine.

The aldehyde group of aldoses is easily oxidized. Thus, **Benedict's reagent** can be used to detect the presence of aldoses, or **reducing sugars**. The aldehyde group is oxidized to a carboxylate group, and Cu^{2+} is reduced to Cu^+.

Although ketones are not readily oxidized, ketoses, such as fructose, react with the Benedict's reagent because they are easily converted to aldoses through the *enediol reaction*.

Benedict's reagent can be used to detect the presence of glucose in the urine. Use of Benedict's reagent to test urine glucose levels has largely been replaced by chemical tests that provide more accurate results. The most common technology is a strip test in which a strip is impregnated with the enzyme glucose oxidase and reagents to produce a color change when the enzyme catalyzes the oxidation of glucose.

Frequently, physicians recommend that diabetics monitor *blood* glucose levels several times a day. This provides a more accurate indication of how well the diabetes is being controlled.

Example 5

Write an equation representing the oxidation of D-glucose by Benedict's solution.

Example 5 continued

Answer

$$
\begin{array}{c}
\text{O} \\
\parallel \\
\text{C}-\text{H} \\
\mid \\
\text{H}-\text{C}-\text{OH} \\
\mid \\
\text{HO}-\text{C}-\text{H} \\
\mid \\
\text{H}-\text{C}-\text{OH} \\
\mid \\
\text{H}-\text{C}-\text{OH} \\
\mid \\
\text{CH}_2\text{OH}
\end{array}
\;+\; 2\,\text{Cu}^{2+} \;+\; 5\,\text{OH}^- \;\longrightarrow\;
\begin{array}{c}
\text{O} \\
\parallel \\
\text{C}-\text{O}^- \\
\mid \\
\text{H}-\text{C}-\text{OH} \\
\mid \\
\text{HO}-\text{C}-\text{H} \\
\mid \\
\text{H}-\text{C}-\text{OH} \\
\mid \\
\text{H}-\text{C}-\text{OH} \\
\mid \\
\text{CH}_2\text{OH}
\end{array}
\;+\; \text{Cu}_2\text{O} \;+\; 3\,\text{H}_2\text{O}
$$

Example 6
Write an equation representing the oxidation of L-glyceraldehyde by Benedict's reagent.

Answer

$$
\begin{array}{c}
\text{O} \\
\parallel \\
\text{C}-\text{H} \\
\mid \\
\text{HO}-\text{C}-\text{H} \\
\mid \\
\text{CH}_2\text{OH}
\end{array}
\;+\; 2\,\text{Cu}^{2+} \;+\; 5\,\text{OH}^- \;\longrightarrow\;
\begin{array}{c}
\text{O} \\
\parallel \\
\text{C}-\text{O}^- \\
\mid \\
\text{HO}-\text{C}-\text{H} \\
\mid \\
\text{CH}_2\text{OH}
\end{array}
\;+\; \text{Cu}_2\text{O} \;+\; 3\,\text{H}_2\text{O}
$$

16.5 Biologically Important Disaccharides

Learning Goal 10: **Draw and name the common disaccharides and discuss their significance in biological systems.**

The α- or β-hydroxyl group of a cyclic monosaccharide can react with a hydroxyl group of another sugar to form an "oxygen bridge" (ether bond) and produce a disaccharide. This reaction is illustrated in Figure 16.8 in the text for two molecules of glucose.

If the disaccharide has an unreacted hemiacetal –OH group at C-1, it is a reducing sugar, because the cyclic structure can break open, forming a free aldehyde. When drawing the structure of a disaccharide, the nonreducing end is written to the left. Most disaccharides are reducing sugars. The single exception is sucrose, which is a **nonreducing sugar.**

Maltose

Maltose is a disaccharide composed of two glucose molecules. Since the C-1 hydroxyl group of one α-D-glucose molecule is attached to C-4 of another glucose molecule (which could be α or β), the bond between the two monosaccharides is called an α(1→4) glycosidic bond. When the reducing end of maltose has a β-hydroxyl group, this disaccharide is called β-maltose.

Lactose

Learning Goal 11:	Describe the difference between galactosemia and lactose intolerance.

Milk sugar, or **lactose**, is a dimer of β-D-galactose and either α- or β-D-glucose. In lactose the C-1 hydroxyl group of β-D-galactose is bonded to the C-4 hydroxyl group of glucose. Thus the bond between these two monosaccharides is called a β(1→4) glycosidic bond.

Lactose, the principal sugar in milk, must be broken down to glucose and galactose before it can be used by the body as an energy source. Glucose can be used directly in energy-harvesting metabolic reactions. However, galactose must be converted into a phosphorylated form of glucose before it can enter the energy harvesting glycolysis pathway. **Galactosemia** is a genetic disease caused by the absence of an enzyme needed for this conversion. Dulcitol, a reduced form of galactose, accumulates to toxic levels, causing severe mental retardation, cataracts, and early death. Galactosemic infants must be provided a diet that does not contain any milk or milk products.

The inability to digest lactose, called **lactose intolerance**, is caused by the absence of the enzyme lactase. The symptoms include intestinal cramping, diarrhea, and dehydration. Because some intestinal bacteria metabolize lactose and release organic acids and CO_2 gas, the individual suffers further discomfort. Avoiding milk and milk products will eliminate these unpleasant symptoms. Lactase is available in tablet form and can be ingested along with dairy products to avoid the symptoms of lactose intolerance.

Sucrose

Sucrose, common table sugar, is a disaccharide of α-D-glucose joined to β-D-fructose. Examine the structure of sucrose in Figure 16.11 in the text. The glycosidic linkage between α-D-glucose and β-D-fructose involves both of the carbons that were previously part of the hemiacetal or hemiketal. This is called an (α1→β2)-glycosidic linkage.

Because both carbons that were previously part of a hemiacetal or a hemiketal have reacted to form this linkage, the ring structure cannot open up. No aldehyde or ketone group can be formed; thus, sucrose is not a reducing sugar.

16.6 Polysaccharides

Most carbohydrates found in nature are **polysaccharides**, high molecular weight polymers of glucose.

Starch

Plants use the energy of sunlight to produce glucose from CO_2 and H_2O. Most plants store glucose in the form of the polysaccharide **starch**.

Starch is composed of the glucose polymers **amylose** (80%) and **amylopectin** (20%). Amylose is a linear polymer of up to 4,000 α-D-glucose molecules connected by α(1→4) glycosidic bonds. Amylose exists as a helix that coils at every sixth glucose unit.

During digestion, amylose is degraded by two enzymes. The enzyme α-amylase cleaves the glycosidic bonds of amylose chains at random along the chain, and β-amylase sequentially removes maltose molecules from the reducing end of the amylose chain.

Amylopectin is a highly branched form of amylose. The main chain consists of α(1→4) glycosidic bonds, and the branches of 20–25 glucose molecules are attached to the C-6 hydroxyl groups by α(1→6) glycosidic bonds.

Glycogen

Glycogen is the principal glucose storage form of animals. It is stored in granules in liver and muscle cells. The structure of glycogen is like that of amylopectin, except that glycogen has more branches, and they are shorter.

The liver regulates the level of blood glucose by the formation and degradation of glycogen. When blood glucose levels are high, liver cells take up glucose from the blood and convert it to glycogen. When the blood glucose levels are too low, the liver breaks down glycogen and releases the glucose into the blood stream.

Cellulose

Cellulose, a polymer of about 3,000 β-D-glucose units linked by β (1→4) glycosidic bonds, is the most abundant organic molecule in the world. Cellulose is an unbranched polymer that forms long fibrils. These long, straight, parallel chains of cellulose form a rigid cage that serves as a structural component of plant cell walls.

We cannot digest cellulose because we lack the enzyme *cellulase*. Cellulose serves as a source of dietary fiber, a necessary component of a healthful diet.

Example 7

Write the "word-reactions" for the hydrolysis of the following carbohydrates:

1. starch 2. maltose 3. sucrose 4. glycogen 5. lactose

Answer

1. starch + many H_2O molecules $\rightarrow$ many glucose molecules
2. maltose + H_2O $\rightarrow$ 2 glucose molecules
3. sucrose + H_2O $\rightarrow$ glucose + fructose
4. glycogen + many H_2O molecules $\rightarrow$ many glucose molecules
5. lactose + H_2O $\rightarrow$ glucose + galactose

For each of the above reactions, acid conditions are required.

Example 8

Why are termites, cows, and goats able to digest cellulose, while humans are unable to do so?

Answer

The termites, cows, and goats have microorganisms within their digestive systems that produce the enzyme *cellulase*, which allows them to break down cellulose. Humans do not have the cellulase enzyme needed for the hydrolysis of cellulose.

Glossary of Key Terms in Chapter Sixteen

aldose (16.2) a sugar that contains an aldehyde (carbonyl) group.

amylopectin (16.6) a highly branched form of amylose; the branches are attached to the C-6 hydroxyl by $\alpha(1\rightarrow6)$ glycosidic linkage; a component of starch.

amylose (16.6) a linear polymer of α-D-glucose in $\alpha(1\rightarrow4)$ glycosidic linkage that is a major component of starch; a polysaccharide storage form.

anomers (16.4) isomers of cyclic monosaccharides that differ from one another in the arrangement of bonds around the hemiacetal carbon.

Benedict's reagent (16.4) a buffered solution of Cu^{2+} ions that can be used to test for reducing sugars.

carbohydrate (16.1) generally sugars and polymers of sugars; the primary source of energy for the cell.

cellulose (16.6) the most abundant organic compound in the world; a polymer of β-D-glucose linked by $\beta(1\rightarrow4)$ glycosidic bonds.

chiral carbon (16.3) a chiral carbon; a carbon bonded to four different groups.

chiral molecule (16.3) molecules capable of existing in mirror image forms.

disaccharide (16.1) a sugar composed of two monosaccharides joined through an oxygen atom bridge.

enantiomers (16.3) stereoisomers that are nonsuperimposable mirror images of one another.

Fischer Projection (16.3) a two-dimensional drawing of a molecule, which shows a chiral carbon at the intersection of two lines with the horizontal lines representing bonds projecting out of the page and the vertical lines representing bonds that project into the page.

fructose (16.4) a ketohexose that is also called levulose and fruit sugar; the sweetest of all sugars, abundant in honey and fruit.

galactose (16.4) an aldohexose that is a component of lactose, milk sugar.

galactosemia (16.5) a human genetic disease caused by the inability to convert galactose to glucose-1-phosphate.

glucose (16.4) an aldohexose, the most abundant monosaccharide; it is a component of many disaccharides, such as lactose and sucrose, and polysaccharides, such as cellulose, starch, and glycogen.

glyceraldehyde (16.3) an aldotriose that is the simplest carbohydrate; phosphorylated forms of glyceraldehyde are important intermediates in cellular metabolic reactions.

glycogen (16.6) the glucose storage form of animals; a linear backbone of α-D-glucose in α(1→4) linkage, with numerous short branches attached to the C-6 hydroxyl group by α(1→6) linkage.

glycosidic bond (16.1) the bond between the hydroxyl group of C-1 carbon of one sugar and a hydroxyl group of another sugar.

Haworth projections (16.4) a means of representing the orientation of substituent groups around a cyclic sugar molecule.

hemiacetal (16.4) the product of the reaction between an aldehyde and an alcohol.

hemiketal (16.4) the product of the reaction between a ketone and an alcohol.

hexose (16.2) a six carbon monosaccharide.

ketose (16.2) a sugar that contains a ketone (carbonyl) group.

lactose (16.5) a disaccharide composed of β-D-galactose and either α- or β-D-glucose in β(1→4) glycosidic linkage; milk sugar.

lactose intolerance (16.5) the inability to produce the enzyme lactase, which degrades lactose to galactose and glucose.

maltose (16.5) a disaccharide composed of α-D-glucose and either α- or β-D-glucose in α(1→4) glycosidic linkage.

monosaccharide (16.1) the simplest type of carbohydrate consisting of a single saccharide unit.

nonreducing sugar (16.5) a sugar that cannot be oxidized by Benedict's reagent.

oligosaccharide (16.1) an intermediate-sized carbohydrate composed of from three to ten monosaccharides.

pentose (16.2) a five carbon monosaccharide.

polysaccharide (16.1) a large, complex carbohydrate composed of long chains of monosaccharides.

reducing sugar (16.4) a sugar that can be oxidized by Benedict's reagent. This includes all monosaccharides and most disaccharides.

ribose (16.4) a five carbon monosaccharide that is a component of RNA and many coenzymes.

saccharide (16.1) a sugar molecule.

stereochemistry (16.3) the study of the spatial arrangement of atoms in a molecule.

stereoisomers (16.3) a pair of molecules having the same structural formulas and bonding patterns, but differing in the arrangement of the atoms in space.

sucrose (16.5) a disaccharide composed of α-D-glucose and β-D-fructose in (α1→β2) glycosidic linkage; table sugar.

tetrose (16.2) a four carbon monosaccharide.

triose (16.2) a three carbon monosaccharide.

Self Test for Chapter Sixteen

1. Classify each of the following as a monosaccharide, disaccharide, or polysaccharide:

   a. deoxyribose

   b. galactose

   c. glucose

   d. starch

   e. sucrose

   f. fructose

2. Classify each of the following as an aldose or a ketose. Also, include in the classification the number of carbon atoms in each.

   a. glucose

   b. fructose

   c. galactose

   d. ribose

   e. deoxyribose

   f. glyceraldehyde

3. What is the term for a pair of stereoisomers that are nonsuperimposable mirror images of one another?

4. Which polysaccharide serves as a structural element in plant cell walls?

5. Which enantiomeric form of glucose is found in humans?

6. Name the simplest carbohydrate.

7. What two enzymes are involved in the degradation of amylose?

8. Name the bond that holds two monosaccharides together in a disaccharide.

9. What is the structural difference between an aldose and a ketose?

10. What is the difference between a triose and a pentose?

11. Determine the D- or L- configurations for each of the following:

   a.

   b.

   c.

   d.

12. List three common names for glucose.

13. Which monosaccharide is the sweetest of all carbohydrates?

14. Which polysaccharide is the major glucose storage form of animals?

15. Give word-reactions for the hydrolysis of the following carbohydrates:

   a. glucose

   b. lactose

   c. sucrose

   d. starch

   e. cellulose

16. Which of the disaccharides will not react with Benedict's solution?

17. What substances are described as either polyhydroxyaldehydes or polyhydroxyketones?

18. Why are humans unable to digest cellulose?

19. What is the term for the condition that results from an inability to digest lactose?

20. Which carbohydrate is found in blood in concentrations up to 0.1% (W/V)?

21. Which form of cyclic D-glucose has the –OH at carbon-1 below the ring?

22. What compound is formed in the reaction between an alcohol and a ketone?

23. What monosaccharide is found in RNA?

24. Give the molecular formula of ribose.

25. What is the name of the human genetic disease in which galactose cannot be converted into a form that can be used in cellular metabolic reactions?

26. What is the term for carbohydrates that can reduce metal ions?

27. What is the term for a sugar composed of two monosaccharide units?

28. What are the major sites of glycogen synthesis in the human body?

29. What test might be used to distinguish between a monosaccharide and polysaccharide?

30. What test might be used to distinguish between a reducing sugar and a nonreducing sugar?

Vocabulary Quiz for Chapter Sixteen

1. _____ is a human genetic disease caused by the inability to convert galactose into glucose-1-phosphate.

2. A(n) _____ is the product of the reaction between an aldehyde and an alcohol.

3. A carbon atom bonded to four different chemical groups is called a(an) _____ carbon.

4. Two stereoisomers that are nonsuperimposable mirror images are called _____.

5. _____ is the glucose storage polymer found in the liver and muscle of animals.

6. The most abundant organic compound in the world, composed of a polymer of β-D-glucose linked by $\beta(1\rightarrow4)$ glycosidic bonds is _____.

7. The bond between the hydroxyl group of one sugar and a hydroxyl group of another sugar is called a _____.

8. A sugar that can be oxidized by Benedict's reagent is a _____.

9. The study of the spatial arrangement of atoms in a molecule is _____.

10. _____ is the disaccharide composed of β-D-galactose and either α or β-D-glucose in $\beta(1\rightarrow4)$ glycosidic linkage. It is also called milk sugar.

17 Lipids and Their Functions in Biochemical Systems

Learning Goals

1 Discuss the physical and chemical properties and biological function of each of the families of lipids.
2 Write the structures of saturated and unsaturated fatty acids.
3 Compare and contrast the structure and properties of saturated and unsaturated fatty acids.
4 Write equations representing the reactions that fatty acids undergo.
5 Describe the functions of prostaglandins.
6 Discuss the mechanism by which aspirin reduces pain.
7 Draw the structure of a phospholipid and discuss its amphipathic nature.
8 Discuss the general classes of sphingolipids and their functions.
9 Draw the structure of the steroid nucleus and discuss the functions of steroid hormones.
10 Describe the function of lipoproteins in triglyceride and cholesterol transport in the body.
11 Draw the structure of the cell membrane and discuss its functions.
12 Discuss passive and facilitated diffusion of materials through a cell membrane.
13 Explain the process of osmosis.
14 Describe the mechanism of action of a Na^+-K^+ ATPase.

Introduction

Lipids are a diverse collection of organic molecules grouped together on the basis of their solubility in nonpolar solvents. The four groups of lipids that will be considered in this chapter are fatty acids, glycerides, nonglyceride lipids, and complex lipids.

Example 1
List two properties that all lipids have in common.

Answer
Most are insoluble in water but very soluble in nonpolar solvents.

17.1 Biological Functions of Lipids

Lipids are involved in a variety of biological processes. They are structural components of cell membranes. They are an energy storage form in the body, stored in adipocytes (fat cells). Some lipids, like the steroids, are hormones. Others are vitamins, required for processes such as blood clotting, proper bone development, and vision.

Example 2

List four important functions of the class of compounds called the lipids.

Answer

1. Cell membranes are composed of lipids, including phospholipids and steroids. The cell membrane creates a barrier between the cell and its environment and provides a means for the controlled passage of materials into and out of the cell.
2. Most of the available stored energy in animals is in the form of lipids known as triglycerides.
3. Several of the lipids have hormonelike properties.
4. Others are lipid-soluble vitamins.

17.2 Fatty Acids

Structure and Properties

Fatty acids are long-chain (12–24 carbons) monocarboxylic acids. The general formula for a **saturated fatty acid** is

$$CH_3(CH_2)_nCOOH$$

where n is an even integer between 10 and 22. In a saturated fatty acid, each carbon in the chain is bonded to the maximum number of hydrogen atoms.

In the case of **unsaturated fatty acids,** there is at least one carbon-carbon double bond, as seen in this example:

$$CH_3(CH_2)_7CH = CH(CH_2)_7COOH$$

As we observed with alkanes, the melting points of saturated fatty acids increase with increasing carbon number. This general trend is also seen for unsaturated fatty acids, except that the melting point also decreases markedly as the number of carbon-to-carbon double bonds increases.

Example 3

List the properties that most fatty acids have in common.

Answer

1. They are long-chain (12–24) monocarboxylic acids.

2. They often contain an even number of carbon atoms.

3. They form continuous chains of carbon atoms with no branches.

4. They are either saturated, with only single bonds in the R group, or unsaturated, with some double bonds in the R group.

Example 4

What physical property clearly distinguishes a saturated and unsaturated fatty acid with the same number of carbon atoms?

Answer

The melting points of saturated fatty acids are much higher. The greater the number of double bonds in the unsaturated fatty acid, the lower its melting point.

Example 5

Is the fatty acid $C_{13}H_{25}COOH$ saturated or unsaturated?

Answer

It is unsaturated, since all saturated fatty acids (similar to alkanes) will fit the general formula of $C_nH_{2n-1}COOH$. If it were saturated, its formula would have to be $C_{13}H_{27}COOH$; thus, the $C_{13}H_{25}COOH$ has two fewer hydrogen atoms than the saturated one. This means that the $C_{13}H_{25}COOH$ contains one double bond.

Chemical Reactions of Fatty Acids

Learning Goal 4: Write equations representing the reactions that fatty acids undergo.

The reactions of fatty acids are similar to those of short-chain carboxylic acids.

Esterification

Esterification is the reaction of a fatty acid with an alcohol to form an ester and water.

Example 6

Complete the following esterification reaction:

$$
\begin{array}{c}
\quad\quad\text{H} \\
\quad\quad| \\
\text{H}-\text{C}-\text{OH} \quad\quad C_{17}H_{35}COOH \\
\quad\quad| \\
\quad\quad\quad\quad\quad\quad\quad\quad\quad\quad\quad\text{H+, heat} \\
\text{H}-\text{C}-\text{OH} \quad + \quad C_{17}H_{35}COOH \quad\longrightarrow \\
\quad\quad| \\
\text{H}-\text{C}-\text{OH} \quad\quad C_{17}H_{35}COOH \\
\quad\quad| \\
\quad\quad\text{H}
\end{array}
$$

Work

A triglyceride is produced by the reaction of three fatty acids and glycerol. Line up the carboxyl group of each fatty acid with the hydroxyl groups (–OH) of the glycerol molecule. Then remove a water molecule between each to form an ester structure at each of the three positions. The resulting triglyceride will always have the following general structure:

$$
\begin{array}{c}
\text{H} \quad\quad\quad \text{O} \\
| \quad\quad\quad\quad || \\
\text{H}-\text{C}-\text{O}-\text{C}-\text{R} \\
| \quad\quad\quad\quad\quad \text{O} \\
\quad\quad\quad\quad\quad\quad || \\
\text{H}-\text{C}-\text{O}-\text{C}-\text{R} \\
| \quad\quad\quad\quad\quad \text{O} \\
\quad\quad\quad\quad\quad\quad || \\
\text{H}-\text{C}-\text{O}-\text{C}-\text{R} \\
| \\
\text{H}
\end{array}
$$

Answer

The R groups will always be the R groups that were found in each specific fatty acid. Notice that there are no hydroxyl or carboxyl groups present; only three ester groups are now holding the glycerol to the three fatty acid.

Example 6 continued

The following triglyceride would be formed, along with three molecules of water:

$$
\begin{array}{c}
H \qquad\qquad O \\
| \qquad\qquad\; || \\
H-C-O-C-C_{17}H_{35} \\
| \qquad\qquad\; O \\
\qquad\qquad\;\; || \\
H-C-O-C-C_{17}H_{35} \\
| \qquad\qquad\; O \\
\qquad\qquad\;\; || \\
H-C-O-C-C_{17}H_{35} \\
| \\
H
\end{array}
$$

Hydrolysis

   Hydrolysis is the addition of water to a fatty acid ester to produce a fatty acid and an alcohol.

Example 7
Give the complete acid hydrolysis of the following triglyceride:

$$
\begin{array}{c}
H \qquad\qquad O \\
| \qquad\qquad\; || \\
H-C-O-C-(CH_2)_{12}CH_3 \\
| \qquad\qquad\; O \\
\qquad\qquad\;\; || \\
H-C-O-C-(CH_2)_{16}CH_3 \\
| \qquad\qquad\; O \\
\qquad\qquad\;\; || \\
H-C-O-C-(CH_2)_{14}CH_3 \\
| \\
H
\end{array}
$$

1-Myristoyl -2-stearoyl-3-palmitoylglycerol

Work
The hydrolysis of a triglyceride requires three water molecules and a catalyst (generally acid). The products formed are glycerol and one molecule of each of the fatty acids used to prepare the triglyceride. The name of the above triglyceride tells us that glycerol plus myristic acid, stearic acid, and palmitic acid were used to make it.

Example 7 continued

Answer

$$
\begin{array}{l}
H-\overset{\overset{\displaystyle H}{|}}{C}-O-\overset{\overset{\displaystyle O}{||}}{C}-(CH_2)_{12}CH_3 \\[2mm]
H-\overset{\overset{\displaystyle |}{C}}{C}-O-\overset{\overset{\displaystyle O}{||}}{C}-(CH_2)_{16}CH_3 \quad +\; 3\,H_2O \\[2mm]
H-\overset{\overset{\displaystyle |}{C}}{C}-O-\overset{\overset{\displaystyle O}{||}}{C}-(CH_2)_{14}CH_3 \\[2mm]
\underset{\displaystyle H}{|}
\end{array}
\xrightarrow{\text{catalyst}}
\begin{array}{l}
H-\overset{\displaystyle H}{\underset{\displaystyle |}{C}}-OH \\[2mm]
H-\overset{\displaystyle |}{C}-OH \\[2mm]
H-\overset{\displaystyle |}{C}-OH \\[2mm]
\underset{\displaystyle H}{|}
\end{array}
\begin{array}{l}
+\quad HO-\overset{\overset{\displaystyle O}{||}}{C}-(CH_2)_{12}CH_3 \\[2mm]
+\quad HO-\overset{\overset{\displaystyle O}{||}}{C}-(CH_2)_{16}CH_3 \\[2mm]
+\quad HO-\overset{\overset{\displaystyle O}{||}}{C}-(CH_2)_{14}CH_3
\end{array}
$$

Example 8
Write the hydrolysis products for each of the following:

1. Tristearoylglycerol $+\ 3H_2O$
2. 1-Lauroyl-2-stearoyl-3-palmitoylglycerol $+\ 3H_2O$
3. phosphatidylcholine $+\ 4H_2O$

Answer
1. glycerol + three molecules of stearic acid
2. glycerol + lauric acid + stearic acid + palmitic acid
3. glycerol + 2 fatty acids + H_3PO_4 + choline

Saponification

Saponification is the base-catalyzed hydrolysis of an ester. The product of this reaction, an ionized salt, is a soap.

Reaction at the Double Bond

Hydrogenation of an unsaturated fatty acid is the addition of hydrogen to a double bond. This is an example of an addition reaction. Hydrogenation is used in the food industry to convert polyunsaturated vegetable oils into solid fats.

Example 9
Show how linoleic acid can be converted in the laboratory into stearic acid.

Example 9 continued

Answer

$$CH_3(CH_2)_4CH=CHCH_2CH=CH(CH_2)_7COOH \quad + \quad 2 \ H_2$$

heat
pressure
Ni

$$CH_3(CH_2)_{16}COOH$$

Stearic acid

Eicosanoids: Prostaglandins, Leukotrienes, and Thromboxanes

Learning Goal 5:	**Describe the functions of prostaglandins in physiological processes.**

Learning Goal 6:	**Discuss the mechanism by which aspirin reduces pain.**

Linolenic acid and linoleic acid are **essential fatty acids**. They must be obtained in the diet because they cannot be synthesized by the body. Linoleic acid is required for the biosynthesis of **arachidonic acid**, the precursor of a class of 20-carbon, hormonelike molecules known as **eicosanoids**. The eicosanoids include *prostaglandins, leukotrienes,* and *thromboxanes.*

The **prostaglandins** are extremely potent biological molecules with hormonelike activity. They are carboxylic acids with a five-carbon ring. All are composed of the basic C_{20} carbon skeleton of prostanoic acid and are grouped under the designations of A, B, E, F, and so on, which indicate the basic carbon skeletal arrangement. Each prostaglandin also has a number designation that indicates the number of carbon-carbon double bonds in the compound. Figure 17.3 in the text provides examples of prostaglandin structure and nomenclature.

Prostaglandins are made in all tissues and exert biological effects on the cells that produce them and on neighboring cells. Their functions are briefly summarized below.

1. *Blood clotting:* Thromboxane A_2, produced by blood platelets, enhances blood clotting. PGI_2 (prostacyclin), produced by the cells lining the blood vessels, inhibits the clotting process. Working together, these molecules ensure that blood clots form only when necessary.
2. *The inflammatory response*: Prostaglandins promote the pain and fever associated with the inflammatory response.

3. *Reproductive system:* PGE$_2$ stimulates smooth muscle contraction, especially uterine contractions. Dysmenorrhea (painful menstruation) is the result of an excess of two prostaglandins. Drugs that inhibit prostaglandin synthesis provide relief from symptoms.
4. *Gastrointestinal tract:* Prostaglandins inhibit the secretion of stomach acid and increase the secretion of a protective mucous layer into the stomach. Aspirin inhibits prostaglandin synthesis; thus, prolonged use may contribute to stomach ulcers.
5. *Kidneys*: Prostaglandins increase the excretion of water and electrolytes.
6. *Respiratory tract:* Leukotrienes promote the constriction of the bronchi associated with asthma. Other prostaglandins have the opposite effect, bronchodilation.

Aspirin relieves pain by inhibiting prostaglandin synthesis. It works by inhibiting the enzyme cyclooxygenase, which catalyzes the first step in prostaglandin synthesis. The acetyl group from aspirin is covalently bound to the enzyme, inhibiting its activity.

17.3 Glycerides

Glycerides are lipids that contain the alcohol glycerol. They may be subdivided into two classes: **neutral glycerides** and **phosphoglycerides**.

Neutral Glycerides

The esterification of glycerol with one, two, or three fatty acids produces a **mono-**, **di-**, or **triglyceride**. These are also referred to as *mono-*, *di-*, or *triacylglycerols*. Although mono- and diglycerides are present in nature, the most common is the triglyceride, the major storage form of lipids found in fat cells. Triglycerides are named by using the "backbone" name, glycerol, as the suffix. The names of the fatty acyl groups are then placed before it.

Neutral glycerides do not dissociate into charged species, because bonding throughout the molecule is covalent and nonpolar. Consequently, they readily stack with one another and are easily stored in adipocytes. In fact, their major function in the body is energy storage. If more nutrients are consumed than are needed for daily metabolic processes, the excess is converted to triglycerides and stored in adipocytes, which form *adipose tissue*. If energy is needed, the triglycerides are broken down, and their stored energy is released for use by the body.

Phosphoglycerides

Learning Goal 7:	**Draw the structure of a phospholipid and discuss its amphipathic nature.**

Phospholipids contain a phosphoryl group (PO$_4^{3-}$). The presence of the phosphoryl group produces an amphipathic molecule with a polar head (the phosphoryl group) and a nonpolar tail (the hydrocarbon chain of the fatty acid).

Phosphoglycerides contain acyl groups derived from long-chain fatty acids esterified at C-1 and C-2 of glycerol-3-phosphate. The simplest, phosphatidate, contains a free phosphoryl group. More complex phosphoglycerides arc formed when the phosphoryl group is bonded to another hydrophilic molecule.

Phosphatidycholine (lecithin), phosphatidylserine, and *phosphatidylethanolamine* (*cephalin*) are commonly found in cell membranes. The structures of these three phospholipids are shown in Figure 17.7 of the text. Each possesses a polar "head" and a nonpolar "tail." The ionic "head" is hydrophilic and interacts with water molecules, and the nonpolar "tail" is hydrophobic and interacts with nonpolar molecules. This bipolar nature is essential to the structure and properties of biological membranes.

Example 10

Draw a phosphatidylcholine that contains the R groups from linolenic acid (position 1) and linoleic acid (position 2), and then label the hydrophobic and hydrophilic ends.

Answer

The general structure of a phosphatidylcholine is shown on the left. The R groups of linolenic and linoleic acids are esterified at positions 1 and 2 on the glycerol molecule, respectively, in the structure on the right.

General Structure

$$H-C-O-C-R_1$$
$$H-C-O-C-R_2$$
$$H-C-O-P-CH_2CH_2-N^+ \quad CH_3 / CH_3$$

Answer

$$H-C-O-C-C_{17}H_{29}$$
$$H-C-O-C-C_{17}H_{31}$$
$$H-C-O-P-CH_2CH_2-N^+-CH_3$$

↑ Hydrophobic   ↑ Hydrophilic

17.4 Nonglyceride Lipids

Sphingolipids

| Learning Goal 8:   Discuss the general classes of sphingolipids and their functions. |

Sphingolipids are lipids that are not derived from glycerol. They are phospholipids derived from the amino alcohol sphingosine.

Sphingomyelin, a sphingolipid, is abundant in the myelin sheath that surrounds and insulates cells of the central nervous system and is essential to proper nerve transmission.

Glycosphingolipids, or *glycolipids*, include the cerebrosides, sulfatides, and gangliosides. They are built on a ceramide backbone structure, a fatty acid amide of sphingosine. The cerebroside galactocerebroside is found almost exclusively in brain cells. The gangliosides are found in most tissues of the body, although they were first isolated from nervous tissue.

Steroids

Learning Goal 9:	Draw the structure of the steroid nucleus and discuss the functions of steroid hormones.

Steroids are an important family of lipids derived from cholesterol. The steroids include the bile salts that aid in the emulsification and digestion of lipids, and the sex hormones testosterone and estrone. They are members of a large, diverse collection of lipids called isoprenoids, all of which are derived from the five-carbon isoprene unit, seen here:

$$\underset{H_2C}{} = \underset{\underset{|}{CH_3}}{C} - CH = CH_2$$

All steroids are structured around the steroid nucleus (steroid carbon skeleton), which consists of four fused rings. Two fused rings share one or more common bonds as part of their ring backbones.

Cholesterol is found in most cell membranes, where it functions to regulate the fluidity of the membrane. A high serum cholesterol concentration is associated with heart disease, especially **atherosclerosis** or hardening of the arteries. Cholesterol and other substances coat the inside of the arteries. This causes the arteries to become narrower, and more pressure is needed to cause blood to flow through them. This results in elevated blood pressure (hypertension), which is also linked to heart disease.

Many steroids play roles in the reproductive cycle. *Progesterone*, the most important hormone associated with pregnancy, is synthesized from cholesterol. Produced in the ovaries and placenta, it prepares the uterine lining to accept the fertilized egg. Progesterone is also involved in fetal development and suppression of further ovulation during pregnancy.

Progesterone is the precursor of *testosterone*, a male sex hormone, and estrone, a female sex hormone. Both these hormones are involved in the development of secondary sexual characteristics.

Development of birth control agents has involved application of steroid chemistry. One of the first effective synthetic birth control agents was 19-norprogesterone. Unfortunately, it had to be taken by injection. Norlutin (17-ethynyl-19-nortestosterone) is equally effective and can be taken orally. These compounds all act by inducing a false pregnancy, which prevents ovulation.

Cortisone is involved in carbohydrate metabolism and is an important drug used to treat rheumatoid arthritis, asthma, gastrointestinal, and skin disorders. Care must be taken in the use of cortisone because of the possible side effects.

Waxes

The chemical composition of **waxes** is highly variable. All are insoluble in water and are solid at room temperature. Paraffin wax is a mixture of solid straight-chain hydrocarbons. Carbowax is a synthetic polyether. The natural waxes are composed of a long-chain fatty acid esterified to a long-chain alcohol.

Examples of naturally occurring waxes include beeswax; lanolin, used in skin creams; carnauba wax, used in automobile polish; and whale oil, once used as a fuel and for candles.

17.5 Complex Lipids

Learning Goal 10:	Describe the function of lipoproteins in triglyceride and cholesterol transport in the body.

Complex lipids are lipids that are bonded to other types of molecules. **Plasma lipoproteins** are complex lipids that transport other lipids through the blood stream. Because lipids are only slightly soluble in water, their movement from organ to organ requires such a transport system. Lipoprotein particles consist of a core of hydrophobic lipids surrounded by a shell of polar lipids and proteins.

There are four major classes of human plasma lipoproteins. **Chylomicrons** carry triglycerides from the intestine to other tissues. **Very low-density lipoproteins (VLDL)** carry triglycerides synthesized in the liver to other tissues for storage. **Low-density lipoproteins (LDL)** carry cholesterol to peripheral tissues and help regulate cholesterol levels. **High-density lipoproteins (HDL)** transport cholesterol from peripheral tissues to the liver.

The path of lipid transport begins with dietary fat, which is emulsified in the small intestine. Triglycerides are hydrolyzed by lipase, releasing fatty acids and monoglycerides that are absorbed by intestinal cells and reassembled into triglycerides. Chylomicrons are produced, which eventually enter the bloodstream for transport to cells throughout the body. Triglycerides and cholesterol synthesized in the liver are transported in lipoproteins. Triglycerides are carried in VLDL particles, and cholesterol is transported in LDL particles.

LDL particles bind to specific protein receptors (LDL receptors) on cell membranes. The LDL-receptor complex is taken into the cell by a process called *receptor-mediated endocytosis*. In this process, the membrane is pulled into the cell, bringing the entire LDL particle into the cytoplasm in a membrane-bound sac called an endosome. Lysosomes, membrane-bound sacs containing digestive enzymes, fuse with the endosome, and the LDL particles are digested by the lysosomal enzymes. Cholesterol is released into the cell cytoplasm, where it inhibits its own biosynthesis, activates an enzyme for its own storage, and inhibits the synthesis of LDL receptors.

A genetic defect in the gene for the LDL receptor causes an accumulation of LDL-cholesterol in the plasma, resulting in atherosclerosis in the sufferer. While high concentrations of LDL in the blood are associated with hardening of the arteries, high levels of HDL in the blood appear to reduce the incidence of atherosclerosis.

17.6 The Structure of Biological Membranes

> **Learning Goal 11:** Draw the structure of the cell membrane and discuss its functions.

Biological membranes are *lipid bilayers* consisting of phospholipids and cholesterol. The hydrophobic, hydrocarbon tails stack in the center of the bilayer, and the ionic head groups are exposed on the surfaces.

Fluid Mosaic Structure of Biological Membranes

Membranes are fluid, having the consistency of olive oil. The degree of fluidity is determined by the amounts of saturated and unsaturated fatty acids and by the length of the fatty acid tails. Shorter, more unsaturated fatty acid tails produce a more fluid membrane. Floating in the sea of phospholipids are many proteins that are critical to normal cellular function. When viewed by electron microscopy, these proteins look like a mosaic. Because of the fluid consistency of the membrane and the presence of numerous proteins, our concept of membrane structure is called the **fluid mosaic theory**.

Peripheral proteins are found only on the surfaces of the membrane. **Transmembrane proteins** are embedded within the membrane and extend completely through it.

Membrane Transport

The cell membrane is responsible for the controlled passage of molecules into and out of the cell. Transport of some molecules across the membrane is regulated by transmembrane transport proteins, while other molecules pass through the membranes unassisted, by **passive transport.**

Passive Diffusion: The Simplest Form of Membrane Transport

> **Learning Goal 12:** Discuss passive and facilitated diffusion of materials through a cell membrane.

Small uncharged molecules, such as O_2 and CO_2, pass through the membrane by **simple diffusion,** the net movement of a substance from an area of high concentration to an area of low concentration. In this case, diffusion involves the movement of solutes across a cellular membrane. Diffusion is a form of passive transport because it requires no energy expenditure on the part of the cell.

Large or highly charged molecules cannot pass through the lipid bilayer directly because the membrane is **selectively permeable**. That is, the membrane allows the diffusion of some molecules, but not others.

Facilitated Diffusion: Specificity of Molecular Transport

Most molecules diffuse across membranes through specific protein carriers called **permeases**. This process, called **facilitated diffusion**, is also a means of passive transport because the direction of movement will depend on the concentration of the substance on each side of the membrane. Facilitated diffusion occurs through pores within the permease. Each pore has a shape complementary to the shape of the molecule to be transported. Only molecules having the right shape can pass through the pore and the rate of diffusion is limited by the number of permease molecules in the membrane as well as the concentration of solute on either side of the membrane.

Osmosis: Passive Movement of a Solvent Across a Membrane.

Learning Goal 13:   Explain the process of osmosis.

Membranes permeable to solvent but not to solute are called **semipermeable membranes**. **Osmosis** is the diffusion of a solvent (water in biological systems) through a semipermeable membrane in response to a water concentration gradient.

The osmotic strength of a solution is referred to as the **osmotic concentration** or **osmolarity**. It depends only on the ratio of the number of solute particles to the number of solvent particles. Thus, only the concentration of a solution and not the chemical nature and size of the solute is important. Blood plasma has an osmolarity equivalent to a 0.30 M glucose solution or a 0.15 M NaCl solution (remember that NaCl dissociates into Na^+ and Cl^- in solution). It is important when infusing solutions into the body that the osmolarity of the solutions is the same as that within the cells of the body. Such a solution is said to be **isotonic**. Since the osmolarity is the same inside and outside, there will be no net movement of water into or out of the red blood cells, and they will remain the same size.

If red blood cells are exposed to a **hypotonic** solution (one having a lower osmolarity than the cell, the cell will swell and burst. If red blood cells are exposed to a **hypertonic** solution (one with a greater osmolarity than the cell), water will pass out of the cells and they will dramatically shrink in size.

The principle of osmolarity is very important when considering the delivery of intravenous (IV) solutions into a patient. Any fluids infused must be isotonic with the cells of the body.

Energy Requirements for Transport

Learning Goal 14: Describe the mechanism of action of a Na^+-K^+ ATPase.

To survive, it is often necessary for cells to accumulate high concentrations of nutrients that are in low concentration in the environment. In this circumstance, the cell must expend energy to move nutrients from an area of low concentration to an area of higher concentration. This process is called **active transport**. Most ions and food molecules are imported by active transport through the cell membrane. An example of active transport is the Na$^+$-K$^+$ pump that maintains a high concentration of Na$^+$ outside the cell and a high concentration of K$^+$ inside the cell.

Glossary of Key Terms in Chapter Seventeen

active transport (17.6) the movement of molecules across a membrane against a concentration gradient.

arachidonic acid (17.2) a fatty acid that is derived from linolenic acid; the precursor of the prostaglandins.

atherosclerosis (17.4) deposition of excess plasma cholesterol and other lipids and proteins on the walls of arteries, resulting in a decreased artery diameter and increased blood pressure.

cholesterol (17.4) a 27-carbon steroid ring structure that serves as the precursor of the steroid hormones.

chylomicron (17.5) a plasma lipoprotein (aggregate of protein and triglycerides) that carries triglycerides from the intestine to all body tissues via the bloodstream.

complex lipid (17.5) a lipid that is bonded to other types of molecules.

diglyceride (17.3) the product of the esterification of glycerol with fatty acids at two positions.

eicosanoid (17.2) any of the derivatives of 20-carbon fatty acids, including the prostaglandins, leukotrienes, and thromboxanes.

emulsifying agent (17.3) a bipolar molecule that aids in the suspension of fats in water.

essential fatty acid (17.2) the fatty acids linoleic acid and linolenic acid, which must be supplied in the diet because they cannot be synthesized by the body.

esterification (17.2) the reaction between a fatty acid and an alcohol to form an ester and water.

facilitated diffusion (17.6) movement of a solute across a membrane from an area of high concentration to an area of low concentration through a transmembrane protein or permease.

fatty acid (17.2) any member of the family of continuous-chain carboxylic acids that generally contain 12-20 carbon atoms; the most concentrated source of energy used by the cell.

fluid mosaic model (17.6) the model of membrane structure that describes the fluid nature of the phospholipid bilayer and the presence of numerous proteins embedded within the membrane.

glyceride (17.3) a lipid that contains glycerol.

high density lipoprotein (17.5) a plasma lipoprotein that transports cholesterol from peripheral tissue to the liver.

hydrogenation (17.2) a reaction in which hydrogen (H$_2$) is added to a double or triple bond.

hypertonic solution (17.6) the more concentrated solution of two separated by a semipermeable membrane.

hypotonic solution (17.6) the more dilute solution of two separated by a semipermeable membrane.

isotonic solution (17.6) a solution that has the same concentration as another solution with which it is being compared; a solution that has the same osmotic pressure as a solution existing within a cell.

lipid (17.1) a member of the group of organic molecules of varying composition that are classified together on the basis of their solubility in nonpolar solvents.

low density lipoprotein (17.5) a plasma lipoprotein that carries cholesterol to peripheral tissues and helps to regulate cholesterol levels in those tissues.

monoglyceride (17.2) the product of the esterification of glycerol with one fatty acid.

neutral glyceride (17.3) the product of the esterification of glycerol with fatty acids at one, two, or three positions.

osmosis (17.6) net flow of a solvent across a semipermeable membrane in response to a concentration gradient.

osmotic pressure (17.6) the net pressure with which water enters a solution through a semipermeable membrane.

passive transport (17.6) the net movement of a solute from an area of high concentration to an area of low concentration.

peripheral membrane protein (17.6) a protein that is bound to either the inner or outer surface of a membrane.

phosphatidate (17.3) the simplest phosphoglyceride; a molecule with fatty acids esterified to C-1 and C-2 of glycerol and a free phosphoryl group esterified at C-3.

phosphoglyceride (17.3) a molecule with fatty acids esterified at the C-1 and C-2 positions of glycerol and a phosphoryl group esterified at C-3. There may be an additional hydrophilic group bonded to the phosphoryl group.

phospholipid (17.3) a lipid containing a phosphoryl group.

plasma lipoprotein (17.5) a complex composed of lipid and protein that is responsible for the transport of lipids throughout the body.

prostaglandin (17.2) a family of hormonelike substances derived from the 20-carbon fatty acid arachidonic acid; produced by many cells of the body, they regulate many body functions.

saponification (17.2) a reaction in which a soap is produced; more generally, the hydrolysis of an ester by an aqueous base.

saturated fatty acid (17.2) a long-chain monocarboxylic acid in which each carbon of the chain is bonded to the maximum number of hydrogen atoms.

semipermeable membrane (17.6) a membrane permeable to the solvent but not to the solute; a material that allows the transport of certain substances from one side of the membrane to the other.

sphingolipid (17.4) a phospholipid that is derived from the amino alcohol sphingosine, rather than from glycerol.

sphingomyelin (17.4) a sphingolipid that is found in abundance in the myelin sheath that surrounds and insulates cells of the central nervous system.

steroid (17.4) a lipid that is derived from cholesterol and composed of one five-sided ring and three six-sided rings; the steroids include sex hormones and anti-inflammatory compounds.

terpene (17.4) the general term for lipids that are synthesized from isoprene units; the terpenes include steroids, bile salts, lipid-soluble vitamins, and chlorophyll.

transmembrane protein (17.6) a protein that is embedded within a membrane and crosses the lipid bilayer, protruding from the membrane both inside and outside the cell.

triglyceride (17.3) triacylglycerol; a molecule composed of glycerol esterified to three fatty acids.

unsaturated fatty acid (17.2) a long-chain monocarboxylic acid having at least one carbon-to-carbon double bond.

very low density lipoprotein (VLDL) (17.5) a plasma lipoprotein that binds triglycerides synthesized by the liver and carries them to adipose tissue for storage.

wax (17.4) a collection of lipids that are generally considered to be esters of long-chain alcohols.

Self Test for Chapter Seventeen

1. What is the simplest type of lipid?

2. Why are most lipids insoluble in water?

3. What is the main energy storage compound in animals?

4. List two properties of all fatty acids.

5. A triglyceride is called 1-stearoyl-2-lauroyl-3-palmitoylglycerol. Name the compounds that would be needed to prepare this triglyceride.

6. What products are formed by hydrolysis of a natural wax?

7. List the four major classes of human plasma lipoproteins.

8. What products are released by the complete hydrolysis of a triglyceride?

9. Evidence indicates that high levels of which plasma lipoprotein help reduce the incidence of atherosclerosis?

10. Patients whose diets are high in saturated fat tend to have high levels of what substance in their blood?

11. In the lipid bilayer of cell membranes, which part of the phospholipid is found in the center of the bilayer?

12. What substance dissolved in the hydrophobic region of a biological membrane helps to regulate membrane fluidity?

13. What features of the fatty acid tails of membrane phospholipids determine the degree of fluidity of a biological membrane?

14. Write an equation for the reaction of lauric acid with sodium hydroxide.

15. What structure is found in all steroids?

16. Name foods in your diet that have a high concentration of cholesterol.

17. Which steroid is responsible for both the successful initiation and completion of pregnancy?

18. Which class of biological compounds has members of varying chemical composition, all of which are grouped together on the basis of their solubility?

19. Saturated fatty acids containing ten or more carbon atoms are found in what physical state at room temperature?

20. What products are formed in the reaction of a fatty acid with an alcohol?

21. List two ions found in hard water.

22. Hydrogenation is used in the food industry to convert polyunsaturated vegetable oils into what?

23. List the three different kinds of eicosanoids.

24. Prostaglandins produced in the kidneys cause what effect on renal blood vessels?

25. What are the properties of neutral glycerides?

26. What two regions characterize phosphoglycerides?

27. Which lipid is found in egg yolks and soybeans and is used as an emulsifying agent in ice cream?

28. Which type of lipoprotein is bound to plasma cholesterol and transports it from the peripheral tissues to the liver?

29. Which steroid is used to suppress the inflammatory response in the treatment of rheumatoid arthritis?

30. What is the purpose of the orally ingested synthetic steroid hormone called norlutin?

Vocabulary Quiz for Chapter Seventeen

1. The 27-carbon steroid ring molecule that is the precursor for sex hormones, is required for the regulation of membrane fluidity, and has been implicated in atherosclerosis and high blood pressure is _____.

2. The current model of the structure of biological membranes is called the _____.

3. The most common lipids found in biological membranes that consist of two fatty acids esterified at the C-1 and C-2 positions of glycerol, a phosphoryl group esterified at the C-3 position, and having another hydrophilic molecule attached to the phosphoryl group are called _____.

4. The most common lipid storage form in the body, which consists of three fatty acids esterified to a glycerol molecule, is called a(n) _____.

5. _____ are long-chain monocarboxylic acids.

6. The sphingolipid found in abundance in the sheath that surrounds and insulates cells of the central nervous system is _____.

7. The plasma lipoprotein that carries triglycerides from the intestine to various tissues throughout the body is the _____.

8. A fatty acid that cannot be synthesized by the body and must be obtained in the diet is called a(n) _____.

9. _____ is the general term for lipids that are synthesized from isoprene units; they include steroids, bile salts, and lipid-soluble vitamins.

10. A(n) _____ is a bipolar molecule that aids in the suspension of lipids in water.

18 *Protein Structure and Function*

Learning Goals

1. List the functions of proteins.
2. Draw the general structure of an amino acid and classify amino acids based on their R groups.
3. Describe the primary structure of proteins and draw the structure of the peptide bond.
4. Draw the structures of small peptides and name them.
5. Describe the types of secondary structure of a protein.
6. Discuss the forces that maintain secondary structure.
7. Describe the structure and function of fibrous proteins.
8. Describe the tertiary and quaternary structure of a protein.
9. List the R group interactions that maintain protein conformation.
10. List examples of proteins that require prosthetic groups and explain the way in which they function.
11. Discuss the importance of the three-dimensional structure of a protein to its function.
12. Describe the roles of hemoglobin and myoglobin.
13. Describe how extremes of pH and temperature cause denaturation of proteins.
14. Explain the difference between essential and nonessential amino acids.

Introduction

Proteins are polymers made up of α-amino acids. Most proteins are enormous molecules, composed of many amino acids, and therefore have very high molecular weights. Proteins carry out all the enzymatic reactions of the cell and are essential structural elements of the body.

18.1 Cellular Functions of Proteins

Learning Goal 1:   **List the cellular functions of proteins.**

Enzymes are proteins that serve as biological catalysts. **Immunoglobulins,** or **antibodies**, are specific protein molecules produced in the immune system in response to foreign **antigens**. **Transport proteins** carry materials from one place to another in the body. They include permeases and **receptors** that move molecules across cell membranes and soluble proteins, such as hemoglobin, which transport oxygen from the lungs to the tissues. **Regulatory proteins** control cell function and include hormones such as insulin and glucagon. **Structural proteins** provide mechanical support and outer covering to animals. **Movement proteins** are necessary for all forms of movement. Our muscles contract and expand through the interaction of actin and myosin proteins. **Nutrient proteins** are sources of amino acids for embryos and infants.

Example 1

Provide a summary of the cellular functions of proteins.

Answer

1. Most *enzymes* are proteins. Enzymes catalyze almost all the chemical reactions that occur in living cells. Reactions that would take days in the laboratory occur in a matter of seconds or minutes in the presence of enzymes.

2. *Antibodies* are proteins that help stop infections by binding specifically to an antigen and then causing its destruction or removal from the body.

3. *Transport proteins* carry materials from one place to another in the body.

4. *Regulatory proteins* control many aspects of cell function, including metabolism and reproduction.

5. *Structural proteins* provide mechanical support to large animals and provide them with their outer coverings.

6. *Proteins of motion* allow a single-celled organism or higher organism to have different types of motility.

7. *Nutrient proteins* are sources of amino acids for infants and embryos.

18.2 The α-Amino Acids

Structure of Amino Acids

Learning Goal 2:	**Draw the general structure of an amino acid and classify amino acids based on their R.**

The general structure of an **α-amino acid** is shown below:

$$\text{H}_3{}^+\text{N} - \underset{\underset{\text{R}}{|}}{\overset{\overset{\text{H}}{|}}{\text{C}}} - \overset{\overset{\text{O}}{\|}}{\text{C}} - \text{O}^-$$

The α-carbon is attached to a carboxylate group ($-CO_2{}^-$), a protonated amino group ($-NH_3{}^+$), a hydrogen, and an R group. The carboxylate group and protonated amino groups are necessary for the covalent binding of amino acids to one another to form a protein. The R groups cause the proteins to fold into precise, three-dimensional configurations that determine their ultimate function.

Stereoisomers of Amino Acids

The α-carbon of all the α-amino acids, except glycine, is attached to four different groups and is therefore chiral. The configuration of all the naturally occurring α-amino acids isolated from proteins is L-. See Figure 18-2 for a comparison of D- and L-glyceraldehyde with D- and L- alanine.

Example 2

Draw and compare the structures of glyceraldehyde and alanine isolated from natural sources.

Answer

In living organisms, we find that glyceraldehyde is in the D-isomer form, but the amino acids exist in the L-isomer form.

D-Glyceraldehyde

L-Alanine

Classes of Amino Acids

The amino acids are grouped according to the polarity of their side chains. The side chains of some amino acids are nonpolar. These amino acids are **hydrophobic** (water-fearing), and they are generally found buried in the interior of proteins. The nonpolar amino acids include alanine, valine, leucine, isoleucine, proline, glycine, cysteine, methionine, phenylalanine, and tryptophan.

Example 3

List the amino acids that contain R groups (side chains) that are hydrophobic. Where are these usually found in a protein molecule?

Answer

Alanine, valine, leucine, isoleucine, proline, glycine, cysteine, methionine, phenylalanine and tryptophan contain hydrophobic R groups. These amino acids are generally found buried in the interior of proteins, where they can associate with one another and remain isolated from interaction with water molecules.

The side chains of the remaining amino acids are polar and therefore are **hydrophilic** (water-loving); they are often found on the surfaces of proteins. These polar amino acids can be subdivided into three classes:

1. Polar, neutral amino acids

2. Negatively charged (acidic) amino acids

3. Positively charged (basic) amino acids

The amino acids are often referred to using an abbreviation of three letters. Most of these abbreviations are the first three letters of their names. (For example, the abbreviation for glutamate is Glu, but glutamine is Gln.)

Example 4
List and categorize the amino acids that have hydrophilic side chains.

Answer
1. Arginine, aspartate, glutamate, histidine, lysine, asparagine, glutamine, serine, threonine, and tyrosine have side chains that are hydrophilic.
2. These polar amino acids are divided into three classes:
   a. Polar, neutral amino acids: These have side chains that have a high affinity for water, but they are not ionic. Serine, threonine, tyrosine, asparagine, and glutamine fall into this category.
   b. Acidic amino acids are those that have R groups that can ionize to form the negatively charged carboxylate ion ($-CO_2^-$ ion). Aspartate and glutamate are in this category.
   c. Basic amino acids are those with R groups that can ionize to form the positively charged amine ion ($-NH_3^+$). These amino acids act like bases in water, since the side chains react with water, picking up a proton and releasing a hydroxide ion. Lysine, arginine, and histidine are in this category.

18.3 The Peptide Bond

Learning Goal 3:   **Describe the primary structure of proteins and draw the structure of the peptide bond.**

Learning Goal 4:   **Draw the structure of small peptides and name them.**

Proteins are polymers of L–amino acids joined by **peptide bonds**. This is the covalent bond formed between the α-carboxylate group of one amino acid and the α-amino group of another amino acid. The amino acid with a free $\alpha-NH_3^+$ group is known as the amino terminal, or simply the **N-terminal amino acid**, and the amino acid with a free $\alpha-CO_2^-$ group is known as the carboxyl, or **C-terminal amino acid**. The N-terminal amino acid is always drawn on the left, and the C-terminal amino acid on the right when depicting a series of covalently linked amino acids.

Example 5

Draw the tripeptide made from alanine + glycine + serine.

Work

1. Begin by drawing the backbone of the tripeptide. Make sure that the N-terminal amino acid is on the left and the C-terminal amino acid is on the right, and that the amino acids are joined correctly by amide bonds.

$$
\begin{array}{c}
\quad\quad H \quad O \quad\quad\quad H \quad O \quad\quad\quad H \quad O \\
\quad\quad | \quad\ || \quad\quad\quad | \quad\ || \quad\quad\quad | \quad\ || \\
H_3^{+}N - C - C - N - C - C - N - C - C - O^{-} \\
\quad\quad | \quad\quad\quad\quad | \quad\ | \quad\quad\quad | \quad\ | \\
\quad\quad R \quad\quad\quad H \quad R \quad\quad\quad H \quad R
\end{array}
$$

N-terminal amino acid   C-terminal amino acid

2. Next, fill in the correct R– groups to correspond with the correct amino acids.

Answer

$$
\begin{array}{c}
\quad\quad H \quad O \quad\quad\quad H \quad O \quad\quad\quad H \quad O \\
\quad\quad | \quad\ || \quad\quad\quad | \quad\ || \quad\quad\quad | \quad\ || \\
H_3^{+}N - C - C - N - C - C - N - C - C - O^{-} \\
\quad\quad | \quad\quad\quad\quad | \quad\ | \quad\quad\quad | \quad\ | \\
\quad\quad CH_3 \quad\quad H \quad H \quad\quad\quad H \quad CH_2 \\
\quad\quad\quad\quad\quad\quad\quad\quad\quad\quad\quad\quad\quad\quad | \\
\quad\quad\quad\quad\quad\quad\quad\quad\quad\quad\quad\quad\quad\quad OH
\end{array}
$$

The names of peptides are derived from the C-terminal amino acid, which receives its entire name. For all other amino acids, the ending *-ine* is changed to *-yl*. Thus, the dipeptide tryptophanyl-leucine has leucine as the C-terminal amino acid and tryptophan as the N-terminal amino acid. Example 18.1 in the text describes the procedure for drawing the structure of a peptide chain.

The lone pair of electrons of the nitrogen atom of the peptide bond interacts with the carbon and oxygen of the carbonyl group. Thus, a resonance structure is formed that gives the peptide bond a partially double bond character. As a result, the peptide bond is planar, and the two adjacent α-carbons lie *trans* to it. The hydrogen of the amide nitrogen is also *trans* to the oxygen of the carbonyl group

Example 6

Show the hydrolysis reaction for alanyl-glycyl-cysteine.

Answer

$$H_3^+N-\underset{\underset{CH_3}{|}}{\overset{\overset{H}{|}}{C}}-\overset{\overset{O}{||}}{C}-N-\underset{\underset{H}{|}}{\overset{\overset{H}{|}}{C}}-\overset{\overset{O}{||}}{C}-N-\underset{\underset{\underset{SH}{|}}{\overset{CH_2}{|}}}{\overset{\overset{H}{|}}{C}}-\overset{\overset{O}{||}}{C}-O^- \quad + \ 2 \ H_2O$$

acid ↓

$$H_3^+N-\underset{\underset{CH_3}{|}}{\overset{\overset{H}{|}}{C}}-COO^- \quad + \quad H_3^+N-\underset{\underset{H}{|}}{\overset{\overset{H}{|}}{C}}-COO^- \quad + \quad H_3^+N-\underset{\underset{\underset{SH}{|}}{CH_2}}{\overset{\overset{H}{|}}{C}}-COO^-$$

Alanine   Glycine   Cysteine

18.4 Primary Structure of Proteins

Learning Goal 3:   **Describe the primary structure of proteins and draw the structure of the peptide bond.**

The **primary protein structure** is the linear sequence of amino acids. This is dictated by the genetic information in the DNA. Over time, mutations occur in the genes. These result in changes in the primary sequence of a protein. The greater the number of amino acid differences between proteins from two different organisms, the greater the evolutionary distance between them. Examination of the amino acid sequences of proteins allows scientists to construct evolutionary trees.

18.5 Secondary Structure of Proteins

Learning Goal 5:   Describe the types of secondary structure of a protein.

Learning Goal 6:   Discuss the forces that maintain secondary structure.

Secondary protein structure results from folding of the chain of covalently linked amino acids into regularly repeating structures. The folding pattern is maintained by numerous hydrogen bonds between the amino nitrogen and the carbonyl oxygen of the peptide chain background.

α-Helix

The most common type of secondary structure is a right-handed helical conformation known as the **α-helix**. There are 3.6 amino acids in each turn of the helix. Each carbonyl oxygen in the helix is hydrogen bonded to an amide hydrogen four amino acids away in the chain, producing an array of hydrogen bonds that are parallel to the long axis of the helix.

The **α-keratins** are α-helical proteins. These **fibrous proteins** form the covering (hair, wool, and fur) of most land animals. The individual α-helices of the keratins coil together in a bundle, producing a three-stranded protofibril that is part of an array known as a microfibril. These molecular pigtails possess great mechanical strength.

β-Pleated Sheet

The second common secondary structure in proteins resembles the pleated folds of drapery, and is known as **β-pleated sheet.** In this secondary structure, the polypeptide chain is nearly completely extended, with all the carbonyl oxygens and amide hydrogens involved in hydrogen bonds. The polypeptide chains in a β-pleated sheet may be *parallel* or *antiparallel*. In the parallel structure, the N-termini are aligned head to head, and in the antiparallel structure, the N-terminus of one chain is aligned with the C-terminus of a second chain (head to tail). Silk fibroin is an example of a protein whose structure is an antiparallel β-pleated sheet.

18.6 Tertiary Structure of Proteins

Most cellular proteins are **globular**. This globular, three-dimensional structure is called the **tertiary protein structure**. The peptide chain with its regions of secondary structure further folds on itself to achieve the tertiary structure.

The forces that maintain the tertiary structure of a protein include the following:

1. *van der Waals attractions* between the R groups of nonpolar amino acids
2. *hydrogen bonds* between the polar R groups of the polar amino acids
3. *ionic bonds* between the R groups of oppositely charged amino acids
4. *covalent bonds* between the thiol-containing amino acids; two cysteines can be oxidized to a dimeric amino acid, cystine, that can be a cross-link between different proteins or hold two segments within a protein together.

The tertiary structure of the protein determines its biological function; therefore, the weak interactions that hold the protein in its correct three-dimensional shape are extremely important.

18.7 Quaternary Structure of Proteins

Learning Goal 8:	Describe the tertiary and quaternary structure of a protein.

Learning Goal 9:	List the R group interactions that maintain protein conformation.

Learning Goal 10:	List examples of proteins that require prosthetic groups and explain the way in which they function.

The active form of some proteins is an aggregate of two or more smaller globular proteins. This is the **quaternary protein structure**. The attractions that hold two or more peptides together are the same as those that maintain tertiary structure: hydrogen bonds, ionic bridges, disulfide bridges, and van der Waals forces. Some proteins must be bound to a nonprotein **prosthetic group** in order to be functional. Hemoglobin is an example of a protein with quaternary structure that is bound to heme groups. **Glycoproteins** are proteins with covalently bonded sugar groups.

Example 7

Summarize the four types of protein structure and their relationship to one another.

Answer

1. The primary structure of a protein is the linear sequence of amino acids bonded to one another through amide (peptide) bonds.

$$H_3{}^+N-\overset{\overset{\displaystyle H}{|}}{\underset{\underset{\displaystyle R}{|}}{C}}-\overset{\overset{\displaystyle O}{\|}}{C}-\overset{\overset{\displaystyle H}{|}}{\underset{\underset{\displaystyle H}{|}}{N}}-\overset{\overset{\displaystyle H}{|}}{\underset{\underset{\displaystyle R'}{|}}{C}}-\overset{\overset{\displaystyle O}{\|}}{C}-\overset{\overset{\displaystyle H}{|}}{\underset{\underset{\displaystyle H}{|}}{N}}-\overset{\overset{\displaystyle H}{|}}{\underset{\underset{\displaystyle R''}{|}}{C}}-\overset{\overset{\displaystyle O}{\|}}{C}-O^-$$

amino acid 1   amino acid 2   amino acid 3

2. The secondary structure involves the hydrogen bonding that can occur between the carbonyl oxygen and the amide hydrogen of the peptide bonds. α-Helix and the β-pleated sheet structures are the most common kinds.

Example 7 continued

3. The tertiary structure involves the gross overall folding of the entire protein molecule. Both noncovalent interactions between the side chains (R groups) and covalent –S–S– bridges play a role in determining the overall tertiary structure.

4. The quaternary structure involves the aggregation of two or more peptide chains with respect to one another. The quaternary structure is maintained by R group interactions.

18.8 An Overview of Protein Structure and Function

Learning Goal 9:   **List the R group interactions that maintain protein conformation.**

Learning Goal 11:   **Discuss the importance of the three-dimensional structure of a protein to its function.**

Primary structure is the linear sequence of amino acids covalently joined by peptide bonds. **Secondary structure** is the result of folding of the peptide chain into an α-helix or a β-pleated sheet. It is maintained by hydrogen bonding. **Tertiary structure** is the further folding of a peptide to produce a globular structure. It is maintained by a variety of noncovalent interactions between the R groups of amino acids, including hydrogen bonding, ionic bonding, and van der Waals forces. **Quaternary structure** is the association of two or more peptides to form the functional protein. It is maintained by the same noncovalent R group interactions that are responsible for tertiary structure.

18.9 Myoglobin and Hemoglobin

Learning Goal 12:   **Describe the roles of hemoglobin and myoglobin.**

Myoglobin and Oxygen Storage

Myoglobin is the oxygen storage protein of skeletal muscle. It is bound to a **heme** group that serves as the site for oxygen binding. Myoglobin has a greater affinity for oxygen than hemoglobin does, and therefore it serves as an efficient molecule to receive and store oxygen in the muscle.

Hemoglobin and Oxygen Transport

Hemoglobin (Hb) is the blood protein responsible for oxygen transport from the lungs to other tissues. It is composed of four peptide subunits, two α subunits and two β subunits, each of which contains a heme group.

Example 8
Explain the general structure and function of hemoglobin.

Answer
Hemoglobin is the oxygen transport protein of the blood. It is composed of four separate peptide subunits. There are two identical α chains and two identical β chains. In addition, each subunit of hemoglobin contains a heme group. A hemoglobin molecule therefore has the ability to bind and carry four molecules of oxygen.

Oxygenation of hemoglobin in the lungs is favored by the high partial pressure of oxygen (pO_2) in the air we breath and the low pO_2 in the blood. Thus, oxygen diffuses from the region of high pO_2 to the region of low pO_2. When the blood reaches actively metabolizing tissues, this situation is reversed and oxygen diffuses into tissues from the blood.

Oxygen Transport from Mother to Fetus

A fetus receives oxygen from its mother by simple diffusion across the placenta. The fetus makes fetal hemoglobin (Hb F), which has a greater affinity for oxygen than the hemoglobin of the mother, Hb A. Thus, the fetus is assured of a constant and adequate supply of oxygen.

Sickle Cell Anemia

Sickle cell anemia is a human genetic disease caused by a mutation in the gene that encodes the β-subunit of hemoglobin. This mutation results in the synthesis of sickle cell hemoglobin (Hb S). There is only a single amino acid difference between normal and sickle cell hemoglobin. That single change causes deoxyhemoglobin to polymerize, causing the red blood cells to sickle. This can result in damage to many organs and death at an early age.

In individuals with sickle cell anemia, both copies of the β-globin gene produce the mutant protein. Individuals with *sickle cell trait* have one mutant and one normal β-globin gene. Typically, they do not suffer the symptoms of the disease; however, they may pass the mutant gene on to their children.

Example 9
Explain what causes sickle cell anemia.

Answer
Sickle cell anemia is a human genetic disease that first appeared in tropical west and central Africa. It afflicts about 0.4% of African Americans. These individuals produce a mutant hemoglobin known as sickle cell hemoglobin (Hb S). Sickle cell anemia receives its name from the sickled appearance of the red blood cells that form in this condition.

Sickle cell hemoglobin differs from normal hemoglobin by a single amino acid that as been replaced in each of the β-chains. A valine has replaced the amino acid glutamic acid in each of the two chains. This substitution provides a basis for the binding of

Example 9 continued

hemoglobin S molecules to one another. When oxyhemoglobin S molecules release their oxygen, they bind to one another as long fibers. These fibers radically alter the shape of the red blood cell, resulting in the sickling effect. The sickled cells are unable to pass through the small capillaries, thereby causing severe medical problems.

18.10 Denaturation of Proteins

Learning Goal 13: **Describe how extremes of pH and temperature cause denaturation of proteins.**

When a protein solution is exposed to temperatures that are too high, the bonds within the protein molecules begin to vibrate violently. The weak interactions that hold the three-dimensional structure of the protein are disrupted, and the protein molecules are **denatured**. These molecules then **coagulate** as they clump together and precipitate out of solution.

If the pH of a solution of proteins becomes too high or too low, the characteristic electric charge on the surface of the protein is neutralized. The **isoelectric point** is the pH at which the net charge on the protein is zero, and it can no longer interact with the surrounding water molecules. This causes the proteins to aggregate with one another, and coagulation occurs.

When the pH rises too high, proteins become polyanions. When the pH is too low, proteins become polycations. In either case, proteins are denatured due to charge repulsion.

If the pH of the blood becomes too acidic or too basic, the proteins dissolved in the blood and the proteins of the cells in the blood can be denatured. Thus it is critical that the body have mechanisms to maintain the pH of the blood within a narrow range.

18.11 Dietary Protein and Protein Digestion

Learning Goal 14: **Explain the difference between essential and nonessential amino acids.**

Proteins are the third major type of energy source in the diet. They are hydrolyzed to amino acids that may be oxidized to provide energy or used directly in protein synthesis. Amino acids are the precursors of a large, diverse group of nitrogen compounds that includes some hormones and the heme group.

Example 10
What are some common dietary sources of proteins?

Answer
Meat is an excellent source of protein. Also, dried beans, peas, and nuts are good sources.

Essential amino acids are those that cannot be synthesized by the body and are required in the diet. **Nonessential amino acids** are those amino acids that can be synthesized by the body and need not be included in the diet.

Proteins are also classified as **complete** or **incomplete**. Animal protein is complete, providing all of the essential and nonessential amino acids in approximately the correct amounts for biosynthesis. Vegetable protein is generally incomplete, lacking a sufficient amount of one or more essential amino acids.

Example 11

How can vegetable proteins be mixed to provide all of the essential amino acids?

Answer

A complete protein provides all of the essential amino acids needed by the human body. Proteins derived from animal sources are generally complete. In contrast, proteins derived from vegetable sources are generally incomplete because they lack a sufficient amount of one or more amino acids. No one vegetable source is a good source of all the essential amino acids, but by mixing many vegetable sources, an adequate amount of each essential amino acid may be obtained.

Protein digestion begins in the stomach. The enzyme pepsin begins the degradation of dietary protein to amino acids. Further digestion occurs in the small intestine. These hydrolysis reactions are carried out by enzymes such as trypsin and chymotrypsin.

Glossary of Key Terms in Chapter Eighteen

α-amino acids (18.2) the basic subunits of proteins; each is composed of an α-carbon bonded to a carboxylate group, a protonated amino group, a hydrogen atom, and a variable "R" group.

antibody (18.1) one of the specific proteins produced by cells of the immune system in response to invasion by infectious agents.

antigen (18.1) any substance able to stimulate the immune system; they are usually proteins or large carbohydrates.

coagulation (18.11) the process by which proteins in solution are denatured and aggregate with one another to produce a solid.

complete protein (18.11) a protein source that contains all the essential and nonessential amino acids.

C-terminal amino acid (18.3) the amino acid at the end of the protein chain that has a free $\alpha–CO_2^-$ group; this is the last amino acid in a peptide.

defense proteins (18.1) proteins that defend the body against infectious diseases; antibodies are defense proteins.

denaturation (18.10) the process by which the organized structure of a protein is disrupted resulting in a completely disorganized, nonfunctional form of the protein.

enzyme (18.1) a protein that serves as a biological catalyst.

essential amino acid (18.11) an amino acid that cannot be synthesized by the body and must therefore be supplied by the diet.

fibrous protein (18.5) a protein composed of peptides arranged in long sheets or fibers.

globular protein (18.6) a protein composed of polypeptide chains that are tightly folded into a compact spherical shape.

glycoprotein (18.7) a protein with sugars as prosthetic groups; often these are receptors on the cell surface.

α-helix (18.5) a right-handed coiled secondary structure maintained by hydrogen bonds between the amide hydrogen of one amino acid and the carbonyl oxygen of an amino acid four residues away.

heme group (18.9) the chemical group found in hemoglobin and myoglobin that is responsible for the ability to carry oxygen.

hemoglobin (18.9) the major protein component of red blood cells; the function of this red, iron-containing protein is transport of oxygen.

hydrophilic amino acid (18.2) "water loving," one of the polar or ionic amino acids that has a high affinity for water.

hydrophobic amino acid (18.2) "water fearing," a nonpolar amino acid that prefers contact with other nonpolar amino acids over contact with water.

incomplete protein (18.11) a protein source that does not contain all of the essential and nonessential amino acids.

isoelectric point (18.10) the pH at which the protein has an equal number of positive and negative charges and therefore has an overall net charge of zero.

α-keratin (18.5) fibrous proteins that form the covering of most land animals; they are major components of fur, skin, beaks, and nails.

movement protein (18.1) a protein involved in any aspect of movement in an organism, for instance, actin and myosin in muscle tissue and flagellin that composes bacterial flagella.

myoglobin (18.9) the oxygen storage protein found in muscle.

nonessential amino acid (18.11) an amino acid that can be synthesized by the body.

N-terminal amino acid (18.3) the amino acid at the end of a protein chain that has a free α-N$^+$H$_3$ group; this is the first amino acid of a peptide.

nutrient protein (18.1) a protein that serves as a source of amino acids for embryos or fetuses.

peptide bond (18.3) the covalent linkage between two amino acids in a peptide chain; an amide formed by a condensation reaction between two amino acids.

β-pleated sheet (18.5) a common secondary structure that resembles the pleats of an oriental fan.

primary structure (of a protein) (18.4) the sequence of amino acids in a protein; this is determined by the genetic information of the gene for each protein.

prosthetic group (18.7) a nonprotein group that is attached to a protein and is essential to the biological activity of that protein; often a complex organic compound.

protein (Intro) a macromolecule whose primary structure is a linear sequence of α-amino acids and whose final structure results from folding of the chain into a specific three-dimensional structure; they have many functions in the organism, including serving as catalysts, structural components, and nutritional elements.

quaternary structure (of a protein) (18.7) aggregation of two or more globular proteins to form a functional protein.

regulatory protein (18.1) a protein that controls cell functions such as metabolism and reproduction.

secondary structure (of a protein) (18.5) folding of the primary structure of a protein into an α-helix or a β-pleated sheet; the forces that maintain the folding are hydrogen bonds between the amide hydrogen and the carbonyl oxygen of the peptide bond.

sickle cell anemia (18.9) a human genetic disease resulting from inheriting mutant hemoglobin genes from both parents.

structural protein (18.1) a protein that provides mechanical support for large plants and animals.

tertiary structure (of a protein) (18.6) the globular, three-dimensional structure of a protein that results from folding the regions of secondary structure; this folding occurs spontaneously as a result of interactions of the side chains or R groups of the amino acids.

transport protein (18.1) a protein that transports materials across the cell membrane or soluble proteins, such as hemoglobin, that carry substances through the body.

Self Test for Chapter Eighteen

1. What elements found in dietary proteins are not found in fats and carbohydrates?

2. Name the simplest amino acid.

3. Which amino acid contains the following structure as its R group?

4. What are enzymes?

5. Where and when are antibodies produced?

6. What two proteins are responsible for the transport and storage of oxygen in higher organisms?

7. What protein is found in hair and fingernails?

8. What is produced by the controlled hydrolysis of proteins?

9. Draw the structure of serine as it appears in a living cell.

10. Which amino acids have negatively charged side chains at physiological pH?

11. Which amino acids have positively charged side chains at physiological pH?

12. Give the three-letter abbreviations used for the following amino acids:

a. alanine   c. glycine   e. lysine

b. cysteine   d. histidine

13. What is the term for the linkage between two amino acids in a peptide?

14. Draw the structure of alanyl-valyl-glycyl-cysteine.

15. What level of protein structure is defined by the sequence of amino acids in the protein?

16. Which level of protein structure is caused by hydrogen bonding between the different peptide bonds?

17. What level of protein structure is defined by the binding of two or more peptides to produce an active functional protein?

18. A polypeptide or protein has equal numbers of positive and negative charges. What is the term used to describe this situation?

19. What kind of secondary structure characterizes the fibrous proteins of muscles?

20. Amino acids in water at pH = 7 exist as dipolar ions. What are these dipolar ions called?

21. What is the term used to describe amino acids that must be acquired from the diet?

22. What dietary source provides complete protein?

23. What enzyme begins the digestion of protein in the stomach?

24. What is the term for a protein that serves as a source of amino acids for infants or embryos?

25. Which amino acid is a secondary amine?

26. What type of secondary structure characterizes silk fibroin?

27. What type of covalent bond is involved in maintaining the tertiary structure of a protein?

28. What is meant by the term *hydrophobic*?

29. What happens to the overall charge on a protein when the pH drops too low?

Vocabulary Quiz for Chapter Eighteen

1. An amino acid that is nonpolar and prefers contact with other nonpolar amino acids over contact with water is described as _____.

2. The chemical group that is attached to a protein and is essential to the biological activity of that protein is called the _____.

3. _____ are proteins that have sugars as prosthetic groups and often are receptors on the cell surface.

4. The process by which the organized structure of a protein is disrupted resulting in a completely disorganized, nonfunctional form of the protein is called _____.

5. The_____ is the globular, three-dimensional structure of a protein that results from folding the regions of secondary structure. This folding occurs spontaneously as a result of interactions of the side chains or R groups of the amino acids.

6. The _____ is the covalent linkage between two amino acids in a peptide chain. It is formed by a condensation reaction.

7. A protein that has an equal number of positive and negative charges and therefore has an overall net charge of zero is said to be _____.

8. _____ is the oxygen storage protein found in muscle.

9. A common secondary structure that resembles the pleats of an oriental fan is called _____.

10. A (n) _____ is an amino acid that cannot be synthesized by the body and must therefore be supplied by the diet.

19 *Enzymes*

Learning Goals

1 Classify enzymes according to the type of reaction catalyzed and the type of specificity.
2 Give examples of the correlation between an enzyme's common name and its function.
3 Describe the effect that enzymes have on the activation energy of a reaction.
4 Explain the effect of substrate concentration on enzyme-catalyzed reactions.
5 Discuss the role of the active site and the importance of enzyme specificity.
6 Describe the difference between the lock-and-key model and the induced fit model of enzyme-substrate complex formation.
7 Discuss the roles of cofactors and coenzymes in enzyme activity.
8 Explain how pH and temperature affect the rate of an enzyme-catalyzed reaction.
9 Describe the mechanisms used by cells to regulate enzyme activity.
10 Discuss the mechanisms by which certain chemicals inhibit enzyme activity.
11 Discuss the role of the enzyme chymotrypsin and other serine proteases.
12 Provide examples of medical uses of enzymes.

Introduction

Enzymes are proteins that function as catalysts in cell processes. They are essential for the thousands of metabolic reactions that allow life to exist. Enzymes are remarkably specific, usually recognizing and binding to only a single type of **substrate**, or reactant, and facilitating its conversion to a **product**.

Example 1
Describe some characteristics of enzymes.

Answer

1. Life depends upon the simultaneous occurrence of hundreds of chemical reactions that must take place rapidly under mild conditions. Enzymes allow these critical reactions to occur under the mild conditions required for life.

2. The enzymes are proteins that speed up biochemical reactions by lowering the energy of activation and increasing the rate of the reaction.

3. An enzyme is very specific. It generally recognizes only one, or occasionally a few, molecules (substrates) upon which it will work its magic.

4. Enzyme-catalyzed reactions often occur from 1 to 100 million times faster than the uncatalyzed reactions.

19.1   Classification and Nomenclature

Classification of Enzymes

> **Learning Goal 1: Classify enzymes according to the type of reaction catalyzed and the type of specificity.**

Enzymes may be classified according to the type of reaction that they catalyze. These six classes are as follows: (1) **oxidoreductases**, (2) **transferases**, (3) **hydrolases**, (4) **lyases**, (5) **isomerases**, and (6) **Ligases.**

Example 2

Explain completely the functions of oxidoreductases.

Answer

These enzymes catalyze oxidation-reduction reactions. They are responsible for the removal of protons and electrons from a substrate to cause its oxidation, or the addition of protons and electrons to a substrate to cause its reduction. Lactate dehydrogenase is a good example. This enzyme transiently binds the coenzyme NAD^+, which accepts a hydride anion ($H:^-$) from the substrate lactate. The product is pyruvate, as shown below:

$$\begin{array}{c}
COO^- \\
| \\
HO-C-H \\
| \\
CH_3
\end{array}
\;+\; \text{enzyme-}NAD^+ \;\longrightarrow\;
\begin{array}{c}
COO^- \\
| \\
C=O \\
| \\
CH_3
\end{array}
\;+\; \text{enzyme-NADH}$$

| lactate (reduced form) | lactate dehydrogenase (oxidized form) | pyruvate (oxidized form) | lactate dehydrogenase (reduced form) |

Example 3

List the six types of enzymes classified according to the type of reaction that they catalyze. Briefly describe the function of each class.

Answer

1. *Oxidoreductases*: these enzymes catalyze electron transfers from one molecule to another.
2. *Transferases*: these enzymes catalyze the transfer of functional groups from one molecule to another.
3. *Isomerases*: these enzymes catalyze the rearrangement of functional groups within a molecule to convert the substrate into a different isomeric form.
4. *Hydrolases*: these enzymes catalyze hydrolysis reactions.

Example 3 continued

5. *Lyases*: these enzymes catalyze the addition of a group to a double bond or the removal of a group to form a double bond.
6. *Ligases*: these enzymes catalyze the condensation or joining of two molecules.

Example 4

To which class does each of the following enzymes belong?

1. pyruvate kinase
2. lipase (hydrolysis of triglycerides)

3. triose isomerase
4. lactate dehydrogenase

Answer
1. transferase
2. hydrolase

3. isomerase
4. oxidoreductase

Nomenclature of Enzymes

Learning Goal 2:   **Give examples of the correlation between an enzyme's common name and its function.**

An enzyme's name often tells us the substrate of the reaction and the nature of the reaction. For instance, the enzyme sucrase hydrolyzes the disaccharide sucrose and the enzyme lactate dehydrogenase removes hydrogen (H^-) from lactate ions. Some enzymes have historical names that do not reveal the nature of the substrate or of the reaction. The substrate and reaction catalyzed by such enzymes as catalase and trypsin must simply be memorized.

Example 5

Explain how the common names of enzymes are derived and give four examples.

Answer
1. The common name of an enzyme is derived from the name of the substrate with which the enzyme interacts, and/or the type of reaction that it catalyzes. Most enzyme names end in -*ase*.
2. The following are specific examples of common names for enzymes:
   a. Urea is the substrate acted on by the enzyme urease.
   b. Lactose is the substrate of lactase.
   c. Dehydrogenase is an enzyme that removes hydrogen from a substrate.
   d. Decarboxylase is an enzyme that removes a carboxyl group from a substrate.

19.2 The Effect of Enzymes on the Activation Energy of a Reaction

Learning Goal 3:   **Describe the effect that enzymes have on the activation energy of a reaction.**

Every chemical reaction is characterized by an equilibrium constant. Enzymes do not alter that equilibrium constant. They do, however, change the path of the reaction, providing a lower energy route for the conversion of the substrate into the product. Thus, enzymes increase the rate of a chemical reaction by lowering the activation energy and therefore increasing the rate at which the reaction reaches equilibrium.

19.3 Effect of Substrate Concentration on Enzyme-Catalyzed Reactions

Learning Goal 4:	Explain the effect of substrate concentration on enzyme-catalyzed reactions.

For an enzyme-catalyzed reaction, the rate of the reaction increases as the substrate concentration increases. However, at a certain substrate concentration, the reaction rate reaches a maximum. At its maximum rate, the active sites of all the enzyme molecules are occupied by substrate molecules. The reaction rate is then limited by the speed with which the substrate is converted to product and product is released.

19.4 The Enzyme-Substrate Complex

Learning Goal 5:	Discuss the role of the active site and the importance of enzyme specificity.

The first step in an enzyme-catalyzed reaction is the binding of the substrate by the enzyme to form the **enzyme-substrate complex**. The groove or pocket in the enzyme that binds to the substrate is the **active site**. The active site of an enzyme is small compared to the overall size of the enzyme. The substrate is held within the active site by weak, noncovalent interactions. The conformation (shape and charge distribution) of the active site is complementary to the conformation of the substrate. **Thus, the conformation of the active site determines the specificity of the enzyme.** Only those substrates that fit into the active site can bind the enzyme.

Example 6
Describe the formation of the enzyme-substrate complex.

Answer
The first step in an enzyme-catalyzed reaction involves the enzyme binding to the substrate to form the enzyme-substrate complex. The portion of the enzyme that is in contact with the substrate is called the active site.

19.5 Specificity of the Enzyme-Substrate Complex

Learning Goal 6:	Describe the difference between the lock-and-key model and the induced-fit model of enzyme-substrate complex formation.

Originally it was thought that the substrate simply snapped into place in the active site, like a piece of a jigsaw puzzle. This view was called the **lock-and-key model**. Our current model describes enzymes as flexible molecules and an active site that is not a rigid pocket into which the substrate fits precisely, The active site is thought to be a flexible pocket that approximates the shape of the substrate. When the substrate binds, the active site molds itself around the substrate. This is called the **induced-fit model**.

Enzymes show a high degree of specificity. For instance, the enzyme urease catalyzes the decomposition of urea to carbon dioxide and ammonia, but it will not accept the related molecule methylurea.

Enzyme specificities differ, and often enzymes are classified into the following four groups based on this property:

1. **Absolute specificity**
2. **Group specificity**
3. **Linkage specificity**
4. **Stereochemical specificity**

Example 7
Describe the enzyme classification based on specificity.

Answer

1. *Absolute specificity*—an enzyme that catalyzes the reaction of only one substrate into product is absolutely specific.

2. *Group specificity*—an enzyme that catalyzes processes involving similar molecules containing the same functional group is group specific.

3. *Linkage specificity*—in this type of specificity, an enzyme always catalyzes the formation or breakage of only one kind of bond.

4. *Stereochemical specificity*—in this type of specificity, an enzyme is capable of catalyzing a reaction involving only one enantiomer.

19.6 The Transition State and Product Formation

The overall process of an enzymatic reaction can be summarized by the following set of reversible reactions:

	Step I		Step II		Step III		Step IV	
E + S	$\rightleftharpoons$	ES	$\rightleftharpoons$	ES*	$\rightleftharpoons$	EP	$\rightleftharpoons$	E + P
Enzyme + Substrate		Enzyme substrate complex		Transition state		Enzyme + Product		Enzyme product complex

In the **transition state**, the shape of the substrate is altered, due to its interaction with the enzyme, into an intermediate form having features of both the substrate and the final product. This favors the conversion of the substrate into the product, which is subsequently released. There are several ways in which an enzyme could cause a reaction to proceed more quickly. In some cases, the enzyme exerts "pressure" on a bond, thereby facilitating bond breakage. An enzyme may simply bring reactants into close proximity and in the proper orientation for reaction to occur. Alternatively, the active site of an enzyme may modify the pH of the microenvironment surrounding the substrate by serving as a donor or acceptor of H+. Remember that the enzyme is unchanged by the chemical reaction. If it transiently loses or gains a proton, it will be converted back to its original form by either regaining or losing a proton, respectively.

Example 8

Summarize the three ways in which an enzyme may lower the activation energy of a reaction.

Answer

1. The enzyme might destroy a bond in a substrate and therefore facilitate bond breakage.

2. An enzyme may facilitate a reaction by bringing two reactant molecules close together and in the proper orientation, so that a reaction easily occurs.

3. The active site of an enzyme may so modify the pH in the microenvironment of the substrate that a reaction will occur quickly.

19.7 Cofactors and Coenzymes

Learning Goal 7:   **Discuss the roles of cofactors and coenzymes in enzyme activity.**

Some enzymes require an additional nonprotein **prosthetic group** in order to function. The protein portion of such an enzyme is called the **apoenzyme,** and the nonprotein prosthetic group is called the **cofactor**. Usually, cofactors are metal ions that bind to the enzyme and help maintain the correct shape of the active site. Cofactors may also be organic compounds or organometallic compounds.

Other enzymes require the transient binding of a **coenzyme**. Coenzymes are organic molecules that generally serve as carriers of electrons or chemical groups. They take part in a reaction by either donating or accepting chemical groups. Most coenzymes contain modified vitamins as part of their structure.

Nicotinamide adenine dinucleotide (NAD^+) is a coenzyme that accepts hydride ions (a hydrogen atom with two electrons) from the substrate that is oxidized. The portion of NAD^+ derived from the vitamin niacin is reduced to produce NADH.

19.8 Environmental Effects

Learning Goal 8:   **Explain how pH and temperature affect the rate of an enzyme-catalyzed reaction.**

Effect of pH

As we learned in the previous chapter, if the pH of a solution becomes too acidic or too basic, a protein is denatured. The pH at which an enzyme functions best is called the **pH optimum**. Enzyme function decreases as the pH rises above or falls below the pH optimum, and at extremes of pH, they are denatured and cease to function.

Most cellular enzymes function optimally at a pH near 7. However, **pepsin**, a proteolytic enzyme found in the stomach where the pH is very low, has a pH optimum of 2. Another proteolytic enzyme, **trypsin**, functions under the conditions of higher pH found in the intestine and has a pH optimum around 8.5.

Effect of Temperature

Enzymes are rapidly denatured if the temperature of the solution rises much above 37°C, but they remain stable at much lower temperatures. For this reason, enzymes used for clinical assays are stored in refrigerators or freezers. Since heating enzymes destroys their activity, cells can't survive extremes of temperatures. Thus, heat is an effective means of sterilizing medical instruments and solutions.

Some bacteria and yeast can survive very high temperatures, living in active volcanoes or in hot springs where the temperature is near the boiling point of water. The proteins of these bacteria have a structure that is stable at these extraordinary temperatures.

19.9   Regulation of Enzyme Activity

> **Learning Goal 9:**   **Describe the mechanisms used by cells to regulate enzyme activity.**

Often enzyme activity is regulated by the cell as a means of controlling metabolic processes. One mechanism of regulation, used by bacteria, is to produce the enzyme only when the substrate is present.

Allosteric Enzymes

A more complex level of enzyme regulation involves enzymes that have more than a single binding site. These are called **allosteric enzymes**. One site, the active site, binds a substrate and catalyzes a biochemical reaction. The second site binds an **effector molecule**. Effector binding alters the shape of the active site. If the active site is converted to an inactive form, **negative allosterism** occurs. If effector binding converts the active site to an active configuration, **positive allosterism** is the result.

Feedback Inhibition

Feedback inhibition usually involves a biosynthetic pathway of many enzymatic steps. If the product of this multistep pathway builds up to high levels, it inhibits the entire pathway by serving as a negative allosteric effector of an enzyme early in the pathway.

Proenzymes

Proenzymes are enzymes produced in an inactive form. They are converted to the active form when they reach the site of their activity. This is a protective mechanism. Most of the proteolytic digestive enzymes are produced in an inactive form so that they do not destroy the cells that make them. These enzymes include elastase, trypsin, chymotrypsin, and pepsin.

Protein Modifications

In **protein modification**, adding or removing a covalently bound group either activates or inactivates an enzyme. The most common type of protein modification is the addition (or removal) of a phosphoryl group to (or from) a free –OH on the R group of a serine, tyrosine, or threonine in the protein chain of the enzyme. The covalent modification of an enzyme is catalyzed by other enzymes, such as *protein kinases* and *phosphatases*. Protein modification is reversible, allowing an enzyme to be turned on or off quickly in response to environmental or physiological conditions.

19.10 Inhibition of Enzyme Activity

Learning Goal 10: Discuss the mechanisms by which certain chemicals inhibit enzyme activity.

Enzyme inhibitors are chemicals that bind to enzymes and either eliminate or drastically reduce their catalytic ability. They are classified on the basis of whether the inhibition is reversible or irreversible, competitive or noncompetitive.

Irreversible Inhibition

Irreversible inhibitors, such as arsenic, bind very tightly, sometimes even covalently, to the enzyme. This binding irreversibly blocks substrate binding or enzyme catalysis.

Reversible, Competitive Inhibitors

Generally, these inhibitors are **structural analogs**, molecules that "look like" the structure of the natural substrate for an enzyme because of similarities in shape and charge distribution. Because of this resemblance, the inhibitor can occupy the enzyme active site, but no reaction can occur. Since it is only bound by weak interactions, the inhibitor is easily removed from the active site, providing an opportunity for the substrate to bind. This is competitive inhibition because the degree of inhibition depends on the relative concentrations of substrate and inhibitor and their relative affinities for the active site.

Reversible, Noncompetitive Inhibitors

These inhibitors bind only weakly to R groups of amino acids or perhaps to the metal ion cofactors. Enzyme activity is restored when the inhibitor dissociates from the enzyme-inhibitor complex.

19.11 Proteolytic Enzymes

Learning Goal 11: Discuss the role of the enzyme chymotrypsin and other serine proteases.

Proteolytic enzymes are enzymes that catalyze the hydrolysis of the peptide bonds that maintain the primary protein structure. **Chymotrypsin** is a proteolytic enzyme found in the small intestine that cleaves peptide bonds on the carbonyl side of tyrosine, tryptophan, and phenylalanine. The specificity of chymotrypsin results from the hydrophobic pocket, a portion of the active site into which the flat aromatic side chains of these amino acids can fit.

Trypsin, **chymotrypsin**, and **elastase** are proteolytic enzymes produced in the pancreas. They all digest protein in the small intestine, have similar primary and tertiary structures, and share the same mechanism of action. They evolved by divergent evolution of a single gene that was probably duplicated. Each copy of the gene underwent mutation, producing many different proteolytic enzymes with different specificities. Chymotrypsin cleaves peptide bonds on the carbonyl side of aromatic amino acids; trypsin cleaves peptide bonds on the carbonyl side of basic amino acids; and elastase cleaves peptide bonds on the carbonyl side of the amino acids glycine and alanine.

19.12 Use of Enzymes in Medicine

Learning Goal 12: Provide examples of medical uses of enzymes.

Enzymes are used as diagnostic tools in medicine. They are also used to treat a variety of human genetic disorders.

The serum concentration of some enzymes is used in disease diagnosis. An enzyme assay is a test that is performed to measure the activity or concentration of an enzyme, expressed in international units. The international unit is equal to the amount of enzyme that will catalyze conversion of one µmole (10^{-6} moles) of substrate to product in one minute at standard conditions of temperature and pH.

Liver disease is indicated by elevated levels of one of the isoenzymes of lactate dehydrogenase (LDH_5), as well as elevated levels of alanine aminotransferase/serum glutamate-pyruvate transaminase (ALT/SGPT) and aspartate aminotransferase/serum glutamate-oxaloacetate transaminase (AST/SGOT) in blood serum. Elevated blood serum concentrations of amylase and lipase are indications of pancreatitis, an inflammation of the pancreas.

In the clinical laboratory, enzymes are valuable analytical reagents. For example, enzymes are used in the BUN test (Blood Urea Nitrogen test). Direct measurement of urea levels is difficult due to the complexity of blood. So the enzyme urease is added, which converts each molecule of urea into two molecules of ammonia (the indicator of urea). Ammonia concentration is easily measured.

The enzyme *glucocerebrosidase* is used to treat Gaucher's disease. Those who suffer this genetic disorder are incapable of producing active glucocerebrosidase and are thus unable to carry out the degradation of the glycolipid called glucocerebroside. Glucocerebroside accumulates in the macrophages of the liver, spleen, and bone marrow. If untreated, these abnormal macrophages will displace normal bone marrow cells, causing severe anemia. Glucocerebrosidase, administered intravenously, degrades glucocerebroside to glucose and ceramide, which are further metabolized.

Glossary of Key Terms in Chapter Nineteen

absolute specificity (19.5) the property of certain enzymes that allows them to bind and catalyze the reaction of only one substrate.

active site (19.4) the cleft in the surface of an enzyme that is the site of substrate binding.

allosteric enzymes (19.9) enzymes that have an effector binding site as well as an active site; effector binding changes the shape of the active site, rendering it either active or inactive.

apoenzyme (19.7) the protein portion of an enzyme that requires a cofactor in order to function in catalysis.

coenzyme (19.7) an organic group required by some enzymes; generally a donor or acceptor of electrons or functional groups in a reaction.

cofactor (19.7) metal ions, organic compounds, or organometallic compounds that must be bound to an apoenzyme to maintain the correct configuration of the active site.

competitive inhibitor (19.10) a structural analog; a molecule that has a structure very similar to the natural substrate of an enzyme, competes with the natural substrate for binding to the enzyme active site, and inhibits the reaction.

enzyme (Intro) a protein that acts as a biological catalyst.

enzyme specificity (19.5) the ability of an enzyme to bind to only one, or a very few, substrates and thus catalyze only a single reaction.

enzyme-substrate complex (19.4) a molecular aggregate formed when the substrate binds to the active site of the enzyme.

feedback inhibition (19.9) when produced in excess, the product of a biosynthetic pathway can turn off the entire pathway for its own synthesis. This is an example of negative allosterism. The product of the pathway is a negative allosteric effector of one of the enzymes early in the pathway.

group specificity (19.5) the property of certain enzymes that allows them to catalyze reactions involving similar substrate molecules having the same functional groups.

holoenzyme (19.7) an active enzyme consisting of an apoenzyme bound to a cofactor.

hydrolase (19.1) an enzyme that catalyzes a hydrolysis reaction.

induced-fit model (19.4) the theory of enzyme-substrate binding that assumes that the enzyme is a flexible molecule and that both the substrate and enzyme change their shapes to accommodate one another as the enzyme-substrate complex forms.

irreversible enzyme inhibitor (19.10) a chemical that binds strongly to the R groups of an amino acid in the active site and eliminates enzyme activity.

isomerase (19.1) an enzyme that catalyzes the conversion of one isomer to another.

ligase (19.1) an enzyme that catalyzes the joining of two molecules.

linkage specificity (19.5) the property of certain enzymes that allows them to catalyze reactions involving only one kind of bond in the substrate molecule.

lock-and-key model (19.4) the theory of enzyme-substrate binding that describes enzymes as inflexible molecules and a substrate that fits into the rigid active site in the same way a key fits into a lock.

lyase (19.1) an enzyme that catalyzes a reaction involving double bonds.

negative allosterism (19.9) effector binding inactivates the active site of an allosteric enzyme.

oxidoreductase (19.1) an enzyme that catalyzes oxidation-reduction reactions.

pancreatic serine proteases (19.11) a family of enzymes, including trypsin, chymotrypsin, and elastase, that hydrolyze dietary proteins in the small intestine.

pII optimum (19.8) the pH at which an enzyme catalyzes the reaction at maximum efficiency.

positive allosterism (19.9) effector binding activates the active site of an allosteric enzyme.

product (19.2) the chemical species that results from an enzyme-catalyzed reaction.

proenzyme (19.9) the inactive form of a proteolytic enzyme

protein modification (19.9) a means of enzyme regulation in which a chemical group is covalently added or removed from a protein; the chemical modification either turns the enzyme on or turns it off.

proteolytic enzyme (19.11) an enzyme that hydrolyzes the peptide bonds between amino acids in a protein chain.

reversible, competitive enzyme inhibitor (19.10) a chemical that resembles the structure and charge distribution of the natural substrate and competes with it for the active site.

reversible, noncompetitive enzyme inhibitor (19.10) a chemical that binds weakly to an amino acid R group or cofactor and inhibits activity. When the inhibitor dissociates, the enzyme is restored to its active form.

stereochemical specificity (19.5) the property of certain enzymes that allows them to catalyze reactions involving only one enantiomer of the substrate.

structural analog (19.10) a chemical having a structure and charge distribution very similar to a natural enzyme substrate.

substrate (19.1) the reactant in a chemical reaction that binds to the enzyme active site and is converted to the product.

temperature optimum (19.8) the temperature at which an enzyme functions optimally and the rate of reaction is maximal.

transferase (19.1) an enzyme that catalyzes the transfer of a functional group from one molecule to another.

transition state (19.6) the unstable intermediate in catalysis in which the enzyme has altered the form of the substrate so that it now shares properties of both the substrate and the product.

vitamin (19.7) an organic substance that is required in the diet in small amounts; water-soluble vitamins are used in the synthesis of coenzymes required for the function of cellular enzymes; lipid-soluble vitamins are involved in calcium metabolism, vision, and blood clotting.

Self Test for Chapter Nineteen

1. How does an enzyme speed up a biological chemical reaction?

2. What is the term for the reactant in an enzyme-catalyzed reaction?

3. Write a balanced equation for the reaction catalyzed by catalase.

4. What is the substrate for each of the following enzymes:

    a. succinate dehydrogenase

    b. sucrase

    c. glycogen phosphorylase

5. Write an equation representing the specific reaction catalyzed by lactate dehydrogenase.

6. What do kinase enzymes do?

7. Which enzyme catalyzes the hydrolysis of triglycerides?

8. What effect does an enzyme have on the equilibrium constant of a reaction?

9. What is the first step in an enzyme-catalyzed reaction?

10. List the different classes of enzyme specificity.

11. What are apoenzymes and cofactors?

12. What is the coenzyme of an enzyme that is a dehydrogenase?

13. Why are the water-soluble vitamins important?

14. What is the term for the pH at which an enzyme has the greatest activity?

15. At extreme pH values, an enzyme may lose the normal three-dimensional shape of the active site. What is the term used to describe this condition?

16. List two classes of enzyme inhibitors.

17. How do the sulfa drugs work as inhibitors?

18. What is the function of the pancreatic enzymes trypsin, chymotrypsin, and elastase?

19. In enzyme-catalyzed reactions, what happens as the concentration of the substrate is increased?

20. At the maximum rate of an enzyme-catalyzed reaction, what may be said about all the enzymes and substrate molecules?

21. What is the term used to describe an enzyme that catalyzes the reaction of only one substrate?

22. Where are the catalytic groups of enzymes located?

23. Most enzymes are rapidly destroyed if the temperature is much higher than what value?

24. What process regulates the activity of allosteric enzymes?

25. In feedback inhibition, which compound causes the first reaction of a pathway to be shut off?

26. What is the term for an enzyme that is first produced in an inactive form?

Vocabulary Quiz for Chapter Nineteen

1. The _____ is the groove or cleft in the surface of an enzyme that is the site of substrate binding.

2. _____ is the property of certain enzymes that allows them to catalyze reactions involving only one enantiomer of the substrate.

3. The _____ is the theory of enzyme-substrate binding that assumes that the enzyme is a flexible molecule and that both the substrate and enzyme change their shapes to accommodate one another as the enzyme-substrate complex forms.

4. _____ is a process in which excess product of a biosynthetic pathway can turn off the entire pathway for its own synthesis.

5. The _____ is the unstable intermediate in catalysis in which the enzyme has altered the form of the substrate so that it now shares properties of both the substrate and the product.

6. A(n) _____ is an enzyme that catalyzes the joining of two molecules.

7. An enzyme that catalyzes the transfer of a functional group from one molecule to another is called a(n) _____.

8. A(n) _____ is the protein portion of an enzyme that requires a cofactor in order to function in catalysis.

9. _____ occurs when effector binding activates the active site of an allosteric enzyme.

10. _____ is the pH at which an enzyme catalyzes the reaction at maximum efficiency.

20 *Introduction to Molecular Genetics*

Learning Goals

1 Draw the general structure of DNA and RNA nucleotides.
2 Describe the structure of DNA and compare it with RNA.
3 Explain DNA replication.
4 List three classes of RNA molecules and describe their functions.
5 Explain the process of transcription.
6 List and explain the three types of post-transcriptional modifications of eukaryotic mRNA.
7 Describe the essential elements of the genetic code and develop a "feel" for its elegance.
8 Describe the process of translation.
9 Define mutation and understand how mutations can cause cancer and cell death.
10 Describe the tools used in the study of DNA and in genetic engineering.
11 Describe the process of polymerase chain reaction and discuss potential uses of the process.
12 Discuss strategies for genome analysis and DNA sequencing.

Introduction

Deoxyribonucleic acid (DNA) carries all the genetic information in the cell. The primary structure of all proteins is dictated by the sequence of nucleotides in the DNA. **Ribonucleic acid (RNA)** molecules are the intermediates that carry out the translation of the genetic code into the structure of a protein.

Any change in the nucleotide sequence of the DNA is a **mutation**. Mutations may be spontaneous or may result from the action of a mutagen. Recombinant DNA technology has allowed a better understanding of human genetic disorders and has produced protein products used for the treatment of some of these diseases.

20.1 The Structure of the Nucleotide

Learning Goal 1:   Draw the general structure of DNA and RNA nucleotides.

Chemical Composition of DNA and RNA

The chemical composition of DNA and RNA is summarized in the following table.

	DNA	RNA
Sugar	2'-Deoxyribose	Ribose
Purine nitrogenous bases	Adenine (A) Guanine (G)	Adenine (A) Guanine (G)
Pyrimidine nitrogenous bases	Cytosine (C) Thymine (T)	Cytosine (C) Uracil (U)

In addition, both DNA and RNA contain phosphoryl groups.

Example 1
Explain the chemical composition of DNA and RNA.

Answer
The treatment of DNA with a strongly acidic solution releases 2'-deoxyribose, phosphoric acid, and the following four heterocyclic bases: adenine, guanine, cytosine, and thymine. The same treatment of RNA releases ribose, phosphoric acid, and the following four heterocyclic bases: adenine, guanine, cytosine, and uracil.

Nucleotide Structure

Nucleotides are made up of a sugar, a nitrogenous base, and at least one phosphoryl group. Because this large structure contains two cyclic molecules, the sugar and the base, the ring atoms of the sugar are designated with a prime to distinguish them from atoms in the nitrogenous base. The covalent bond between the sugar and the phosphoryl group is a phosphoester bond. The bond between the base and the sugar is called a β-*N*-glycosidic linkage.

20.2 The Structure of DNA and RNA

Learning Goal 2: Describe the structure of DNA and compare it with RNA.

DNA Structure: The Double Helix

A strand of DNA is a polymer of 2'-deoxyribonucleotide units bonded to one another by 3'-5' phosphodiester bonds. The backbone of the polymer is called the sugar-phosphate backbone because it is composed of alternating deoxyribose and phosphoryl groups in phosphodiester linkage. A nitrogenous base is bonded to each sugar by a β-*N*-glycosidic linkage.

Example 2
Summarize the primary structure of DNA.

Answer
The primary structure of DNA is the linear sequence of its 2'-deoxyribonucleotides. The nucleotides are linked by 3'-5' phosphodiester bonds. Each single strand of DNA is characterized by a backbone of alternating deoxyribose and phosphoryl groups in phosphodiester linkages. Each deoxyribose is also linked to one of the nitrogenous bases by a β-*N*-glycosidic bond.

DNA is a **double helix** of two strands of DNA held together by hydrogen bonds. The sugar-phosphate backbone spirals around the outside of the helix, and the nitrogenous bases extend into the center at right angles to the axis of the helix. Adenine forms hydrogen bonds with thymine, and cytosine forms hydrogen bonds with guanine. The hydrogen-bonded bases are called **base pairs**. It is the hydrogen bonds between the base pairs that hold the two strands of the double helix together. The two strands of DNA are complementary, because the sequence of bases on one strand automatically determines the sequence of bases on the opposite strand. There are ten base pairs in each turn of the helix. The two strands of the double helix are **antiparallel** to one another; they proceed in opposite directions.

Example 3
Summarize the double helix structure of DNA.

Answer
DNA consists of two strands wound around each other. Each strand has a helical conformation, and the resultant structure is called a double helix. The major properties of the DNA double helix are as follows:
1. The sugar phosphate backbone winds around the outside of the bases like the handrails of a spiral staircase.
2. Each purine base is hydrogen bonded to a pyrimidine base in the interior of the double helix. Adenine is always paired with thymine, and guanine is always paired with cytosine. Each base pair lies at nearly right angles to the long axis of the helix, like the stairs of the spiral staircase. Due to the base pairing of A to T and G to C, the two strands are complementary to one another. Therefore, the sequence of bases on one strand automatically determines the sequence of the other strand.
3. The two strands of the DNA double helix are antiparallel. One strand advances in the $5' \rightarrow 3'$ direction, and the second strand advances in the $3' \rightarrow 5'$ direction.
4. The double helix of DNA completes one full turn every ten nucleotides.

Chromosomes

Chromosomes are pieces of DNA that carry the genetic instructions, or genes, of an organism. The complete set of genetic information in all the chromosomes of an organism is called the **genome**.

The genome of **prokaryotes** is circular and supercoiled, and consists of a single chromosome. Chromosomes in **eukaryotes** are much more complex. The first level of eukaryotic chromosome structure is the **nucleosome**, which consists of a strand of DNA wrapped around a small disk made up of histone proteins. At this level the DNA looks like beads along a string. This string of beads then coils into larger, more complex structures, the *30 nm fiber* and the *200 nm fiber*.

RNA Structure

RNA molecules are single stranded. The sugar-phosphate backbone of RNA consists of ribonucleotides linked by 3'-5' phosphodiester bonds. The sugar in RNA is ribose rather than 2'-deoxyribose, and uracil (U) replaces thymine.

20.3 DNA Replication

Learning Goal 3:   **Explain DNA replication.**

DNA must be replicated before cell division so that each daughter cell inherits a copy of each gene. The mechanism of faithful replication of DNA involves the enzyme **DNA polymerase III,** which "reads" each parental strand, also called the template, and catalyzes the polymerization of a complementary daughter strand. Thus, each daughter DNA molecule consists of one parental strand and one newly synthesized daughter strand. This mode of DNA replication is called **semiconservative replication**.

Bacterial DNA Replication

Bacterial chromosomes are circular DNA molecules. Replication begins at a **replication origin** and proceeds bidirectionally around the circular chromosome. The site at which the new DNA is being synthesized is called the **replication fork**.

The first step in DNA replication is the separation of the two strands of DNA by the enzyme *helicase. Topoisomerase* relieves the stress of the supercoiling, and *single-stranded binding protein* binds to each strand. *Primase* catalyzes the synthesis of a short RNA primer to which *DNA polymerase III* will add DNA nucleotides to begin synthesis of the daughter strand.

Because the two strands of DNA are antiparallel, and replication can proceed in only the 5'→3' direction, the precise mechanism for replication on the two strands is different. The *leading strand* is replicated continuously from a single RNA primer; and the *lagging strand* is replicated discontinuously from many RNA primers. Each of these primers is removed and replaced with DNA. The small fragments produced in this way are joined together by *DNA ligase*.

Eukaryotic DNA Replication

In eukaryotes, DNA replication begins at many replication origins and proceeds bidirectionally along each chromosome.

Example 4
Summarize DNA replication.

Answer
DNA replication occurs each time a cell divides; thus, all the genetic information is passed from one generation to the next. As a result of specific base pairing, the sequence of bases along each strand of DNA automatically specifies the sequence of bases in the complementary strand. DNA replication begins at the replication origin, and each parental strand is copied.

20.4 Information Flow in Biological Systems

The genetic information in the DNA must be expressed to produce the proteins that actually carry out the work of the cell. Gene expression involves two steps. First, DNA is transcribed to produce a variety of RNA molecules. This process is called transcription. Then the RNA molecules participate in translation, a process in which proteins are produced. This unidirectional expression of the genetic information is called the **central dogma** of molecular biology and can be summarized as follows:

$$DNA \rightarrow RNA \rightarrow PROTEIN$$

Example 5
Explain the following terms: central dogma, transcription, and translation.

Answer
The *central dogma* of molecular biology states that, in cells, the flow of genetic information contained in DNA is a one-way street that leads from DNA to RNA to protein synthesis. The process by which a single strand of DNA serves as a template for the synthesis of an RNA molecule is called *transcription*. In this process, part of the information of DNA is copied into a strand of RNA. The process by which the message is converted into protein is called *translation*.

Classes of RNA Molecules

Learning Goal 4: **List three classes of RNA molecules and describe their functions.**

There are three classes of RNA found in the cell, and they are classified by their function in gene expression.

Messenger RNA (mRNA) carries the genetic information for a protein from DNA to the ribosomes. It is a complementary RNA copy of a gene on the DNA.

Ribosomal RNA (rRNA) is a structural and functional component of the ribosomes, which are "platforms" on which protein synthesis occurs.

Transfer RNA (tRNA) translates the genetic code of the mRNA into the primary sequence of amino acids in the protein. In addition to the primary structure, tRNA molecules have a cloverleaf-shaped secondary structure resulting from base pair hydrogen bonding (traditional A–U and G–C, as well as other unusual base pairings), and a tertiary structure. The sequence CCA is found at the 3'-end of the tRNA. The 3'-adenosine of this sequence can be covalently attached to an amino acid. Three nucleotides at the base of the cloverleaf structure form the **anticodon**. This triplet of bases forms hydrogen bonds to a **codon** (complementary sequence of bases) in a messenger RNA (mRNA) molecule on the surface of a ribosome during protein synthesis. This interaction assures that the correct amino acid is brought to the site of protein synthesis at the appropriate location in the growing peptide chain.

Example 6
Describe the different classes of RNA molecules.

Answer
An RNA molecule is classified by its cellular location and by its function.

1. *Messenger RNA (mRNA)* carries the genetic information for a protein.
2. *Ribosomal RNA (rRNA)* is a structural and functional component of the ribosomes.
3. *Transfer RNA (tRNA)* is responsible for translating the genetic code of the mRNA.

Transcription

Learning Goal 5:	Explain the process of transcription.

Transcription is catalyzed by the enzyme **RNA polymerase**. RNA polymerase binds to a specific nucleotide sequence, the **promoter**, at the beginning of a gene. It then separates the two strands of DNA so that it can "read" the sequence of the DNA. Chain elongation begins as the RNA polymerase reads the DNA template strand and catalyzes the polymerization of a complementary RNA copy. The final stage of transcription is termination. The RNA polymerase finds a termination sequence at the end of the gene and releases the newly formed RNA molecule.

Post-Transcriptional Processing of RNA

Learning Goal 6:	List and explain the three types of post-transcriptional modifications of eukaryotic mRNA.

In eukaryotes, transcription produces a **primary transcript** that undergoes extensive **post-transcriptional modification** before it is exported from the nucleus for translation.

The first modification is the addition of a **cap structure** to the 5' end of the RNA. This facilitates efficient translation.

The second modification is the addition of a **poly(A) tail** to the 3' end of the RNA. This protects the mRNA from enzymatic degradation.

The third modification is **RNA splicing**. Bacterial genes are continuous, and all the nucleotide sequences of the gene are found in the mRNA. Eukaryotic genes are discontinuous; there are extra DNA sequences within the genes. The initial mRNA, or primary transcript, carries both the protein-coding sequences and these extra sequences, which are termed intervening sequences or introns. The **introns** are removed by the process of RNA splicing. Splicing is carried out with the assistance of ribonucleoprotein particles called **spliceosomes**. They aid in the recognition of specific sequences at the splice boundaries and stabilize the splicing complex.

20.5 The Genetic Code

> **Learning Goal 7:** Describe the essential elements of the genetic code and develop a "feel" for its elegance.

The **genetic code** is a triplet code. Each code word (codon) consists of three nucleotides. The three-letter genetic code contains 64 codons, but there are only 20 amino acids. Thus, there are 44 more codons than are required. Three of the codons (UAA, UAG, and UGA) are translation termination signals, leaving 41 additional codons. The genetic code is said to be highly **degenerate**. Methionine and tryptophan are the only amino acids encoded by only a single codon. All others are encoded by at least two codons, and serine and leucine each have six codons. As a result of this, the genetic code is quite resistant to mutation. For those amino acids that have multiple codons, the first two bases define the amino acid, and the third position is variable. A **point mutation** (the change of a single base) in the third position, therefore, often has no effect upon the amino acid that is incorporated into a protein.

20.6 Protein Synthesis

> **Learning Goal 8:** Describe the process of translation.

Protein synthesis, or **translation**, occurs on ribosomes. Ribosomes are complexes of ribosomal RNA and proteins consisting of two subunits, a small and a large ribosomal subunit.

Many ribosomes simultaneously translate each mRNA. These structures are called polyribosomes or **polysomes**.

The Role of Transfer RNA

The molecule that decodes the information on the mRNA molecule into the primary structure of a protein is transfer RNA (tRNA). In order to do this, the tRNA must be covalently linked to one specific amino acid. The enzyme that binds the specific tRNA to its specific amino acid is an **aminoacyl tRNA synthetase**. The product is called an **aminoacyl tRNA**.

The tRNA recognizes the appropriate codon on the mRNA by codon-anticodon hydrogen bonding.

The Process of Translation

Initiation

Initiation factors mediate the formation of a translation initiation complex composed of an mRNA molecule, the small and large ribosomal subunits, and the initiator tRNA. The initiator tRNA specifically recognizes the **initiation codon**, AUG, on the mRNA. The ribosome has two sites for binding tRNA molecules. The first site is called the **peptidyl tRNA binding site (P-site)**. It holds the initiator tRNA in the initiation complex and then carries the tRNA bound to the growing peptide during the remainder of protein synthesis. The second site, called the **aminoacyl tRNA binding site (A-site)**, holds the aminoacyl tRNA carrying the next amino acid to be added to the peptide chain.

Chain Elongation

This occurs in three steps that are repeated until protein synthesis is complete. It involves the interaction of several **elongation factors.**

1. Binding of an aminoacyl tRNA molecule to the empty A-site.
2. Peptidyl transferase catalyzes the formation of the peptide bond, and the peptide chain is shifted to the tRNA that occupies the A-site.
3. The uncharged tRNA molecule is discharged, and the ribosome changes positions (**translocation**), so that the next codon on the mRNA occupies the A-site, shifting the new peptidyl tRNA from the A-site to the P-site.

Termination

When a stop, or **termination codon** (UAA, UAG, or UGA), is encountered, translation is terminated. A **release factor** finds the empty A-site and causes release of the newly formed peptide chain and the ribosomal subunits.

The newly synthesized protein folds into its characteristic three-dimensional shape, and if it has quaternary structure, it may associate with other protein subunits. Some proteins are further modified following protein synthesis by the addition of carbohydrate or lipid molecules.

20.7 Mutations, Ultraviolet Light, and DNA Repair

Learning Goal 9: Define mutation and understand how mutations cause cancer and cell death.

The Nature of Mutations

Any change in the nucleotide sequence of a DNA molecule is called a **mutation**. Mutations can arise from errors during DNA replication or may be the result of chemicals, called **mutagens**, which damage the DNA.

Mutations are classified by the kind of change that occurs in the DNA. The substitution of a single nucleotide for another is called a **point mutation**. Loss of one or more nucleotides is a **deletion mutation**. Addition of one or more nucleotides to a DNA sequence is an **insertion mutation**.

The Results of Mutations

Some mutations are **silent**; that is, they cause no change in the organism. Often, however, the result of a mutation has a negative effect on the health of the organism. The effect of a mutation depends on how it alters the genetic code for a protein. There are approximately 4000 human genetic disorders that are known to be caused by mutations, including sickle cell anemia, hemophilia, cystic fibrosis, and color-blindness.

Mutagens and Carcinogens

Often mutagens are also **carcinogens**, cancer-causing chemicals. Most cancers result from mutations in a single normal cell. These mutations result in the loss of normal growth control, causing the abnormal cell to proliferate. If that growth is not controlled or destroyed, it may result in the death of the individual.

Ultraviolet Light Damage and DNA Repair

Ultraviolet (UV) light causes damage to DNA by inducing the formation of **pyrimidine dimers.** Such dimers consist of adjacent pairs of pyrimidine bases covalently bonded to one another. This interferes with normal hydrogen bonding between these pyrimidines and the complementary bases on the opposite strand. As a result, this region of DNA cannot be replicated or transcribed and the cell dies.

Bacteria have four different mechanisms to repair ultraviolet light damage. Mutations occur when the repair system makes an error and causes a change in the nucleotide sequence of the DNA.

Ultraviolet lights (germicidal lamps) are used in hospitals to kill bacteria in the air and on environmental surfaces, such as in a vacant operating room. In addition, excessive exposure to UV light (as during sun tanning) has been correlated with an increased incidence of skin cancer.

Consequences of Defects in DNA Repair

The human repair system for pyrimidine dimers requires at least five enzymes. A mutation in the gene for one of these enzymes results in the genetic skin disorder *xeroderma pigmentosum*. Individuals with this disorder are extremely sensitive to the UV rays of sunlight and usually develop multiple skin cancers before the age of twenty.

20.8 Recombinant DNA

Learning Goal 10:	Describe the tools used in the study of DNA and in genetic engineering.

Tools Used in the Study of DNA

Many of the techniques and tools used to study DNA and to clone genes were developed or discovered during basic studies on bacterial DNA replication and gene expression. These include many enzymes that catalyze reactions of DNA molecules, gel electrophoresis, cloning vectors, and hybridization techniques.

Restriction Enzymes

Restriction enzymes are bacterial enzymes that "cut" the sugar-phosphate backbone of DNA molecules at specific nucleotide sequences. Often these enzymes leave short single-stranded stretches at the ends of the DNA molecules. These are called "sticky ends" because they can reassociate with one another by hydrogen bonding. This property of the DNA fragments generated by restriction enzymes is very important to gene cloning.

These enzymes are used to digest large DNA molecules into smaller fragments of specific size. DNA from an individual will produce a reproducible set of DNA fragments when digested with a particular restriction enzyme. This is essential for the study or cloning of DNA from any source.

Agarose Gel Electrophoresis

One means of studying the DNA fragments produced by restriction enzyme digestion is agarose gel electrophoresis. The digested DNA sample is placed in a sample well in the gel and an electric current is applied. DNA fragments move through the gel away from the negative electrode (cathode) and toward the positive electrode (anode). Smaller DNA fragments move more rapidly than larger ones and, as a result, the DNA fragments end up distributed throughout the gel according to their size.

Hybridization

Hybridization is a technique used to identify the presence of a gene on a particular DNA fragment. This technique is based on hydrogen bonding of complementary DNA or RNA strands to one another. It doesn't matter whether the single-stranded sequences come from the same organism or from two different kinds of organisms. If the nucleotide sequences are complementary, they will hybridize with one another.

One technique, called *Southern blotting*, involves hybridization of DNA fragments from an agarose gel. This technique allows the detection of specific nucleotide sequences on a DNA fragment of known size.

DNA Cloning Vectors

A **cloning vector** is a piece of DNA having its own replication origin, so that it can be replicated inside a host cell. Often the bacterium *Escherichia coli* serves as the host cell in which the vector carrying the cloned DNA is replicated in abundance.

There are two major kinds of cloning vectors: *phage vectors* and *plasmid vectors*. Phage vectors are specially modified bacterial viruses and plasmids are extra pieces of circular DNA found in most kinds of bacteria. Both can carry a cloned DNA fragment into a bacterial cell and carry out replication that allows the amplification of the vector and the cloned DNA fragment.

Genetic Engineering

To clone a gene, vector DNA and target DNA (the source of the gene to be cloned) are digested with the same restriction enzyme. They are then mixed together under conditions that allow the sticky ends of the target and vector DNA to hybridize with one another. DNA ligase is added to covalently join the ends of the DNA molecules.

Finally, these recombinant DNA molecules are introduced into a bacterial cell. This can be done by transformation, a process in which bacterial cells are specially treated to favor the entry of DNA into the cells. Antibiotic selection and hybridization can be used to detect those clones carrying the target gene.

Genetic engineers have had to overcome many obstacles to clone eukaryotic genes of special interest medically. For instance, if the researcher desires to produce the protein product of a eukaryotic gene, he or she cannot simply clone cellular DNA because there are introns within eukaryotic genes. Molecular biologists found that a DNA copy of a eukaryotic mRNA could be made using the enzyme reverse transcriptase of a class of viruses called retroviruses. This DNA copy of the mRNA carries all the protein-coding sequences of a gene, but none of the intron sequences. Thus bacteria are able to transcribe and translate the cloned DNA and produce valuable products for use in medicine and other applications.

Great progress has been made by applying genetic engineering to genetic disorders, but most of these products represent only a treatment, not a cure. The ultimate dream for the future is to be able to replace mutated genes.

20.9 Polymerase Chain Reaction

Learning Goal 11:	Describe the process of polymerase chain reaction and discuss potential uses of the process.

Polymerase chain reaction is a sensitive technique for the amplification of specific DNA sequences. A primer, which is a short piece of single-stranded DNA that will hybridize to a region of the gene of interest, is required. The DNA, primer, four DNA nucleotides, and a heat stable DNA polymerase (Taq polymerase) are mixed and placed in a thermocycler at 94-96°C to melt the two strands of DNA. The temperature is dropped to 50-56°C to allow the primers to hybridize to the template DNA. Finally, the temperature is raised to 72°C to allow Taq polymerase to synthesize a daughter strand. These three steps are repeated many times, doubling the amount of target DNA with each cycle.

20.10 The Human Genome Project

> **Learning Goal 12:   Discuss strategies for genome analysis and DNA sequencing.**

Genetic Strategies for Genome Analysis

The goals of the Human Genome Project (HGP) were to identify and map the approximately 30,000 genes of the human genome and to determine the complete DNA sequence of each of the chromosomes. To do this, genomic libraries were generated and the DNA sequences of the clones were determined. To map the sequences along each chromosome, chromosome walking was used. The initial goal of the HGP was to complete this sequence by the year 2005. Technological advances developed as a result of the project allowed its completion by 2003.

DNA Sequencing

DNA sequencing involves reactions in which DNA polymerase copies specific DNA sequences. Nucleotide analogues that cause chain termination (dideoxynucleotides) are incorporated randomly into the growing DNA chain. This generates a family of DNA fragments that differ in size by one nucleotide. DNA sequencing gels separate these fragments and provide DNA sequence data.

Glossary of Key Terms in Chapter Twenty

aminoacyl tRNA (20.6) the transfer RNA covalently linked to its cognate amino acid.

aminoacyl tRNA binding site of ribosome (A-site) (20.6) a pocket on the surface of a ribosome that holds the aminoacyl tRNA during translation.

aminoacyl tRNA synthetase (20.6) an enzyme that recognizes one tRNA and covalently links the appropriate amino acid to it.

anticodon (20.4) a sequence of three ribonucleotides on a tRNA that are complementary to a codon on the mRNA; codon-anticodon binding results in the tRNA bringing the correct amino acid to the site of protein synthesis.

antiparallel strands (20.2) a term describing the polarities of the two strands of the DNA double helix; on one strand, the sugar-phosphate backbone advances in the $5' \rightarrow 3'$ direction while on the opposite, complementary strand, the sugar phosphate backbone advances in the $3' \rightarrow 5'$ direction.

base pair (20.2) a hydrogen-bonded pair of bases within the DNA double helix; the standard base pairs always involve a purine and a pyrimidine; in particular, adenine always base pairs with thymine and cytosine with guanine.

bioinformatics (20.10) an interdisciplinary field that uses computer information sciences and DNA technology to devise methods for understanding, analyzing, and applying DNA sequence information.

cap structure (20.4) a 7-methylguanosine unit covalently bonded to the 5' end of a mRNA by a 5'-5' triphosphate bridge.

carcinogen (20.7) any chemical or physical agent that causes mutations in the DNA that lead to uncontrolled cell growth or cancer.

central dogma (20.4) a statement of the directional transfer of the genetic information in cells: DNA $\rightarrow$ RNA $\rightarrow$ Protein.

chromosome (20.2) a piece of DNA that carries all the genetic instructions, or genes, of an organism.

cloning vector (20.8) a DNA molecule that can carry a cloned DNA fragment into a cell and that has a replication origin that allows the DNA to be replicated abundantly within the host cell.

codon (20.4) a group of three ribonucleotides on the mRNA that specifies the addition of a specific amino acid onto the growing peptide chain.

complementary strands (20.2) the opposite strands of the double helix are hydrogen bonded to one another such that adenine and thymine or guanine and cytosine are always paired.

degenerate code (20.5) a term used to describe the fact that different triplet codons may be used to specify a single amino acid.

deletion mutation (20.7) a mutation that results in the loss of one or more nucleotides from a DNA sequence.

deoxyribonucleic acid (DNA) (20.1) the nucleic acid molecule that carries all of the genetic information of an organism; the DNA molecule is a double helix composed of two strands, each of which is composed of phosphate groups, 2'-deoxyribose, and the nitrogenous bases thymine, cytosine, adenine, and guanine.

deoxyribonucleotide (20.1) a nucleotide composed of a nitrogenous base in β-*N*-glycosidic linkage to the 1' carbon of the sugar 2'-deoxyribose and with one, two, or three phosphoryl groups esterified at the hydroxyl of the 5' carbon.

DNA polymerase III (20.3) the enzyme that catalyzes the polymerization of daughter DNA strands using the parental strand as template.

double helix (20.2) the spiral staircaselike structure of the DNA molecule characterized by two sugar-phosphate backbones wound around the outside and nitrogenous bases extending into the center.

elongation factor (20.6) a protein that facilitates the elongation phase of translation.

eukaryote (20.2) an organism having cells containing a true nucleus enclosed by a nuclear membrane and having a variety of membrane-bound organelles that segregate different cellular functions into different compartments.

exon (20.4) the protein-coding sequences of a gene that are found on the final mature mRNA.

genome (20.2) the complete set of genetic information in all the chromosomes of an organism.

hybridization (20.8) a technique for identifying DNA or RNA sequences that is based on specific hydrogen bonding between a radioactive probe and complementary DNA or RNA.

initiation factors (20.6) proteins that are required for formation of the translation initiation complex, which is composed of the large and small ribosomal subunits, the mRNA, and the initiator tRNA.

insertion mutation (20.7) a mutation that results in the addition of one or more nucleotides to a DNA sequence.

intron (20.4) a noncoding sequence within a eukaryotic gene; it must be removed from the primary transcript to produce a functional mRNA.

lagging strand (20.3) the DNA strand that is replicated discontinuously from many RNA primers.

leading strand (20.3) the DNA strand that is replicated continuously from a single RNA primer.

messenger RNA (mRNA) (20.4) an RNA species produced by transcription, which specifies the amino acid sequence for a protein.

mutagen (20.7) any chemical or physical agent that causes changes in the nucleotide sequence of a gene.

mutation (20.7) any change in the nucleotide sequence of a gene.

nucleosome (20.2) the first level of chromosome structure consisting of a strand of DNA wrapped around a small disk of histone proteins.

nucleotide (20.1) a molecule composed of a nitrogenous base, a five-carbon sugar, and one, two, or three phosphoryl groups.

peptidyl tRNA binding site of ribosome (P-site) (20.6) a pocket on the surface of the ribosome that holds the tRNA bound to the growing peptide chain.

point mutation (20.7) the substitution of a single base in a codon; this may or may not alter the genetic code of the mRNA resulting in the substitution of one amino acid in the protein.

poly(A) tail (20.4) a tract of 100-200 adenosine monophosphate units covalently attached to the 3' end of a eukaryotic mRNA molecule.

polysome (20.6) complexes of many ribosomes all simultaneously translating a single mRNA.

post-transcriptional modification (20.4) alterations of the primary transcripts produced in eukaryotic cells; these include addition of the poly(A) tail to the 3' end of the transcript, addition of the cap structure to the 5' end of the mRNA, and RNA splicing.

primary transcript (20.4) the RNA product of transcription in eukaryotic cells, before post-transcriptional modifications are carried out.

prokaryote (20.2) an organism with simple cellular structure in which there is no true nucleus enclosed by a nuclear membrane and there are no true membrane-bound organelles in the cytoplasm.

promoter (20.4) the sequence of nucleotides immediately before a gene that is recognized by the RNA polymerase and signals the start point and direction of transcription.

purine (20.1) a family of nitrogenous bases that are components of DNA and RNA and consist of a six-sided ring fused to a five-sided ring. The common purines in nucleic acids are adenine and guanine.

pyrimidine (20.1) a family of nitrogenous bases that are components of nucleic acids and consist of a single six- sided ring. The common pyrimidines of DNA are cytosine and thymine; the common pyrimidines of RNA are cytosine and uracil.

pyrimidine dimer (20.7) two adjacent pyrimidine bases in a DNA strand become covalently linked to one another; as a result of this ultraviolet light induced damage, there can be no hydrogen bonding to the opposite, complementary strand.

release factor (20.6) a protein that binds to the termination codons in the empty A-site and causes the peptidyl transferase to hydrolyze the bond between the peptide and the peptidyl tRNA.

replication fork (20.3) the point at which new nucleotides are added to the growing daughter DNA strand.

replication origin (20.3) the region of a DNA molecule where DNA replication always begins.

restriction enzyme (20.8) a bacterial enzyme that recognizes specific nucleotide sequences on a DNA molecule and cuts the sugar-phosphate backbone of the DNA at or near that site.

ribonucleic acid (RNA) (20.1) single-stranded nucleic acid molecules that are composed of phosphate groups, ribose, and the nitrogenous bases uracil, cytosine, adenine, and guanine.

ribonucleotide (20.1) a nucleotide composed of a nitrogenous base in β-*N*-glycosidic linkage to the 1' carbon of the sugar ribose and with one, two, or three phosphoryl groups esterified at the hydroxyl of the 5' carbon of the ribose.

ribosomal RNA (rRNA) (20.4) the RNA species that are a structural and functional component of the small and large ribosomal subunits.

ribosome (20.6) an organelle composed of a large and a small subunit, each of which is made up of ribosomal RNA and proteins; it functions as a platform on which translation can occur and has the enzymatic activity that forms peptide bonds.

RNA polymerase (20.4) the enzyme that catalyzes the synthesis of RNA molecules using DNA as a template.

RNA splicing (20.4) removal of portions of the primary transcript that do not encode protein sequences.

semiconservative replication (20.3) DNA polymerase reads each parental strand of DNA and produces a complementary daughter strand; thus, all newly synthesized DNA molecules consist of one parental and one daughter strand.

silent mutation (20.7) a mutation that changes the sequence of the DNA but does not alter the amino acid sequence of the protein encoded by the DNA.

termination codon (20.6) a triplet of ribonucleotides with no corresponding anticodon on a tRNA. As a result, translation will end, since there is no amino acid to transfer to the peptide chain.

transcription (20.4) the synthesis of RNA from a DNA template.

transfer RNA (tRNA) (20.4) small RNAs that bind to a specific amino acid at the 3' end and mediate its addition at the appropriate site in a growing peptide chain; this is accomplished by recognition of the correct codon on the mRNA by the complementary anticodon on the tRNA.

translation (20.4) the synthesis of a protein from the genetic code carried on the mRNA; the process occurs on ribosomes and the code of the mRNA is decoded by the anticodon of the tRNA.

translocation (20.6) movement of the ribosome along the mRNA during translation; each time an amino acid is added to the chain, the ribosome moves to the next codon on the mRNA.

Self Test for Chapter Twenty

1. Which molecule in living things is the carrier of genetic information?

2. In which type of nucleic acid is thymine usually found?

3. Give the correct abbreviations for the following compounds:
    a. adenine + deoxyribose + triphosphate unit
    b. cytosine + ribose + triphosphate unit
    c. thymine + deoxyribose + monophosphate unit

4. State the central dogma of molecular biology.

5. What two men elucidated the structure of DNA in 1953?

6. In naming the ring atoms of a pentose, what designation is used to distinguish them from atoms in the base?

7. What type of bonding helps to hold the two strands of DNA together as a double helix?

8. Each base pair in the DNA molecule lies at what angle to the long axis of the helix?

9. Which base in one strand of DNA is always hydrogen-bonded to adenine in the other strand?

10. One strand of the DNA molecule has a $5' \rightarrow 3'$ orientation. What is the orientation of the opposite strand?

11. As a result of specific base pairing in the DNA molecule, what is the relationship of the sequence of bases on one strand to the sequence of bases on the other strand?

12. What is the name of the DNA damage that is caused by ultraviolet light?

13. What is the first event to occur during replication of DNA?

14. DNA replication in *E. coli* begins at a unique sequence on the circular chromosome. What is the term for this site?

15. What is the term for the site at which the new deoxyribonucleotides are added to the growing daughter strand?

16. What are the two major functions of the enzyme DNA polymerase?

17. What disease is caused by the accumulation of mutations that cause uncontrolled cell growth and division?

18. List the classes of different RNA molecules.

19. Which form of RNA carries the needed amino acids to the ribosomes for use in protein synthesis?

20. Name the enzyme that catalyzes the transcription of DNA.

21. What sequence on DNA indicates the site at which the RNA polymerase should begin transcription?

22. In eukaryotic cells, the initial mRNA (primary transcript) carries the sequences for a protein but also contains noncoding sequences. What is the term for these noncoding sequences?

23. What is the term for the coding sequences that remain after mRNA splicing?

24. The three-letter genetic code contains 64 words. What is the term for these genetic words?

25. Who first proposed that different triplet codons may serve as code words for the same amino acid?

26. Which two amino acids have only a single codon?

27. When the abnormal gene for the formation of the β-chain in hemoglobin is translated into protein, a valine is substituted for which amino acid?

28. What genetic skin disorder is caused by a mutation in the repair endonuclease gene or in other genes in the repair pathway?

29. What are the terms used to describe the two subunits of the ribosome?

30. What is the name of the complexes of many ribosomes along a single mRNA?

31. What is the term used to describe a mutation that results in no change in the organism?

32. What is the term for an enzyme that recognizes a specific DNA nucleotide sequence and "cuts" the sugar-phosphate backbone at that site?

33. List two types of cloning vectors.

34. What is the technique that relies on specific complementary base pairing to identify a particular gene on a DNA fragment?

35. List five medically important proteins that are currently produced by genetic engineering.

Vocabulary Quiz for Chapter Twenty

1. The _____ is composed of a 5-carbon sugar, a nitrogenous base, and one, two, or three phosphoryl groups.

2. The term _____ describes the double stranded spiral nature of the DNA molecule.

3. The unidirectional flow of biological information (DNA → RNA → Protein) is called _____.

4. The class of RNA that carries the genetic code for a protein is _____.

5. The process of protein synthesis is called _____.

6. The platform on which protein synthesis occurs is called the _____.

7. Any change in the sequence of DNA is called a _____.

8. The enzyme that is responsible for transcription is _____.

9. An enzyme that cuts DNA molecules at specific DNA sequences is _____.

10. A mutation that results in no change in the sequence of amino acids in the protein is called a _____.

21 Carbohydrate Metabolism

Learning Goals

1 Discuss the importance of ATP in cellular energy transfer processes.
2 Describe the three stages of catabolism of dietary proteins, carbohydrates, and lipids.
3 Discuss glycolysis in terms of its two major segments.
4 Looking at an equation representing any of the chemical reactions that occur in glycolysis, describe the kind of reaction that is occurring and the significance of that reaction to the pathway.
5 Describe the mechanism of regulation of the rate of glycolysis. Discuss particular examples of that regulation.
6 Discuss the practical and metabolic roles of fermentation reactions.
7 List several products of the pentose phosphate pathway that are required for biosynthesis.
8 Compare glycolysis and gluconeogenesis.
9 Summarize the regulation of blood glucose levels by glycogenesis and glycogenolysis.

Introduction

Cells need a constant supply of energy to maintain essential life processes such as *active transport*, *biosynthesis*, and *mechanical work*. A supply of energy-rich food molecules, carbohydrates, protein, and fats is required to provide this needed cellular energy. Carbohydrates are the most readily used energy source, and glycolysis is the pathway for the first stages of carbohydrate degradation.

21.1 ATP: The Cellular Energy Currency

Learning Goal 1:	Discuss the importance of ATP in cellular energy transfer processes.

Catabolism is the degradation of fuel molecules. The energy released in catabolism is stored as *chemical bond energy*. The molecule used for the storage of chemical energy, and often called the universal energy currency, is **adenosine triphosphate (ATP)**.

ATP is a **nucleotide** composed of the nitrogenous base adenine bonded in *N*-glycosidic linkage to the sugar ribose. Ribose is bonded to one (AMP), two (ADP), or three (ATP) phosphoryl groups. The molecule is a high-energy compound because the phosphoanhydride bonds holding the terminal phosphoryl groups are high-energy bonds.

This means that a large amount of energy is released when these bonds are hydrolyzed. Such high-energy bonds are indicated as squiggles (~).

ATP has a higher energy content than the compounds to which it donates energy, but, as must be the case, it contains less energy than the compounds that are involved in forming it. The secret to the success of ATP as the energy currency is that both the reactions that produce ATP and the hydrolysis of ATP to provide energy for cellular work are energetically favored reactions. ATP is an ideal "go-between," shuttling energy from exothermic reactions to endothermic reactions.

Example 1
What is the function of ATP?

Answer
Adenosine triphosphate (ATP) is a "go-between" molecule that can store or release chemical energy. The secret to the function of ATP as a go-between lies in its chemical structure. The molecule is a high-energy compound because of the phosphoanhydride bonds holding the terminal phosphoryl groups. When these bonds are hydrolyzed, they release a large amount of energy that can be used for cellular work. Then other pathways allow the resynthesis of ATP by providing the energy needed for the reaction between $ADP + P_i$ which produces ATP.

Example 2
Show how the hydrolysis of ATP is coupled with the synthesis of glucose-6-phosphate.

Answer
In the first step of glycolysis, there is a transfer of a phosphoryl group ($-PO_3^{2-}$) from ATP to the C-6 hydroxyl group of a glucose molecule. This reaction is catalyzed by the enzyme hexokinase. This reaction can be "dissected" to reveal the role of ATP as a source of energy. The first thing that happens is the hydrolysis of ATP to ADP and an inorganic phosphate group, abbreviated P_i. This reaction releases 7 kcal/mole of energy:

$$\textbf{ATP} + \textbf{H}_2\textbf{O} \rightarrow \textbf{ADP} + \textbf{P}_i + \textbf{7 kcal/mole}$$

Next, the synthesis of glucose-6-phosphate from glucose and phosphate requires 3.0 kcal/mole:

$$\textbf{3.0 kcal/mole} + \textbf{glucose} + \textbf{P}_i \rightarrow \textbf{glucose-6-phosphate} + \textbf{H}_2\textbf{O}$$

Thus, by the coupling of a reaction needing energy with the hydrolysis of ATP to release energy, the cell can undergo hundreds of reactions.

21.2 Overview of Catabolic Processes

Learning Goal 2: Describe the three stages of catabolism of dietary proteins, carbohydrates, and lipids.

The first stage of catabolism is the hydrolysis of dietary macromolecules into small subunits. Large polymeric molecules are degraded into their constituent subunits, which are taken into the cells. Polysaccharides are hydrolyzed to monosaccharides. Proteins are hydrolyzed to oligopeptides and amino acids, and fats are hydrolyzed into fatty acids, glycerol, and monoglycerides. All these molecules are then absorbed by the cells lining the intestine. The second stage of catabolism is the conversion of monomers into a form that can enter one of the catabolic pathways and be degraded to yield energy. The final stage of catabolism is the complete oxidation of nutrients and the production of ATP.

21.3 Glycolysis

| Learning Goal 3: | Discuss glycolysis in terms of its two major segments. |

General Considerations

Glycolysis, the first stage of carbohydrate catabolism, was probably the first successful pathway for energy generation that evolved on earth. It is an anaerobic process, requiring no oxygen, and it is carried out by enzymes in the cytoplasm of the cell. The degradation of glucose by glycolysis yields chemical energy in the form of ATP. Four ATP molecules are produced by **substrate-level phosphorylation**, a reaction in which high-energy phosphoryl groups from one of the substrates in glycolysis are transferred to ADP to form ATP. However, the net yield of ATP is only 2, because 2 ATP are used in the early stages of glycolysis. Chemical energy in the form of reduced **nicotinamide adenine dinucleotide,** NADH, is also produced. Under aerobic conditions, the electrons are donated to an electron transport system for the generation of ATP by **oxidative phosphorylation**; but under anaerobic conditions, NADH is used as a source of electrons in fermentation reactions. The final product of glycolysis is two pyruvate molecules. Under anaerobic conditions, pyruvate is used as a substrate in fermentation reactions.

Example 3
Describe the first stage of carbohydrate metabolism.

Answer
The pathway called glycolysis is the beginning stage for the use of glucose as a source of chemical energy. This pathway is anaerobic, which means that no oxygen is required. The glycolysis pathway releases only about 2% of the potential energy of the glucose molecule. In the pathway, glucose is finally transformed after many steps into two molecules of pyruvate.

Reactions of Glycolysis

| Learning Goal 4: | Looking at an equation representing any of the chemical reactions that occur in glycolysis, describe the kind of reaction that is occurring and the significance of that reaction to the pathway. |

The structures of the intermediates of glycolysis can be found on pages 721-725 of the text (Section 21.3). The following is a brief description of those reactions.

Reaction 1
Glucose is phosphorylated by the enzyme *hexokinase*. A phosphoryl group from ATP is transferred to C-6 of glucose, and the product is glucose-6-phosphate.

Reaction 2
The enzyme *phosphoglucose isomerase* isomerizes the glucose-6-phosphate to produce fructose-6-phosphate.

Reaction 3
The enzyme *phosphofructokinase* transfers a phosphoryl group from ATP to the C-1 hydroxyl of fructose-6-phosphate, producing fructose-1, 6-bisphosphate.

Reaction 4
Aldolase splits the fructose-1, 6-bisphosphate into glyceraldehyde-3-phosphate (G3P) and dihydroxyacetone phosphate (DHAP).

Reaction 5
Only G3P can be used in glycolysis, and the DHAP is isomerized by the enzyme *triose phosphate isomerase*. This produces a second molecule of G3P.

Reaction 6
Glyceraldehyde-3-phosphate dehydrogenase catalyzes the oxidation of the aldehyde group of G3P to a carboxyl group. The coenzyme in this reaction is NAD^+, which is reduced to NADH. Next, an inorganic phosphate group is transferred to the carboxylate group to produce 1,3-bisphosphoglycerate.

Reaction 7
Phosphoglycerate kinase catalyzes the transfer of a phosphoryl group from 1,3-bisphosphoglycerate to ADP. This is the first substrate-level phosphorylation of glycolysis to produce ATP.

Reaction 8
Phosphoglycerate mutase catalyzes the isomerization of 3-phosphoglycerate to 2-phosphoglycerate.

Reaction 9
Enolase catalyzes the dehydration of 2-phosphoglycerate, producing the energy-rich product phosphoenolpyruvate.

Reaction 10
Pyruvate kinase catalyzes the last substrate-level phosphorylation, in which a phosphoryl group from phosphoenolpyruvate is transferred to ADP to produce ATP. The final product of glycolysis is pyruvate.

Reactions 5 through 10 occur twice per glucose molecule; thus, the final products of glycolysis are 2 NADH, 2 pyruvate, and 4 ATP. However, since 2 ATP/glucose were invested early in glycolysis, the net yield is 2 ATP/glucose molecule.

Example 4

List the steps in glycolysis in which ATP molecules are hydrolyzed to provide energy for specific reactions.

Answer

1. Reaction 1 is a coupled reaction in which the enzyme *hexokinase* catalyzes the transfer of a phosphoryl group from ATP to glucose. The product is glucose-6-phosphate.
2. Reaction 3 is a coupled reaction in which the enzyme *phosphofructokinase* catalyzes the transfer of a phosphoryl group from ATP to fructose-6-phosphate. The product is fructose-1,6-bisphosphate.

Example 5

List the two steps of the anaerobic glycolysis pathway that result in the production of ATP molecules.

Answer

1. Reaction 7: This reaction is the first step of the pathway in which energy is harvested in the form of ATP. One of the phosphoryl groups of 1,3-bisphosphoglycerate is transferred to an ADP molecule in the first substrate-level phosphorylation of glycolysis. This reaction is catalyzed by the enzyme *phosphoglycerate kinase*.
2. Reaction 10: The final substrate-level phosphorylation in glycolysis involves the transfer of a phosphoryl group from phosphoenolpyruvate to ADP to form ATP. This reaction is catalyzed by the enzyme *pyruvate kinase*.

Regulation of Glycolysis

Learning Goal 5: **Describe the mechanism of regulation of the rate of glycolysis. Discuss particular examples of that regulation.**

Energy-harvesting pathways are responsive to the energy needs of the cell. Reactions of the pathway speed up when there is a demand for ATP. They slow down when there is abundant ATP to meet the energy requirements of the cell.

One of the major mechanisms for the control of the rate of the glycolytic pathway is the use of allosteric enzymes. The allosteric effectors that indicate the energy needs of the cell include molecules such as ATP. If the ATP concentration is high, the cell must have a sufficient energy reserve. ADP and AMP indicate that the cell is in need of ATP.

Hexokinase is allosterically inhibited by the product of the reaction it catalyzes, glucose-6-phosphate. ATP is an allosteric inhibitor of phosphofructokinase, while AMP, ADP, and citrate are allosteric activators.

Pyruvate kinase is activated by fructose-1,6-bisphosphate.

21.4 Fermentation

<table>
<tr><td>Learning Goal 6:</td><td>Discuss the practical and metabolic roles of fermentation reactions.</td></tr>
</table>

In order for glycolysis to continue to degrade glucose and produce ATP, the NADH must be reoxidized, and the pyruvate must be utilized and removed. Under anaerobic conditions, these two requirements are met through fermentation reactions.

Example 6

Explain how the NADH produced by glycolysis in reaction 6 is regenerated into NAD⁺.

Answer
1. If the cell is functioning under aerobic conditions, the NADH will be reoxidized, and pyruvate will be completely oxidized by aerobic respiration.
2. Under anaerobic conditions, cells of different types employ a variety of fermentation reactions to accomplish the conversion of NADH to NAD⁺.

Lactate Fermentation

Some cells, such as muscle cells and dairy bacteria, utilize lactate fermentation. Under anaerobic conditions, the enzyme *lactate dehydrogenase* reduces pyruvate to lactate. NADH is the reducing agent for this reaction, and thus NAD⁺ is regenerated, while the pyruvate is used up.

This reaction occurs in muscle cells during strenuous exercise, when the body cannot provide sufficient oxygen for aerobic respiration to provide enough ATP. Lactate eventually builds up in the muscle to the extent that glycolysis can no longer proceed, and the muscle cells can no longer function. This point of exhaustion is called the **anaerobic threshold**.

When exercise ceases, the liver takes up the lactate from the blood and converts it back to pyruvate, and the ATP stores are replenished by aerobic energy-harvesting reactions, such as the citric acid cycle, which will be discussed in Chapter 22.

Lactate fermentation is used by bacteria that produce yogurt and some cheeses. Lactate is one of the molecules that gives these dairy products their tangy flavor.

Alcohol Fermentation

Under anaerobic conditions, yeast cells ferment sugars produced by fruit and grains to produce ethanol. First, *pyruvate decarboxylase* removes CO_2 from (decarboxylates) the pyruvate, producing CO_2 and acetaldehyde. Then, *alcohol dehydrogenase* catalyzes the reduction of acetaldehyde to ethanol, reoxidizing NADH in the process. Eventually, the stable fermentation end product, ethanol, builds up to a concentration that kills the yeast cells. It is characteristic of fermentations that the end product eventually builds up to levels that inhibit the cells carrying out the reaction.

Example 7

Summarize the lactate and ethanol fermentation pathways.

Answer

1. *Lactate fermentation*: This pathway occurs when you have exercised beyond the capacity of your lungs and circulatory system to deliver enough oxygen to the working muscles. The aerobic energy-generating pathways will no longer be able to supply enough ATP, but the muscles still demand energy. Under anaerobic conditions, pyruvate can be reduced by lactate dehydrogenase to form lactate. Simultaneously, NADH is oxidized to NAD^+, which is required for the continued anaerobic functioning of the glycolysis pathway.

2. *Alcohol fermentation:* Under anaerobic conditions in yeast (usually not humans!), the pyruvate formed by glycolysis is converted into ethanol and CO_2 in two enzyme-catalyzed steps. First, pyruvate is converted into acetaldehyde as CO_2 is released. Then the acetaldehyde is reduced by NADH. This produces ethanol and NAD^+ that allows anaerobic glycolysis to continue.

21.5 The Pentose Phosphate Pathway

Learning Goal 7: **List several products of the pentose phosphate pathway that are required for biosynthesis.**

The **pentose phosphate pathway** is an alternative pathway for glucose degradation that is particularly abundant in the liver and adipose tissue. It can be divided into three stages, each of which provides a unique biosynthetic requirement for the cell.

In the reactions of the *oxidative stage*, glucose-6-phosphate is oxidized and decarboxylated to produce ribulose-5-phosphate, NADPH, and CO_2. NADPH is the reducing agent required for many biosynthetic pathways. The second stage involves isomerization reactions that convert ribulose-5-phosphate into other pentose phosphates. Among these is ribose-5-phosphate, which is required for nucleotide biosynthesis. In the final stage, a complex series of reactions occurs that involve carbon-carbon bond breakage and formation. The products are two molecules of fructose-6-phosphate and one molecule of glyceraldehyde-3-phosphate. Among the intermediates of the third stage is erythrose-4-phosphate, a precursor in the biosynthesis of the aromatic amino acids.

21.6 Gluconeogenesis: The Synthesis of Glucose

Learning Goal 8: **Compare glycolysis and gluconeogenesis.**

During starvation and following extended exercise, the body must make glucose. **Gluconeogenesis** is the process by which glucose is produced from noncarbohydrate precursors. Lactate, all amino acids, except leucine and lysine, and glycerol from fats all serve as precursors for glucose biosynthesis. Gluconeogenesis, shown in Figure 21.10 in the text, appears to be the reverse of glycolysis. Certainly most of the intermediates of the two pathways are identical. However, steps 1, 3, and 10 of glycolysis are irreversible and must be bypassed by other enzymes. Glucose-6-phosphate is dephosphorylated by the enzyme glucose-6-phosphatase, found in the liver but not in muscle. Similarly, the phosphorylation of fructose-6-phosphate by phosphofructokinase is irreversible. Fructose bisphosphatase catalyzes the removal of the phosphoryl group from fructose-1,6-bisphosphate. Finally, step 10 of glycolysis is bypassed by two enzymes. Pyruvate carboxylase carboxylates pyruvate by the addition of atmospheric CO_2, producing oxaloacetate. Then phosphoenolpyruvate carboxykinase removes the CO_2 and adds a phosphoryl group, producing phosphoenolpyruvate. The phosphoryl group donor is **guanosine triphosphate, GTP**.

In the Cori Cycle, lactate produced by working muscle is transported to the liver by the bloodstream. In the liver, the lactate is converted to pyruvate, which may be used to produce glucose by gluconeogenesis. The glucose produced in the liver may be degraded for energy or stored as glycogen.

21.7 Glycogen Synthesis and Degradation

Learning Goal 9:	**Summarize the regulation of blood glucose levels by glycogenesis and glycogenolysis.**

The liver helps regulate the blood glucose level. One of the ways this is accomplished is through the uptake and storage of excess glucose as glycogen (**glycogenesis**). Alternatively, liver cells may degrade glycogen (**glycogenolysis**) and release glucose into the blood. **Glycogen** is a highly branched glucose polymer. The primary chain is linked by $\alpha(1{\rightarrow}4)$ glycosidic bonds. The branches are linked to the primary chain by $\alpha(1{\rightarrow}6)$ glycosidic bonds. Glycogen is stored in the cytoplasm of liver and muscle cells as **glycogen granules**. These granules are complexes of glycogen and the enzymes that carry out glycogenesis and glycogenolysis.

Glycogenolysis: Glycogen Degradation

Glycogenolysis is controlled by two hormones, **glucagon** and **epinephrine**. Glucagon is released from the pancreas in response to low blood glucose, and epinephrine is released from the adrenal glands in response to a threat or a stress. Both hormones regulate the activity of glycogen phosphorylase and glycogen synthase. Glycogen phosphorylase, involved in glycogenolysis, is activated; glycogen synthetase, involved in glycogenesis, is inactivated. The steps in glycogenolysis are summarized below:

Step 1: The enzyme *glycogen phosphorylase* catalyzes phosphorolysis of a glucose unit at one end of glycogen, producing glucose-1-phosphate.

Step 2: The enzyme *α(1→6) glycosidase*, sometimes called the debranching enzyme, hydrolyzes the *α*(1→6) glycosidic bond at a branch point, thereby removing the branches. Branches are further degraded by glycogen phosphorylase to produce glucose-1-phosphate.

Step 3: *Phosphoglucomutase* converts glucose-1-phosphate to glucose-6-phosphate, which can be degraded by glycolysis or dephosphorylated for transport into the bloodstream.

Example 8
Explain why glycogenolysis is important.

Answer
Glucose is the sole source of energy for mammalian red blood cells and the major source of energy for the brain. Neither of these can store glucose; thus, a constant supply must be available as blood glucose. This is provided by dietary glucose and by the production of glucose either by gluconeogenesis or by glycogenolysis, the degradation of glycogen molecules. Glycogen is a long, branched-chain polymer of glucose that is stored in the liver and skeletal muscles. Breakdown of glycogen in the liver mobilizes the glucose when hormonal signals register a need for increased levels of blood glucose.

Glycogenesis: Glycogen Synthesis

The hormone **insulin**, produced by the pancreas in response to high blood glucose levels, stimulates glycogenesis. It accelerates the uptake of glucose by most cells of the body. In the liver, insulin promotes glycogenesis by inhibiting glycogen phosphorylase, thus inhibiting glycogenolysis and stimulating glycogen synthase and glucokinase, two enzymes involved in glycogen synthesis. The following is a summary of the steps of glycogenesis:

Step 1: *Glucokinase* phosphorylates glucose, using ATP as a phosphoryl group donor and forming glucose-6-phosphate.

Step 2: *Phosphoglucomutase* isomerizes glucose-6-phosphate to glucose-1-phosphate.

Step 3: *Pyrophosphorylase* catalyzes bond formation between the C-1 phosphoryl group of glucose and the α-phosphoryl group of uridine triphosphate (UTP) to produce UDP-glucose. A pyrophosphate group (PP_i) is released in the reaction.

Step 4: *Glycogen synthase* breaks the phosphoester linkage of UDP-glucose and forms an $\alpha(1 \rightarrow 4)$ glycosidic bond between the glucose and the growing glycogen chain, releasing UDP.

Step 5: Finally, the branches are added. The *branching enzyme* removes sections of the linear $\alpha(1 \rightarrow 4)$-linked glycogen and reattaches them elsewhere in the chain by $\alpha(1 \rightarrow 6)$ glycosidic linkages.

Compatibility of Glycogenesis and Glycogenolysis

Glycogenesis and glycogenolysis are regulated by hormonal controls. When blood glucose levels are too high (**hyperglycemia**), insulin stimulates glucose uptake. It also stimulates glucokinase and glycogen synthase activity and inhibits the first enzyme in glycogen degradation, glycogen phosphorylase. The net effect is the removal of glucose from the blood and its conversion into glycogen in the liver.

Glucagon is produced in response to low blood glucose levels (**hypoglycemia**). Glycogenolysis is stimulated because the activity of glycogen phosphorylase, which catalyzes the first stage of glycogenolysis, is accelerated.

Glossary of Key Terms in Chapter Twenty-One

adenosine triphosphate (ATP) (21.1) a nucleotide composed of the purine adenine, the sugar ribose, and a triphosphate group; it serves as the major energy storage form of the cell.

anaerobic threshold (21.4) the point at which the level of lactate in the exercising muscle inhibits glycolysis; when deprived of energy the muscle ceases to function.

catabolism (21.1) the degradation of fuel molecules and harvest of energy for cellular functions.

Cori Cycle (21.6) a metabolic pathway in which the lactate produced by working muscle is converted back to glucose by gluconeogenesis in the liver.

fermentation (21.4) anaerobic catabolic reactions that occur with no net oxidation; pyruvate or an organic compound produced from pyruvate is reduced as NADH is oxidized.

glucagon (21.7) a peptide hormone released by the pancreas in response to low blood glucose; it promotes glycogenolysis and thereby increases the concentration of blood glucose.

glycogen (21.7) a long, branched polymer of glucose stored in the liver and in muscles.

glycogen granule (21.7) a core of glycogen surrounded by enzymes responsible for glycogen synthesis and degradation.

glycogenesis (21.7) the metabolic pathway that results in the addition of glucose to growing glycogen polymers when blood glucose levels are high.

glycogenolysis (21.7) the biochemical pathway that results in the removal of glucose molecules from glycogen polymers when blood glucose levels are low.

glycolysis (21.3) the enzymatic pathway that converts a glucose molecule into two molecules of pyruvate; this anaerobic process harvests energy in the form of 2 ATP and 2 NADH.

hyperglycemia (21.7) blood glucose levels that are higher than normal.

hypoglycemia (21.7) blood glucose levels that are lower than normal.

insulin (21.7) a hormone released from the pancreas in response to high blood glucose levels; it promotes glycogenesis and fat storage.

nicotinamide adenine dinucleotide, NAD$^+$ (21.3) a molecule synthesized from the vitamin niacin and the nucleotide ATP, which serves as a carrier of hydride anions.

nucleotide (21.1) a molecule composed of a nitrogenous base, a five carbon sugar, and a triphosphate group.

oxidative phosphorylation (21.3) production of ATP using the energy of electrons harvested during biological oxidation-reduction reactions.

pentose phosphate pathway (21.5) an alternative pathway for glucose degradation that provides the cell with reducing power in the form of NADPH.

substrate-level phosphorylation (21.3) the production of ATP by the transfer of a phosphoryl group from the substrate of a reaction to ADP.

uridine triphosphate (21.6) a nucleotide composed of the pyrimidine uracil, the sugar ribose, and a triphosphate group, which serves as a carrier of glucose-1-phosphate in glycogenesis.

Self Test for Chapter Twenty-One

1. Which pathway is thought to be the first successful energy-generating pathway available to early organisms?

2. What molecule in a cell couples exothermic and endothermic reactions?

3. How many kcal/mole of energy are released by the hydrolysis of ATP?

4. Describe glycolysis in terms of its oxygen requirements.

5. What is the net yield of ATP produced by anaerobic glycolysis?

6. Which coenzyme must be reoxidized so that glycolysis can continue?

7. What is the final fermentation end product in yeast cells?

8. What is the final fermentation end product produced in muscle cells under anaerobic conditions?

9. Of the three classes of food molecules, which one is the most readily used by the human body?

10. Which pathway of carbohydrate degradation is thought to be the most ancient, having existed for at least 3.5 billion years?

11. What subunits are released by digestion of proteins?

12. Which aspect of metabolism involves the degradation of absorbed food molecules to produce energy for the body?

13. Of the several high-energy compounds produced in cells, which one is the principal energy storage compound?

14. What type of chemical reaction is involved in the process of digestion?

15. What happens in the first step of glycolysis that causes glucose to be trapped within the cell?

16. Phosphofructokinase converts what substrate into fructose-1,6-bisphosphate?

17. Where in the cell does glycolysis occur?

18. During glycolysis, 1,3-bisphosphoglycerate is converted into 3-phosphoglycerate. Simultaneously, an ATP is produced. What is the general term for this kind of reaction?

19. What compound is hydrolyzed to produce adenosine-5'-diphosphate, P_i, and energy?

20. Provide the missing product in the following reaction:

    ATP + water + glucose $\rightarrow$ ADP + 4 kcal/mole + ?

21. What is the name of the enzyme found in saliva that begins the hydrolysis of starch?

22. What enzyme hydrolyzes milk sugar?

23. What enzyme in the stomach starts the breakdown of proteins?

24. Where does the digestion of triglycerides begin?

25. What is the first portion of the small intestine called?

26. Hexokinase is used to convert what substrate into glucose-6-phosphate?

    a. glucose

    b. fructose

    c. glycogen

    d. pyruvate

    e. lactate

27. The enzyme triose phosphate isomerase converts dihydroxyacetone phosphate into what compound?

    a. pyruvate

    b. fructose-1,6-bisphosphate

    c. glucose-1,6-bisphosphate

    d. lactate

    e. glyceraldehyde-3-phosphate

28. Following strenuous exercise, the liver takes up excess lactate from the blood. Into what compound is this lactate converted?

    a. oxaloacetate

    b. glucose

    c. pyruvate

    d. acetyl CoA

    e. glycogen

29. True or false: Gluconeogenesis means the production of glycogen from a noncarbohydrate source.

30. True or false: The liver is the only tissue that contains glucose-6-phosphatase.

Vocabulary Quiz for Chapter Twenty-One

1. _____ is a nucleotide composed of the purine adenine, the sugar ribose, and a triphosphate group. It serves as the major energy storage form of the cell.

2. The enzymatic pathway that converts a glucose molecule into two molecules of pyruvate is called _____. This anaerobic process generates energy in the form of 2 ATP and 2 NADH.

3. _____ is the synthesis of glucose from noncarbohydrate precursors.

4. _____ is the production of ATP by the transfer of a phosphoryl group from the substrate of a reaction to ADP.

5. The _____ is an alternative pathway for glucose degradation that provides the cell with reducing power in the form of NADPH.

6. The condition in which blood glucose levels are higher than normal is called
_____.

7. The degradation of fuel molecules and production of energy for cellular functions is called _____.

8. _____ is a long, branched polymer of glucose stored in the liver and in muscles.

9. _____ is a molecule synthesized from the vitamin niacin and the nucleotide ATP, which serves as a carrier of hydride anions.

10. _____ is the production of ATP using the energy of electrons harvested during biological oxidation-reduction reactions.

22 *Aerobic Respiration and Energy Production*

Learning Goals

1 Name the regions of the mitochondria and the function of each region.
2 Describe the reaction that results in the conversion of pyruvate to acetyl CoA, describing the location of the reaction and the components of the pyruvate dehydrogenase complex.
3 Summarize the reactions of aerobic respiration.
4 Looking at an equation representing any of the chemical reactions that occur in the citric acid cycle, describe the kind of reaction that is occurring and the significance of that reaction to the pathway.
5 Explain the mechanisms for the control of the citric acid cycle.
6 Describe the process of oxidative phosphorylation.
7 Describe the conversion of amino acids to molecules that can enter the citric acid cycle.
8 Explain the importance of the urea cycle and describe its essential steps.
9 Discuss the cause and effect of hyperammonemia.
10 Summarize the role of the citric acid cycle in catabolism and anabolism.

Introduction

The energy-harvesting reactions that produce the greatest energy yield are aerobic reactions that occur in the **mitochondria**. Mitochondria are the organelles that serve as the cellular "powerplants." An organelle is a membrane-enclosed compartment within the cytoplasm that has a specialized function. In the mitochondria, the enzymes of the citric acid cycle strip electrons from substrates in the pathway. These are passed through the **electron transport system**, and the energy of the electrons is used to produce ATP. This process is **oxidative phosphorylation**.

Example 1
What is the function of the mitochondria?

Answer
The oxidative reactions of metabolism are responsible for most cellular ATP production. These reactions occur in metabolic pathways located in the mitochondria. This organelle in the cytoplasm is the power plant of the cell, producing most of the ATP for cellular processes. The mitochondria are responsible for the final oxidation of the acetyl group of acetyl CoA from glycolysis, fatty acid degradation, and amino acid catabolism and ATP production.

22.1 Mitochondria

Learning Goal 1:	Name the regions of the mitochondria and the function of each region.

Structure and Function

Mitochondria are bounded by an **outer mitochondrial membrane** and an **inner mitochondrial membrane**. The region between the two membranes is known as the **intermembrane space,** and the region enclosed by the inner membrane is known as the **matrix space**.

The outer membrane is freely permeable to substances of molar mass less than 10,000 g/mol. Thus, metabolites to be oxidized via the citric acid cycle easily enter the intermembrane space through channel proteins.

The inner membrane is a highly folded, continuous structure. The individual folds are known as **cristae**. The inner mitochondrial membrane is virtually impermeable to most substances and contains three types of proteins. *Transport proteins* allow the transport of metabolites across the inner mitochondrial membrane into the matrix. *Electron transport system proteins* are involved in electron transfers, for which O_2 serves as the terminal electron acceptor. The third protein is a very large multiprotein complex known as **ATP synthase**, which is responsible for phosphorylation of ADP.

The enzymes of the citric acid cycle and those for the oxidation of fatty acids and amino acids are located in the matrix space.

Example 2

Describe the structure of the mitochondria.

Answer

Mitochondria are football-shaped organelles that are roughly the size of bacteria. This organelle has both an outer and an inner membrane. The region between the two membranes is called the intermembrane space, and the region enclosed by the inner membrane is known as the matrix space. The outer membrane is freely permeable to substances of less than 10,000 g/mol. This is because of the presence of a large number of transport proteins that form pores in the membrane.

The inner membrane is a continuous structure that is highly folded. These folds of the inner membrane are called cristae. This membrane is virtually impermeable to most substances. The inner membrane contains transport proteins to allow the transport of metabolites across the membrane. Protein complexes for electron transport, ATP synthesis, the citric acid cycle, and fatty acid oxidation are also located in the inner membrane or matrix space.

Origin of the Mitochondria

Mitochondria are roughly the size of a bacterium, and are thought to have evolved from free-living bacteria that were captured by eukaryotic cells. Some of the evidence for this is that they have their own DNA and protein-synthesizing system. In addition, they are self-replicating, growing and dividing to produce new mitochondria.

22.2 Conversion of Pyruvate to Acetyl CoA

Learning Goal 2:	Describe the reaction that results in the conversion of pyruvate to acetyl CoA, describing the location of the reaction and the components of the pyruvate dehydrogenase complex.

Under aerobic conditions, cells use oxygen and completely oxidize glucose to CO_2 in a metabolic pathway called the **citric acid cycle**. **Acetyl CoA** is the molecule that carries two-carbon fragments (acetyl groups) produced from pyruvate into the citric acid cycle.

The coenzyme A portion of acetyl CoA is derived from ATP and the vitamin pantothenic acid. The acetyl groups are linked to the thiol group of coenzyme A by a high-energy thioester bond. Acetyl CoA is an "activated" form of the acetyl group.

The reaction that converts pyruvate to acetyl CoA is carried out by the *pyruvate dehydrogenase complex*. Pyruvate is decarboxylated and oxidized and acetyl CoA is formed.

Acetyl CoA plays a central role in cellular metabolism. It is the product of the degradation of glucose, fatty acids, and some amino acids. It carries acetyl groups to the citric acid cycle for complete oxidation and the ultimate production of large amounts of ATP. It is also used for *anabolic* or biosynthetic reactions to produce cholesterol and fatty acids. Thus, acetyl CoA is the intermediate through which all energy sources (fats, proteins, and carbohydrates) are interconvertible.

22.3 An Overview of Aerobic Respiration

Learning Goal 3:	Summarize the reactions of aerobic respiration.

Aerobic respiration is the oxygen-requiring degradation of food molecules and production of ATP. Acetyl groups derived from the breakdown of sugars, amino acids, or lipids are completely oxidized to CO_2 in the reactions of the citric acid cycle. The electrons harvested in these oxidation reactions are used to reduce 3 NAD^+ and 1 FAD, producing 3 NADH and 1 $FADH_2$.

These electrons are then passed through an electron transport system that simultaneously pumps protons (H^+) into the H^+ reservoir in the mitochondrial intermembrane space. The energy of the H^+ reservoir is used by ATP synthase to produce ATP. The entire process is called **oxidative phosphorylation** because the energy of electrons from the oxidation of substrates is used to phosphorylate ADP and produce ATP.

22.4 The Citric Acid Cycle (The Krebs Cycle)

Reactions of the Citric Acid Cycle

Learning Goal 4:	Looking at an equation representing any of the chemical reactions that occur in the citric acid cycle, describe the kind of reaction that is occurring and the significance of that reaction to the pathway.

The following is a summary of the reactions of the **citric acid cycle**.

1. The acetyl group of acetyl CoA is transferred to oxaloacetate by the enzyme *citrate synthase*, forming citrate.

2. *Aconitase* catalyzes the dehydration of citrate, producing *cis*-aconitate. This same enzyme then adds a water molecule to the *cis*-aconitate, converting it to isocitrate. The result of these two steps is the isomerization of citrate to isocitrate.

3. *Isocitrate dehydrogenase* catalyzes the first oxidative reaction of the citric acid cycle. This is a complex reaction in which three things happen. First, the hydroxyl group of isocitrate is oxidized to a ketone; then carbon dioxide is released; finally NAD^+ is reduced to NADH. The product of this oxidative decarboxylation reaction is α-ketoglutarate.

4. The *α-ketoglutarate dehydrogenase* complex, an enzyme complex very similar to the pyruvate dehydrogenase complex, mediates the next reaction. Once again, three chemical events occur. First, α-ketoglutarate loses a carboxylate group as CO_2; next, NAD^+ is reduced to NADH; and finally, coenzyme A combines with the product to form succinyl CoA. The bond between succinate and coenzyme A is a high-energy thioester linkage.

5. Succinyl CoA is converted to succinate by the enzyme *succinyl CoA synthase*, which removes the CoA group and uses the energy of the thioester bond to add an inorganic phosphate group to GDP, producing GTP. *Dinucleotide diphosphokinase* then shifts the phosphoryl group from GTP to ADP, producing ATP.

6. *Succinate dehydrogenase* then catalyzes the oxidation of succinate to fumarate. The oxidizing agent, flavin adenine dinucleotide (FAD), is reduced in this step.

7. *Fumarase* catalyzes the addition of H_2O to the double bond of fumarate, producing malate.

8. Finally, *malate dehydrogenase* reduces NAD^+ to NADH and oxidizes malate to oxaloacetate. Since the citric acid cycle began with the addition of an acetyl group to oxaloacetate, we have come full circle.

Summary of the Energy Yield

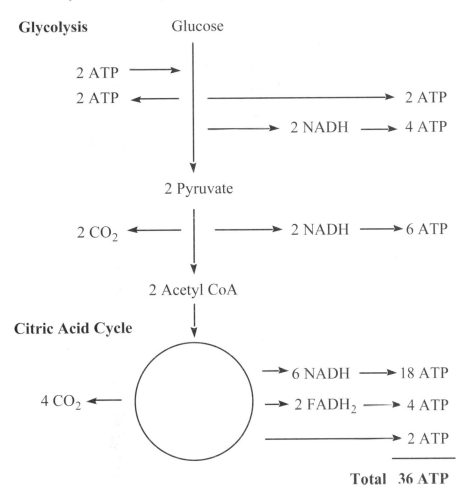

Glycolysis Glucose

2 ATP ⟶

2 ATP ⟵ ⟶ 2 ATP

⟶ 2 NADH ⟶ 4 ATP

2 Pyruvate

2 CO$_2$ ⟵ ⟶ 2 NADH ⟶ 6 ATP

2 Acetyl CoA

Citric Acid Cycle

⟶ 6 NADH ⟶ 18 ATP

4 CO$_2$ ⟵ ⟶ 2 FADH$_2$ ⟶ 4 ATP

⟶ 2 ATP
_____

Total 36 ATP

Example 3

Write the specific reactions in the citric acid cycle in which the coenzymes NAD$^+$ and FAD act as oxidizing agents.

Answer

1. The first oxidative step, Step 3, is catalyzed by isocitrate dehydrogenase:
   isocitrate + NAD$^+$ ⟶ α-ketoglutarate + CO$_2$ + NADH

2. The next step involves the oxidation of α-ketoglutarate into succinate:
   α-ketoglutarate + NAD$^+$ ⟶ succinate + CO$_2$ + NADH

3. In Step 6, succinate is oxidized to fumarate:
   succinate + FAD ⟶ fumarate + FADH$_2$

4. Then in the final step of the citric acid cycle, malate is oxidized to produce oxaloacetate:
   malate + NAD$^+$ ⟶ oxaloacetate + NADH

22.5 Control of the Citric Acid Cycle

Learning Goal 5:   Explain the mechanisms for the control of the citric acid cycle.

The citric acid cycle is regulated so that cellular energy demand determines the rate of cellular energy production. This regulation occurs because some of the enzymes of the citric acid cycle are allosteric enzymes. Binding of an effector, such as ATP, ADP, or NADH, alters the shape of the active site. Effector binding may turn the enzyme on (positive allosterism) or it may inhibit the enzyme (negative allosterism). The pyruvate dehydrogenase complex is inhibited by high concentrations of ATP, acetyl CoA, and NADH. Citrate synthase is an allosteric enzyme inhibited by ATP. Isocitrate dehydrogenase is an allosteric enzyme that is stimulated by ADP binding and is also inhibited by high levels of NADH and ATP. The α-ketoglutarate dehydrogenase complex is inhibited by high levels of NADH, succinyl CoA, and ATP.

22.6 Oxidative Phosphorylation

Learning Goal 6:   Describe the process of oxidative phosphorylation.

Oxidative phosphorylation is a series of reactions that couples the oxidation of NADH and $FADH_2$ to the phosphorylation of ADP to produce ATP.

Electron Transport Systems and the Hydrogen Ion Gradient

Embedded within the mitochondrial inner membrane is a series of electron carriers called the **electron transport system**. Prominent among these electron carriers are the cytochromes, which carry a heme group. These molecules are arranged within the membrane so that they pass electrons from one to the next. Some of these electron carriers are able to carry hydrogen atoms, while others can carry only electrons. Those that do not pass protons deposit them within the intermembrane space.

NADH carries electrons to the first carrier of the electron transport system, *NADH dehydrogenase*. There it is oxidized to NAD^+, donating a pair of hydrogen atoms, and returns to the site of the citric acid cycle to be reduced again. The pair of electrons is passed to the next electron carrier, but the protons are pumped into the intermembrane compartment. The electrons are passed sequentially through the electron transport system and, at two additional points, protons from the matrix are pumped into the intermembrane compartment. In aerobic organisms, the **terminal electron acceptor** is molecular oxygen, O_2, and the product is water. $FADH_2$ donates its electrons to a carrier later in the electron transport system, so only four protons are pumped into the intermembrane space.

As the electron transport system continues to function, a high concentration of hydrogen ions accumulates in the intermembrane space, creating a concentration gradient across the inner membrane. Thus, the system generates a high concentration of protons in the intermembrane space; a lower concentration of protons is found in the matrix.

ATP Synthase and the Production of ATP

The last component needed for oxidative phosphorylation is a multiprotein complex called **ATP synthase**, or F_0F_1 **complex**. The F_0 portion provides a channel in the membrane through which protons may pass. The F_1 portion phosphorylates ADP to produce ATP, using the energy of the proton gradient.

ATP synthase harvests the energy of this gradient and uses it to produce ATP. As the protons pass into the matrix through F_0, some of their energy is used by the enzymatic portion of ATP synthase (F_1) to catalyze the phosphorylation of ADP to ATP.

Example 4

What are the two different ways that ATP can be synthesized by the cell? Where are these pathways located, and how efficient are they?

Answer

1. The two ways that ATP can be synthesized by the cell are substrate-level phosphorylation and oxidative phosphorylation.
2. Substrate-level phosphorylation occurs in the cytoplasm of the cell and produces only a few ATP molecules per glucose molecule.
3. Oxidative phosphorylation occurs in the mitochondria and produces a large number of ATP molecules per glucose molecule.

Example 5

Explain how the electron transport system converts NADH back into NAD^+, so that it may be used again by the citric acid cycle.

Answer

The NADH carries a hydride anion, which was originally from glucose or other food molecules, to the first carrier of the electron transport system. There it is oxidized to NAD^+ as the first carrier, NADH dehydrogenase is reduced. NAD^+ returns to be used again in the citric acid cycle.

Example 6

Write an equation to represent the last step of the electron transport system.

Answer

$$2\,H^+ + 2\,e^- + {}^1\!/_2\,O_2 \rightarrow H_2O$$

22.7 The Degradation of Amino Acids

Learning Goal 7: Describe the conversion of amino acids to molecules that can enter the citric acid cycle.

Amino acids obtained from the degradation of dietary protein may also be oxidized as energy sources. Amino acid degradation occurs mainly in the liver and takes place in two stages: (1) the removal of the α-amino group and (2) the degradation of the remaining carbon skeleton. In land mammals, the amino group is excreted in the urine as urea. The carbon skeletons of amino acids can be converted into a variety of compounds.

Example 7
When is body protein used for energy?

Answer
Carbohydrates are not the only source of energy. Dietary protein is digested to amino acids, which also may be used as an energy source. Most of the amino acids used for energy come from dietary protein. Only under starvation conditions, when stored glycogen and lipids have been greatly depleted, does the body begin to use its own muscle protein for energy.

Removal of α-Amino Groups: Transamination

A **transaminase** catalyzes the transfer of the α-amino group from an α-amino acid to an α-keto acid. This general reaction is summarized below.

$$
\overset{+}{\underset{\underset{R}{|}}{\underset{|}{H-C-COO^-}}} \overset{NH_3}{} \quad + \quad \overset{O}{\underset{\underset{R}{|}}{\underset{||}{C-COO^-}}} \quad \underset{\longrightarrow}{\overset{\text{transaminase}}{\rightleftharpoons}} \quad \overset{O}{\underset{\underset{R}{|}}{\underset{||}{C-COO^-}}} \quad + \quad \overset{+}{\underset{\underset{R}{|}}{\underset{|}{H-C-COO^-}}} \overset{NH_3}{}
$$

Donor   Acceptor   Carbon   New amino acid
amino   keto acid   skeleton of
acid   amino acid

The α-amino groups of many amino acids are transferred to α-ketoglutarate to produce the amino acid glutamate. This glutamate family of transaminases is especially important because the α-keto acid corresponding to glutamate is the citric acid cycle intermediate, α-ketoglutarate. The *glutamate transaminases* thus provide a direct link between amino acid degradation and the citric acid cycle. *Alanine transaminase* and *aspartate transaminase* are also important members of this family.

All transaminases require the prosthetic group pyridoxal phosphate, a coenzyme derived from vitamin B_6 (pyridoxine). During transamination, the α-amino group is first transferred to pyridoxal phosphate and then from pyridoxal phosphate to the α-keto acid.

Example 8
Explain how the *transamination* of amino acids provides a direct linkage between amino acid degradation and the citric acid cycle.

Example 8 continued

Answer

Transaminases catalyze the transfer of the α-amino group from an α-amino acid to an α-keto acid. The α-amino groups of a great many of the amino acids are transferred to α-ketoglutarate to produce the amino acid glutamate. The glutamate family of transaminases is especially important because the α-keto acid corresponding to glutamate is the citric acid cycle intermediate, α-ketoglutarate. Thus, the glutamate transaminases, as well as others, provide a direct link to the citric acid cycle.

Removal of α-Amino Groups: Oxidative Deamination

Ammonium ion is now liberated from the glutamate by oxidative deamination, catalyzed by *glutamate dehydrogenase*:

$$
\begin{array}{l}
COO^- \\
| \\
CH-NH_3^+ \\
| \\
CH_2 \quad + \quad NAD^+ \quad + \quad H_2O \\
| \\
CH_2 \\
| \\
COO^-
\end{array}
\rightleftharpoons
NH_4^+ \quad + \quad
\begin{array}{l}
COO^- \\
| \\
C=O \\
| \\
CH_2 \\
| \\
CH_2 \\
| \\
COO^-
\end{array}
\quad + \quad NADH
$$

Glutamate   α-Ketoglutarate

The Fate of Amino Acid Carbon Skeletons

The carbon skeletons produced by these and other deamination reactions enter the energy-harvesting pathways at many steps, as seen in Figure 22.8 in the text.

Oxidative deamination of glutamate and deamination of other amino acids produce considerable quantities of ammonium ion. This must be incorporated into a biological molecule and removed from the body so that it does not reach toxic levels.

22.8 The Urea Cycle

Learning Goal 8: Explain the importance of the urea cycle and describe its essential steps.

It is critically important to the survival of the organism to be able to excrete ammonium ions, regardless of the energy required. In humans and most terrestrial vertebrates, the means of ammonium ion removal is the urea cycle, which occurs in the liver. This cycle keeps excess ammonium ion out of the blood and allows the excretion of the excess in the form of urea.

Reactions of the Urea Cycle

The reactions of the urea cycle, summarized below, are shown in Figure 22.12 of the text.

Step 1. CO_2 and NH_4^+ react to form carbamoyl phosphate. This reaction requires ATP and H_2O.

Step 2. Carbamoyl phosphate condenses with the amino acid ornithine to produce the amino acid citrulline.

Step 3. Citrulline now condenses with aspartate to produce argininosuccinate. This reaction requires energy released by the hydrolysis of ATP.

Step 4. Argininosuccinate is cleaved to produce the amino acid arginine and the citric acid cycle intermediate fumarate.

Step 5. Finally, arginine is hydrolyzed, producing urea, which will be excreted, and ornithine, the original reactant in the cycle. Note that one of the amino groups of urea is the ammonium ion, and the second is derived from the amino acid aspartate.

Learning Goal 9: **Discuss the cause and effect of hyperammonemia.**

A deficiency of urea cycle enzymes causes an elevation of the concentration of NH_4^+, a condition known as **hyperammonemia**. If there is a complete deficiency of one of the enzymes of the urea cycle, the result is death in early infancy. A partial deficiency results in less severe symptoms and can be treated with a low-protein diet.

Example 9

Give a summary of the reactions of the urea cycle.

Answer

1. *Step 1*: The first step involves the reaction of the waste product CO_2 with ammonium ions from the amino acids to form carbamoyl phosphate.

$$CO_2 + NH_4^+ + 2\,ATP \longrightarrow H_2N-\overset{\overset{\displaystyle O}{\|}}{C}-O-\overset{\overset{\displaystyle O}{\|}}{\underset{\underset{\displaystyle O^-}{|}}{P}}-O^- + 2\,ADP + P_i + 3\,H^+$$

2. *Step 2*: The carbamoyl phosphate then condenses with the amino acid ornithine to produce citrulline (another amino acid).

3. *Step 3*: Citrulline condenses with aspartate to produce argininosuccinate.

4. *Step 4*: The argininosuccinate is cleaved to produce arginine and fumarate.

5. *Step 5*: Finally, arginine is hydrolyzed to generate urea and ornithine. The ornithine can then return to be used again in Step 2.

Example 10

Describe the genetic disorder that results from the deficiency of an enzyme in the urea cycle.

Answer

A deficiency of a urea cycle enzyme causes an elevation of the concentration of ammonium ion, a condition known as *hyperammonemia*. If there is a complete deficiency of one of these enzymes, the result is death in early infancy. If, however, there is a partial deficiency of one of the cycle's enzymes, the result can be retardation, convulsions, and vomiting. If the disorder is caught early, a diet low in protein may lead to less severe clinical symptoms.

22.9 Overview of Anabolism: The Citric Acid Cycle as a Source of Biosynthetic Intermediates

Learning Goal 10: Summarize the role of the citric acid cycle in catabolism and anabolism.

The citric acid cycle plays a key role in biosynthetic reactions, as well as ATP production. Glycolysis and the citric acid cycle provide precursors for the biosynthesis of amino acids, lipids, and the nitrogenous bases found in DNA. They also generate precursors for heme, the prosthetic group required for hemoglobin, myoglobin, cytochrome c, and catalase. Metabolic pathways that function in both **anabolism** and **catabolism** are called **amphibolic pathways**. Cellular energy demand sometimes exceeds the supply of citric acid intermediates. Mammalian cells can produce more oxaloacetate by the carboxylation of pyruvate, a reaction that is also important in gluconcogenesis. This reaction is called an **anaplerotic reaction**, meaning "to fill up or replenish."

Glossary of Key Terms in Chapter Twenty-Two

acetyl CoA (22.2) a molecule composed of coenzyme A and an acetyl group; this intermediate provides acetyl groups for complete oxidation by aerobic respiration.

aerobic respiration (22.3) the oxygen requiring degradation of food molecules and production of ATP.

amphibolic pathways (22.9) metabolic pathways that function both in anabolism and catabolism.

anabolism (22.9) energy requiring biosynthetic pathways.

anaplerotic reactions (22.9) a reaction that replenishes a substrate needed for a biochemical pathway.

ATP synthase (22.6) a multiprotein complex within the inner mitochondrial membrane that uses the energy of the proton gradient to produce ATP.

catabolism (22.9) energy harvesting degradative pathways.

citric acid cycle (22.4) a cyclic biochemical pathway that is the final stage of degradation of carbohydrates, fats, and amino acids; it results from the complete oxidation of acetyl groups derived from these dietary fuels.

coenzyme A (22.2) a molecule derived from ATP and the vitamin pantothenic acid. Coenzyme A functions in the transfer of acetyl groups.

cristae (22.1) the highly folded inner membrane of the mitochondria.

electron transport system (22.6) the series of electron transport proteins embedded in the inner mitochondrial membrane that accept high-energy electrons from NADH and $FADH_2$ and transfer them in stepwise fashion to molecular oxygen (O_2).

F_0F_1 complex (22.6) an alternate term for the ATP synthase, the multiprotein complex in the inner mitochondrial membrane that uses the energy of the proton gradient to produce ATP.

hyperammonemia (22.8) a genetic disorder in one of the enzymes of the urea cycle that results in toxic or even fatal elevations of the concentration of ammonium ions in the body.

inner mitochondrial membrane (22.1) the highly folded, impermeable membrane within the mitochondrion, which is the location of the electron transport system and ATP synthase.

intermembrane space (22.1) the region between the outer and inner mitochondrial membranes, which is the location of the proton reservoir that drives ATP synthesis.

matrix space (22.1) the region of the mitochondrion within the inner membrane; it is the location of the enzymes that carry out the reactions of the citric acid cycle.

mitochondria (22.1) the cellular "power plants" in which the reactions of the citric acid cycle, the electron transport system, and ATP synthase function to produce ATP.

outer mitochondrial membrane (22.1) the membrane that surrounds the mitochondrion and separates it from the contents of the cytoplasm; it is highly permeable to small "food" molecules.

oxidative deamination (22.7) an oxidation-reduction reaction in which NAD^+ is reduced and the amino acid is deaminated.

oxidative phosphorylation (22.6) a series of reactions that couples the oxidation of NADH and $FADH_2$ to the phosphorylation of ADP to produce ATP.

pyridoxal phosphate (22.7) a coenzyme derived from vitamin B_6 that is required for all transamination reactions.

pyruvate dehydrogenase complex (22.2) a complex of all the enzymes and coenzymes required for the synthesis of CO_2 and acetyl CoA.

terminal electron acceptor (22.6) the final electron acceptor in an electron transport system that removes the low-energy electrons from the system; in aerobic organisms, the terminal electron acceptor is molecular oxygen.

transaminase (22.7) an enzyme that catalyzes the transfer of an amino group from one molecule to another.

transamination (22.7) a reaction in which an amino group is transferred from one molecule to another.

urea cycle (22.8) a cyclic series of reactions that detoxifies ammonium ions by incorporating them into urea, which is excreted from the body.

Self Test for Chapter Twenty-Two

1. The oxidative reactions of metabolism provide most cellular energy. In what specific part of the cell do these reactions occur?

2. The number of mitochondria in a eukaryotic cell varies widely. What need is reflected by this variation?

3. To what compound is the acetyl group transferred in the first step of the citric acid cycle?

4. Malate dehydrogenase reduces NAD^+ to NADH and oxidizes malate into what product in the final step of the citric acid cycle?

5. How many molecules of CO_2 are produced by the complete oxidation of an acetyl group by the citric acid cycle?

6. How many molecules of NADH are formed by the complete oxidation of an acetyl group by the citric acid cycle?

7. How many molecules of $FADH_2$ are produced by the complete oxidation of an acetyl group by the citric acid cycle?

8. In glycolysis, how many NADH are generated per glucose molecule?

9. In what organelle do the reactions of the Krebs cycle, fatty acid oxidation, and oxidative phosphorylation occur?

10. What is the term for the series of electron and proton carriers embedded within the mitochondrial inner membrane?

11. Which of the electron carriers in the electron transport system contain a heme unit?

12. When NADH dehydrogenase oxidizes NADH in the mitochondrial matrix, it can pass the electrons to the next carrier. What ion cannot be transferred?

13. Spanning the inner mitochondrial membrane is a protein complex (F_0) that serves as a channel protein. What specific particles pass through this channel?

14. Protruding into the mitochondrial matrix is a spherical protein complex (F_1). What is the enzymatic activity of F_1?

15. NADH carries electrons, originally from glucose, to the first carrier of the electron transport system. What compound is formed by the oxidation of NADH at this site?

16. After NADH is oxidized to NAD^+ in the first step of the electron transport system, what happens to the NAD^+?

17. In the last step of the electron transport system, the electrons have too little energy to accomplish any more work. In aerobic organisms, what is the terminal electron acceptor?

18. What proteins of the electron transport system are poisoned by cyanide?

19. How many ATP are produced for each NADH generated in the mitochondria?

20. How many ATP are produced for each $FADH_2$ generated in the mitochondria?

21. What is the source of most of the amino acids that are used by the human body?

22. Where in the human body does the degradation of amino acids primarily occur?

23. The α-amino groups of a great many amino acids are transferred to α-ketoglutarate. What amino acid is produced in this reaction?

24. The glutamate transaminases provide a direct link between amino acid degradation and which other process?

25. One of the most important transaminases is aspartate transaminase, which catalyzes the transfer of the α-amino group of aspartate to α-ketoglutarate. What are the two new compounds produced in this reaction?

26. Where in the human body does the conversion of ammonium ions to urea occur?

27. Provide the missing product in the following reaction:

$$CO_2 + NH_4^+ + 2ATP + H_2O \rightarrow 2ADP + P_i + 3H^+ + \,?$$

28. What is the term used to describe metabolic pathways that function in both anabolism and catabolism?

29. Where in the mitochondrion do the reactions of the citric acid cycle occur?

30. What is the first step of amino acid catabolism?

Vocabulary Quiz for Chapter Twenty-Two

1. A(n) _____ is an enzyme that catalyzes the transfer of an amino group from one molecule to another.

2. Metabolic pathways that function both in anabolism and catabolism are called _____.

3. _____ is the series of reactions that couples the oxidation of NADH and $FADH_2$ to the phosphorylation of ADP to produce ATP.

4. The _____ is the series of electron transport proteins embedded in the inner mitochondrial membrane that accept high energy electrons from NADH and $FADH_2$.

5. The cellular "power plants" in which the reactions of the citric acid cycle, the electron transport system, and ATP synthase function to produce ATP are called _____.

6. The _____ is the final electron acceptor in an electron transport system that removes the low energy electrons from the system. In aerobic organisms, it is molecular oxygen.

7. _____ is the multiprotein complex within the inner mitochondrial membrane that uses the energy of the proton gradient to produce ATP.

8. The _____ is the region of the mitochondrion within the inner membrane. It is the location of the enzymes that carry out the reactions of the citric acid cycle.

9. The _____ is a cyclic series of reactions that detoxifies ammonium ions by incorporating them into urea, which is excreted from the body.

10. The _____ is the cyclic biochemical pathway that is the final stage of degradation of carbohydrates, fats, and amino acids. It results in the complete oxidation of acetyl groups derived from these dietary fuels.

23 Fatty Acid Metabolism

Learning Goals

1 Summarize the digestion and storage of lipids.
2 Describe the degradation of fatty acids by β-oxidation.
3 Explain the role of acetyl CoA in fatty acid metabolism.
4 Understand the role of ketone body production in β-oxidation.
5 Compare β-oxidation of fatty acids and fatty acid biosynthesis.
6 Describe the regulation of lipid and carbohydrate metabolism in relation the liver, adipose tissue, muscle tissue, and the brain.
7 Summarize the antagonistic effects of glucagon and insulin.

Introduction

Acetyl CoA is the key molecule of lipid metabolism. Fatty acids are degraded to acetyl CoA, which is oxidized by the citric acid cycle. On the other hand, acetyl CoA is also the precursor for the biosynthesis of fatty acids and cholesterol.

23.1 Lipid Metabolism in Animals

Learning Goal 1:   Summarize the digestion and storage of lipids.

Digestion and Absorption of Dietary Fats

Fats are hydrophobic molecules and must be extensively processed before they can be digested and absorbed. The enzymes that hydrolyze lipids are called **lipases**. The most effective lipid digestion occurs in the small intestine, where **bile** causes formation of small lipid micelles. **Micelles** are aggregations of molecules having a polar region facing the aqueous exterior and an internal nonpolar region that dissolves lipid. Bile is produced in the liver and stored in the gall bladder. The presence of lipids in the small intestine stimulates secretion of bile into the duodenum.

Triglycerides are the major lipids in the micelles. A protein called **colipase** binds to the surface of the micelles and facilitates hydrolysis of the triglycerides by pancreatic lipases. The fatty acids and monoglycerides produced by hydrolysis are absorbed by cells of the intestinal epithelium.

Example 1
Explain the digestion and absorption of dietary fats.

In intestinal cells the monoglycerides and fatty acids are reassembled into triglycerides and combined with protein, producing **chylomicrons**. These chylomicrons eventually enter the blood, and the lipids are once again hydrolyzed to products that can be absorbed by cells of the body.

Example 2
Explain what happens to lipid digestion products after absorption.

Answer
The monoglycerides and fatty acids are reassembled into triglycerides, which are combined with protein to produce the class of plasma lipoproteins called chylomicrons. These are secreted into lymphatic vessels and eventually arrive in the bloodstream. Here triglycerides are again hydrolyzed, and the products are absorbed by the cells of the body.

Lipid Storage

Fatty acids are stored as triglycerides within **adipocytes** (fat cells) that compose **adipose tissue**. When the body demands energy, these triglycerides can be hydrolyzed, and the fatty acids are oxidized to generate ATP.

Example 3
Explain lipid storage in the human body.

Answer
Fatty acids are stored as triglycerides within fat droplets in the cytoplasm of adipocytes. These cells contain a large fat droplet, which accounts for nearly the entire volume of the cell.

23.2 Fatty Acid Degradation

Overview of Fatty Acid Degradation

Fatty acids are degraded by a pathway called **β-oxidation**, which consists of a set of five reactions. Each trip through the β-oxidation pathway releases acetyl CoA and returns a fatty acyl CoA molecule that contains two fewer carbons. One $FADH_2$ (2 ATP) and one

NADH (3 ATP) are produced for each cycle of β-oxidation. Acetyl CoA can enter the citric acid cycle, resulting in production of twelve more ATP. The remaining fatty acyl CoA cycles through steps 2-5 of β-oxidation until the fatty acid is completely converted to acetyl CoA.

Example 4
Give an overview of fatty acid degradation.

Answer
The β-oxidation cycle for fatty acid degradation occurs in the mitochondrial matrix. The fatty acids are acted upon by an enzyme system known as the β-oxidation cycle until the fatty acids are all converted into acetyl coenzyme A units.

The Reactions of β-Oxidation

Learning Goal 2:   Describe the degradation of fatty acids by β-oxidation.

Learning Goal 3:   Explain the role of acetyl CoA in fatty acid metabolism.

Special transport mechanisms bring fatty acid molecules into the mitochondrial matrix, where β-oxidation occurs. The reactions of β-oxidation are summarized below.

1. *Activation*: A fatty acyl CoA molecule is formed between coenzyme A and the fatty acid. This reaction requires ATP, which is cleaved to AMP and pyrophosphate. The bond between the fatty acid and coenzyme A is a high-energy thioester bond.
2. *Oxidation*: A pair of hydrogen atoms are removed from the fatty acid, and FAD is reduced to $FADH_2$.
3. *Hydration*: The double bond produced in Step 2 undergoes a hydration reaction, and the β-carbon is hydroxylated.
4. *Oxidation*: The hydroxyl group of the β-carbon is now dehydrogenated, and NAD^+ is reduced to form NADH.
5. *Thiolysis*: A molecule of coenzyme A attacks the β-carbon, releasing acetyl CoA and a fatty acyl CoA that is two carbons shorter than the original fatty acid.

The fatty acyl CoA is further oxidized by cycling through Steps 2 through 5 until the fatty acid carbon chain is completely degraded to acetyl CoA. The acetyl CoA is completely oxidized in the citric acid cycle. The balance sheet for ATP production in the complete oxidation of a C16-fatty acid, palmitic acid, is found in Figure 23.7 of the text.

Example 5
Compare ATP production by β-oxidation of a fatty acid like palmitic acid to that produced by oxidation of an equivalent amount of glucose.

Example 5 continued

Answer
The complete oxidation of the C16-fatty acid palmitic acid produces 129 ATP molecules. This is 3.5 times more energy than results from the complete oxidation of the same amount of glucose.

23.3 Ketone Bodies

Learning Goal 4:   Understand the role of ketone body production in β-oxidation.

If there is not a sufficient supply of citric acid cycle intermediates to allow the complete oxidation of the acetyl CoA produced in β-oxidation, acetyl CoA is converted to **ketone bodies**, such as β-hydroxybutyrate, acetoacetate, and acetone.

Example 6
Explain why ketone bodies are produced in the human body.

Answer
The β-oxidation of fatty acids produces a steady supply of acetyl CoA molecules. If glycolysis and the β-oxidation cycle are functioning at the same rate, there will be a steady supply of pyruvate, which can be converted into oxaloacetate. However, if the supply of oxaloacetate is too low to allow all of the acetyl CoA to enter the citric acid cycle, the acetyl CoA molecules are converted into ketone bodies. The ketone bodies are β-hydroxybutyrate, acetoacetate, and acetone.

Ketosis

Ketosis is the abnormal elevation of blood ketone body concentration. This may occur due to starvation, a diet that is extremely low in carbohydrates, or uncontrolled **diabetes mellitus**. In diabetes, this can lead to **ketoacidosis** because the ketone acids are relatively strong acids and dissociate to release H^+. This causes the pH of the blood to become acidic.

Ketogenesis

The first step in the pathway for ketone body synthesis is the fusion of two molecules of acetyl CoA to produce acetoacetyl CoA. This reaction is catalyzed by the enzyme thiolase. Acetoacetyl CoA reacts with a third acetyl CoA to yield β-hydroxy-β-methylglutaryl CoA (HMG-CoA). HMG-CoA is cleaved to yield acetoacetate and acetyl CoA. While some acetoacetate spontaneously loses carbon dioxide, producing acetone, most of it undergoes NADH-dependent reduction to produce β-hydroxybutyrate.

Acetoacetate and β-hydroxybutyrate, produced in the liver, are circulated to other tissues through the blood. There they may be reconverted to acetyl CoA and used to produce ATP.

Example 7
Explain ketogenesis.

Answer
The pathway for the production of ketone bodies begins with the reversal of the last step of the β-oxidation cycle. When oxaloacetate levels are low, the enzyme that mediates the last reaction of β-oxidation, thiolase, now mediates the fusion of two acetyl CoA molecules to produce acetoacetyl CoA. The acetoacetyl CoA reacts with another molecule of acetyl CoA to produce HMG-CoA. The HMG-CoA is cleaved to yield *acetoacetate* and acetyl CoA. Some of the acetoacetate spontaneously loses carbon dioxide to produce *acetone*. However, most of the acetoacetate is reduced to form *β-hydroxybutyrate*.

23.4 Fatty Acid Synthesis

Learning Goal 5:   Compare β-oxidation of fatty acids and fatty acid biosynthesis.

All organisms possess the ability to synthesize fatty acids. Fatty acid synthesis appears to be simply the reverse of β-oxidation. In fact, the fatty acid chain is constructed by the sequential addition of two carbon acetyl groups, as seen in Figure 23.10 of the text. However, there are several differences between the two processes, including intracellular location, acyl group carriers, enzymes involved, and the electron carriers used. Fatty acid biosynthesis occurs in the cytoplasm. The acyl group carrier is **acyl carrier protein (ACP)**. The portion of ACP that binds the fatty acid is the **phosphopantetheine** group also found in coenzyme A. A multienzyme complex, *fatty acid synthase*, carries out fatty acid biosynthesis; NADPH is the reducing agent.

Example 8
Compare fatty acid synthesis with fatty acid degradation.

Answer
On first examination of fatty acid synthesis, it appears that it is simply the reverse of β-oxidation. Although fatty acid synthesis and breakdown are similar, there are several major differences:

1. The enzymes responsible for fatty acid biosynthesis are located in the cytoplasm of the cells of vertebrates, whereas those responsible for the degradation of fatty acids are found in the mitochondria.
2. The activated intermediates of fatty acid synthesis are bound to a thiol group of acyl carrier protein. Thus, the thioester intermediates of fatty acid synthesis are not derivatives of coenzyme A.
3. The seven steps of fatty acid biosynthesis are carried out by the multienzyme complex known as fatty acid synthase. The enzymes responsible for fatty acid degradation are not associated with one another as a complex.

Example 8 continued

4. NADH and FADH$_2$ are produced by fatty acid oxidation, whereas NADPH is the reducing agent for fatty acid biosynthesis.

23.5 The Regulation of Lipid and Carbohydrate Metabolism

Learning Goal 6:	**Describe the regulation of lipid and carbohydrate metabolism in relation to the liver, adipose tissue, muscle tissue, and brain.**

Fatty acid and carbohydrate metabolism occur at different levels in different organs. The regulation of these two processes is of great physiological importance.

The Liver

The liver plays a central role in the regulation of blood glucose concentration, providing a steady supply of glucose for muscle and brain. It also plays a central role in lipid metabolism. When excess fuel is available, the liver synthesizes fatty acids, which are transported to adipose tissue and stored. During starvation, the liver converts fatty acids to ketone bodies that are exported to other organs. Yet the liver obtains most of its metabolic energy from the carbon skeletons of amino acids.

Example 9

Explain the central role of the liver in lipid metabolism.

Answer

When excess fuel is available, the liver synthesizes fatty acids. These are esterified to produce triglycerides that are transported to the adipose tissues by very low density lipoprotein complexes. This transport is very active when more calories are eaten than are burned. During fasting or starvation conditions, however, the liver converts fatty acids to acetoacetate and other ketone bodies. These ketone bodies are transported to other organs for energy usage.

Adipose Tissue

Adipose tissue is the major storage depot of fatty acids. Triglycerides produced in the liver are transported through the blood in very low density lipoprotein complexes. The triglycerides are hydrolyzed, and the fatty acids are absorbed by adipose tissue. In adipocytes, triglycerides are resynthesized from the absorbed fatty acids and glycerol-3-phosphate, which is produced from intermediates of glycolysis.

Muscle Tissue

The energy demand of resting muscle is generally supplied by the β-oxidation of fatty acids. The heart muscle actually prefers ketone bodies over glucose. Working muscle obtains energy by degradation of glycogen.

The Brain

Normally, the brain uses glucose as its sole source of metabolic energy. However, under starvation conditions, the brain converts to the use of ketone bodies produced by the liver. This is possible because ketone bodies can cross the *blood-brain barrier*.

Example 10
Which compounds can be used by the brain for production of ATP?

Answer
Under normal conditions, the brain uses glucose as its sole source of metabolic energy. If conditions of starvation occur, the glycogen stores drop sharply, and blood glucose concentrations drop. The ketone bodies, acetoacetate and β-hydroxybutyrate, derived from fatty acid degradation, then provide an alternative source of metabolic energy.

23.6 The Effects of Insulin and Glucagon on Cellular Metabolism

Learning Goal 7: Summarize the antagonistic effects of glucagons and insulin.

Insulin is a polypeptide hormone secreted by the β-cells of the islets of Langerhans in the pancreas in response to an increase in the blood glucose level. Insulin lowers the concentration of blood glucose by stimulating its storage as glycogen or triglycerides. In general, insulin activates biosynthetic processes and inhibits catabolism.

Insulin acts only on those *target cells* that possess a specific insulin receptor protein in their membranes. The major target cells for insulin are liver, adipose, and muscle cells. Insulin lowers blood glucose levels by altering cellular carbohydrate, protein, and lipid metabolism.

A second hormone, **glucagon**, is secreted by the α-cells of the islets of Langerhans in response to decreased blood glucose levels. The effects of glucagon are generally the opposite of the effects of insulin.

The effects of insulin and glucagon on a variety of metabolic reactions are summarized in the following table:

Actions	Insulin	Glucagon
Cellular glucose transport	Increased	No effect
Glycogen synthesis	Increased	Decreased
Glycogenolysis in liver	Decreased	Increased
Gluconeogenesis	Decreased	Increased
Amino acid uptake and protein synthesis	Increased	No effect
Inhibition of amino acid release and protein degradation	Decreased	No effect
Lipogenesis	Increased	No effect
Lipolysis	Decreased	Increased
Ketogenesis	Decreased	Increased

Glossary of Key Terms in Chapter Twenty-Three

acyl carrier protein (ACP) (23.4) the protein that forms a thioester linkage with fatty acids during fatty acid synthesis.

adipocyte (23.1) a fat cell.

adipose tissue (23.1) fatty tissue that stores most of the body lipids.

bile (23.1) micelles of lecithin, cholesterol, bile salts, protein, inorganic ions, and bile pigments that aid in lipid digestion by emulsifying fat droplets.

chylomicron (23.1) an aggregate of protein and triglycerides that carries triglycerides from the intestine to all body tissues via the blood stream.

colipase (23.1) a protein that aids in lipid digestion by binding to the surface of lipid droplets and facilitating binding of pancreatic lipase.

diabetes mellitus (23.3) the appearance of glucose in the urine caused by high blood glucose levels; the disease frequently results from the inability to produce insulin.

glucagon (23.6) a peptide hormone synthesized by the α-cells of the islets of Langerhans in the pancreas in response to low blood glucose levels.

insulin (23.6) a hormone released from the pancreas in response to high blood glucose levels; insulin stimulates glycogenesis, fat storage, and cellular uptake and storage of glucose from the blood.

ketoacidosis (23.6) a drop in the pH of the blood caused by elevated ketone levels.

ketone bodies (23.6) acetone, acetoacetone, and β-hydroxybutyrate produced from fatty acids in the liver.

ketosis (23.3) an abnormal rise in the level of ketone bodies in the blood.

lipase (23.1) an enzyme that hydrolyzes the ester linkage between glycerol and the fatty acids of triglycerides.

micelle (23.1) an aggregation of molecules having a nonpolar and a polar region; the nonpolar regions of the molecules aggregate leaving the polar regions facing the surrounding water.

β-oxidation (23.2) the biochemical pathway that results in the oxidation of fatty acids and the production of acetyl CoA.

phosphopantetheine (23.4) the portion of coenzyme A and the acyl carrier protein that is derived from the vitamin pantothenic acid; it forms a thioester linkage with fatty acids.

triglyceride (23.1) (also called triacylglycerol) a molecule composed of glycerol esterified to three fatty acids.

Self Test for Chapter Twenty-Three

1. The metabolism of fatty acids and lipids revolves around which specific compound?

2. Which compound is the starting material for the synthesis of half of the amino acids?

3. In what form do we find dietary fat when it arrives in the duodenum?

4. What mixture is composed of micelles of lecithin, cholesterol, protein, bile salts, inorganic ions, and bile pigments?

5. What is the term for an aggregate of molecules having polar and nonpolar regions?

6. Name the two major bile salts. (Refer to the textbook.)

7. Which protein molecule binds to the surface of lipid droplets and helps pancreatic lipases to adhere to the surface and hydrolyze the ester linkages?

8. Monoglycerides and fatty acids are reassembled into triglycerides and combined with protein in the cells of the small intestine. What class of lipoproteins is formed in this process?

9. What tissue stores most of the body's triglyceride molecules?

10. How many NADH and $FADH_2$ are produced in each round of β-oxidation?

11. What is another term for a fat cell?

12. The complete oxidation of a molecule of palmitic acid, using the β-oxidation cycle and the other oxidation pathways, produces how many molecules of ATP?

13. What is ketosis?

14. List three conditions that can produce ketosis.

15. In patients with uncontrolled diabetes, the very high concentration of ketone acids in the blood leads to what condition?

16. Where in the body are acetoacetate and β-hydroxybutyrate produced?

17. Which muscle derives more of its energy from the metabolic use of ketone bodies than it does from glucose?

18. What is the name of the multienzyme complex used for biosynthesis of fatty acids?

19. As a general rule, NADH is produced by catabolic reactions, but what acts as the reducing agent for biosynthetic reactions?

20. Which organ provides a steady supply of glucose for muscles and brain and plays a major role in the regulation of blood glucose concentration?

21. Adipose tissue is unable to make glycerol-3-phosphate. What pathway must it use for its supply of this precursor?

22. What lipoprotein complex transports triglycerides produced in the liver through the bloodstream?

23. What is the preferred fuel of liver tissue?

24. What compound is required for the synthesis of triglycerides in adipose tissue?

25. What compounds are oxidized to supply the energy demand of resting muscles?

26. What compound is produced in the first step of β-oxidation of fatty acids?

27. What compound is released in the final step of β-oxidation of fatty acids?

28. List the names of the two ketone bodies that are also acids.

29. Muscle tissues produce lactate under what conditions?

30. What are the major target cells for insulin?

Vocabulary Quiz for Chapter Twenty-Three

1. _____ is fatty tissue that stores most of the body lipids.

2. A(n) _____ is a molecule composed of glycerol esterified to three fatty acids.

3. _____ consists of micelles of lecithin, cholesterol, bile salts, protein, inorganic ions, and bile pigments that aid in lipid digestion by emulsifying fat droplets.

4. A(n) _____ is an aggregate of protein and triglycerides that carries triglycerides from the intestine to all body tissues via the blood stream.

5. The biochemical pathway that results in the oxidation of fatty acids and the production of acetyl CoA is called _____.

6. The molecule that serves as a carrier of acetyl groups in lipid and carbohydrate metabolism is _____.

7. _____ is a peptide hormone synthesized by the α-cells of the islets of Langerhans in the pancreas in response to low blood glucose levels.

8. The peptide hormone synthesized by the β-cells of the islets of Langerhans in the pancreas in response to high blood glucose levels is _____.

9. A(n) _____ is an enzyme that hydrolyzes the ester linkage between glycerol and the fatty acids of triglycerides.

10. Acetone, acetoacetone, and β-hydroxybutyrate produced from fatty acids in the liver via acetyl CoA are referred to as _____.

APPENDIX A

Chapter 1
Chemistry: Methods and Measurement
Solutions to the Odd-Numbered Problems

In-Chapter Questions and Problems

1.1   a. physical property
       b. chemical property
       c. physical property
       d. physical property
       e. physical property

1.3   a. pure substance
       b. heterogeneous mixture
       c. homogeneous mixture
       d. pure substance

1.5   a. $1.0 \text{ L} \times \dfrac{10^3 \text{ mL}}{\text{L}} = 1.0 \times 10^3 \text{ mL}$

       b. $1.0 \text{ L} \times \dfrac{10^6 \text{ }\mu\text{L}}{\text{L}} = 1.0 \times 10^6 \text{ }\mu\text{L}$

       c. $1.0 \text{ L} \times \dfrac{1 \text{ kL}}{10^3 \text{ L}} = 1.0 \times 10^{-3} \text{ kL}$

       d. $1.0 \text{ L} \times \dfrac{10^2 \text{ cL}}{\text{L}} = 1.0 \times 10^2 \text{ cL}$

       e. $1.0 \text{ L} \times \dfrac{10^{-1} \text{ daL}}{\text{L}} = 1.0 \times 10^{-1} \text{ daL}$

1.7   a. $0.50 \text{ in} \times \dfrac{2.54 \text{ cm}}{1 \text{ in}} \times \dfrac{1 \text{ m}}{10^2 \text{ cm}} = 1.3 \times 10^{-2} \text{ m}$

       b. $0.75 \text{ qt} \times \dfrac{0.946 \text{ L}}{1 \text{ qt}} = 0.71 \text{ L}$

c. $56.8 \text{ g} \times \dfrac{1 \text{ lb}}{454 \text{ g}} \times \dfrac{16 \text{ oz}}{1 \text{ lb}} = 2.00 \text{ oz}$

d. $1.5 \text{ cm}^2 \times \left(\dfrac{1 \text{ m}}{10^2 \text{ cm}}\right)^2 = 1.5 \text{ cm}^2 \times \dfrac{1 \text{ m}^2}{10^4 \text{ cm}^2} = 1.5 \times 10^{-4} \text{ m}^2$

1.9   a.   Three (all non-zero digits)

    b.   Three (all non-zero digits)

    c.   Four (zeros between non-zero digits are significant)

    d.   Two (trailing zero significant due to decimal)

    e.   Three (leading zeros are not significant)

1.11   a. 2.4×10^{-3}

    b. 1.80×10^{-2}

    c. 2.24×10^2

1.13   a.   8.09 (3 significant figures)

    b.   5.9 (2 significant figures)

    c.   20.19 (4 significant figures)

1.15   a.   51 (2 significant figures)

    b. 8.0×10^1 (2 significant figures)

    c. 1.6×10^2 (2 significant figures)

1.17   a.   61.404 rounds to 61.4

    b.   6.1714 rounds to 6.17

    c.   0.066494 rounds to 0.0665 (or, 6.65×10^{-2})

1.19   a. $^\circ C = \dfrac{^\circ F - 32}{1.8} = \dfrac{32 - 32}{1.8} = 0 ^\circ C$

    b. $K = {^\circ}C + 273 = 0 + 273 = 273 \text{ K}$

1.21   g alcohol = 30.0 mL alcohol x $\dfrac{0.789 \text{ g alcohol}}{1 \text{ mL alcohol}}$ = 23.7 g alcohol

---

End-of-Chapter Questions and Problems

1.23   a.   Chemistry is the study of matter and the changes that matter undergoes.
         b.   Matter is the material component of the universe.
         c.   Energy is the ability to do work.

1.25   a.   Potential energy is stored energy, or energy due to position or composition.
         b.   Kinetic energy is the energy resulting from motion of an object
             $[\text{K.E.} - {}^{1}/_{2}(\text{mass})(\text{velocity}^2)]$.
         c.   Data are a group of facts resulting from an experiment.

1.27   a.   gram (or kilogram)
         b.   liter
         c.   meter

1.29   Weight is the force exerted on a body by gravity; mass is a quantity of matter. Mass is an independent quantity, and weight is dependent on gravity, which may differ from location to location.

1.31   Density is mass per volume. Specific gravity is the ratio of the density of a substance to the density of water at 4°C.

1.33   The scientific method is an organized way of doing science. It uses carefully planned experimentation to study our surroundings.

1.35   The statement "stem-cell research has the potential to provide replacement 'parts' for the human body" is, at the time this book was written, a theory. Pay particular attention to the word "potential" in the statement. The meaning and the scientific status of this statement would be quite different if the words "has the potential" were replaced by the word "can."

1.37   A physical property is a characteristic of a substance that can be observed without the substance undergoing a change in chemical composition.

1.39   Chemical properties of matter include flammability and toxicity.

1.41   A pure substance has constant composition with only a single substance, whereas a mixture is composed of two or more substances.

1.43   Mixtures are composed of two or more substances. A homogeneous mixture has uniform composition, and a heterogeneous mixture has nonuniform composition.

1.45   a.   chemical reaction
       b.   physical change
       c.   physical change

1.47   a.   physical property
       b.   chemical property

1.49   a.   pure substance
       b.   pure substance
       c.   mixture

1.51   a.   homogeneous
       b.   homogeneous
       c.   homogeneous

1.53   A gas is made up of particles that are widely separated. A gas will expand to fill any container and it has no definite shape or volume.

1.55   a.   extensive property
       b.   extensive property
       c.   intensive property

1.57   An element is a pure substance that cannot be changed into a simpler form of matter by any chemical reaction. An atom is the smallest unit of an element that retains the properties of that element.

1.59   a.   an element - iron, oxygen, carbon are just a few of the more than 100 possible elements
       b.   a pure structure - sodium chloride, water, sucrose, ethyl alcohol

1.61   a. $2.0 \text{ lb} \times \dfrac{16 \text{ oz}}{1 \text{ lb}} = 32 \text{ oz}$

       b. $2.0 \text{ lb} \times \dfrac{1 \text{ t}}{2000 \text{ lb}} = 1.0 \times 10^{-3} \text{ t}$

       c. $2.0 \text{ lb} \times \dfrac{454 \text{ g}}{1 \text{ lb}} = 9.1 \times 10^2 \text{ g}$

       d. $2.0 \text{ lb} \times \dfrac{454 \text{ g}}{1 \text{ lb}} \times \dfrac{10^3 \text{ mg}}{1 \text{ g}} = 9.1 \times 10^5 \text{ mg}$

       e. $2.0 \text{ lb} \times \dfrac{454 \text{ g}}{1 \text{ lb}} \times \dfrac{1 \text{ da}}{10^1 \text{ g}} = 9.1 \times 10^1 \text{ da}$

1.63　a.　$3.0 \text{ g} \times \dfrac{1 \text{ lb}}{454 \text{ g}} = 6.6 \times 10^{-3} \text{ lb}$

b.　$3.0 \text{ g} \times \dfrac{1 \text{ lb}}{454 \text{ g}} \times \dfrac{16 \text{ oz}}{1 \text{ lb}} = 1.1 \times 10^{-1} \text{ oz}$

c.　$3.0 \text{ g} \times \dfrac{1 \text{ kg}}{10^3 \text{ g}} = 3.0 \times 10^{-3} \text{ kg}$

d.　$3.0 \text{ g} \times \dfrac{10^2 \text{ cg}}{1 \text{ g}} = 3.0 \times 10^2 \text{ cg}$

e.　$3.0 \text{ g} \times \dfrac{10^3 \text{ mg}}{1 \text{ g}} = 3.0 \times 10^3 \text{ mg}$

1.65　a.　$^\circ C = \dfrac{^\circ F - 32}{1.8} = \dfrac{50.0 - 32}{1.8} = 10.0 ^\circ C$

b.　$K = ^\circ C + 273 = 10.0 + 273.15 = 283.2 \text{ K}$

1.67　a.　$K = ^\circ C + 273 = 20.0 + 273.15 = 293.2 \text{ K}$

b.　If $^\circ C = \dfrac{^\circ F - 32}{1.8}$, then $1.8 ^\circ C = ^\circ F - 32$

and $^\circ F = 1.8 ^\circ C + 32$

$^\circ F = [(1.8)(20.0)] + 32 = 68.0 ^\circ F$

1.69　$9 \text{ pt} \times \dfrac{1 \text{ qt}}{1 \text{ pt}} \times \dfrac{0.946 \text{ L}}{1 \text{ qt}} = 4.257 \text{ L} \approx 4 \text{ L}$ (1 significant figure)

1.71　If $^\circ C = \dfrac{^\circ F - 32}{1.8}$, then $1.8 ^\circ C = ^\circ F - 32$

and $^\circ F = 1.8 ^\circ C + 32$

$^\circ F = [(1.8)(38.5)] + 32 = 101 ^\circ F$

1.73　In order to compare the magnitude of the two lengths, we must convert them to a common unit.

$5.0 \text{ cm} \times \dfrac{1 \text{ in}}{2.54 \text{ cm}} = 2.0 \text{ in}$

Therefore, 5 cm is shorter than 5 in.

1.75   In order to compare the magnitude of the two masses, we must convert them to a common unit.

$$5.0 \text{ mg} \times \frac{10^{-3} \mu g}{1 \text{ mg}} = 5 \times 10^{-3} \text{ } \mu g$$

Therefore, 5.0 μg is smaller that 5.0 mg.

1.77   a. 3
       b. 3
       c. 3
       d. 4
       e. 4
       f. 3

1.79   a. 3.87×10^{-3}

       b. 5.20×10^{-2}

       c. 2.62×10^{-3}

       d. 2.43×10^{1}

       e. 2.40×10^{2}

       f. 2.41×10^{0}

1.81   a. Precision is a measure of the agreement of replicate results.
       b. Accuracy is the degree of agreement between the true value and measured value.

1.83   a. $(23)(657) = 1.5 \times 10^{4}$

       b. $0.00521 + 0.236 = 2.41 \times 10^{-1}$

       c. $\dfrac{18.3}{3.0576} = 5.99$

       d. $1157.23 - 17.812 = 1139.42$

       e. $\dfrac{(1.987)(298)}{0.0821} = 7.21 \times 10^{3}$

1.85   a. 1.23×10^{1}

b. 5.69×10^{-2}

c. -1.527×10^3

d. 7.89×10^{-7}

e. 9.2×10^7

f. 5.280×10^{-3}

g. 1.279×10^0

h. -5.3177×10^2

1.87   a. 3,240

b. 0.000150

c. 0.4579

d. −683,000

e. −0.0821

f. 299,790,000

g. 1.50

h. 602,200,000,000,000,000,000,000

1.89 $d = \dfrac{m}{V} = \dfrac{3.00 \times 10^2 \text{ g}}{50.0 \text{ mL}} = 6.00 \text{ g/mL}$

1.91 $1.50 \times 10^2 \text{ mL} \times \dfrac{7.20 \text{ g}}{1 \text{ mL}} = 1.08 \times 10^3 \text{ g}$

1.93 $d = \dfrac{m}{V} = \dfrac{98 \text{ g}}{1.00 \times 10^2 \text{ cm}^3} = 0.98 \text{ g/cm}^3$, therefore, teak

1.95   Specific gravity of alcohol $= \dfrac{\text{density of alcohol}}{\text{density of water at } 4°C}$

Specific gravity of alcohol $= \dfrac{0.789 \text{ g/mL}}{1.00 \text{ g/mL}} = 0.789$

1.97 $d_{lead} = \dfrac{5.0 \times 10^{1}\,g}{6.36\,cm^{3}} = 7.9\,g/cm^{3}$

$d_{uranium} = \dfrac{75\,g}{3.97\,cm^{3}} = 19\,g/cm^{3}$

$d_{platinum} = \dfrac{m}{V} = \dfrac{2140\,g}{1.00 \times 10^{2}\,cm^{3}} = 21.4\,g/cm^{3}$

Lead has the lowest density and platinum has the greatest density.

1.99 $d = \dfrac{m}{V}$

Thus $V = \dfrac{m}{d} = \dfrac{10.0\,g}{0.791\,g/mL}$

$V = 12.6\,mL$

Chapter 2
The Composition and Structure of the Atom
Solutions to the Odd-Numbered Problems

In-Chapter Questions and Problems

2.1   a.   16 protons and 16 electrons (atomic number = 16)
               32 - 16 = 16 neutrons (mass number - atomic number)
        b.   11 protons and 11 electrons (atomic number = 11)
               23 - 11 = 12 neutrons (mass number - atomic number)

2.3   Step 1. Convert each percentage to a decimal fraction.

$$90.48\,\% \;\; {}^{20}_{10}\text{Ne} \; \text{x} \; \frac{1}{100\%} \; = \; 0.9048 \; {}^{20}_{10}\text{Ne}$$

$$0.27\,\% \;\; {}^{21}_{10}\text{Ne} \; \text{x} \; \frac{1}{100\%} \; = \; 0.0027 \; {}^{21}_{10}\text{Ne}$$

$$9.25\,\% \;\; {}^{22}_{10}\text{Ne} \; \text{x} \; \frac{1}{100\%} \; = \; 0.0925 \; {}^{22}_{10}\text{Ne}$$

Step 2.

$$\begin{array}{c}\text{contributions to} \\ \text{atomic mass by } {}^{20}_{10}\text{Ne}\end{array} = \begin{pmatrix}\text{fraction of all Ne atoms} \\ \text{that are } {}^{20}_{10}\text{Ne}\end{pmatrix} \text{x} \begin{pmatrix}\text{mass of a} \\ {}^{20}_{10}\text{Ne atom}\end{pmatrix}$$

$$= \;\;\; 0.9048 \;\; \text{x} \;\; 19.99 \text{ amu}$$

$$= \;\;\; 18.087 \text{ amu}$$

$$\begin{array}{c}\text{contributions to} \\ \text{atomic mass by } {}^{21}_{10}\text{Ne}\end{array} = \begin{pmatrix}\text{fraction of all Ne atoms} \\ \text{that are } {}^{21}_{10}\text{Ne}\end{pmatrix} \text{x} \begin{pmatrix}\text{mass of a} \\ {}^{21}_{10}\text{Ne atom}\end{pmatrix}$$

$$= \;\;\; 0.0027 \;\; \text{x} \;\; 20.99 \text{ amu}$$

$$= \;\;\; 0.057 \text{ amu}$$

$$\begin{array}{c}\text{contributions to} \\ \text{atomic mass by } {}^{22}_{10}\text{Ne}\end{array} = \begin{pmatrix}\text{fraction of all Ne atoms} \\ \text{that are } {}^{22}_{10}\text{Ne}\end{pmatrix} \text{x} \begin{pmatrix}\text{mass of a} \\ {}^{22}_{10}\text{Ne atom}\end{pmatrix}$$

$$= \;\;\; 0.9025 \;\; \text{x} \;\; 21.99 \text{ amu}$$

$$= \quad 2.034 \text{ amu}$$

Step 3. The weighted average is:

atomic mass of naturally occurring neon = (contribution of Ne-20)
+ (contribution of Ne-21)
+ (contribution of Ne-22)

$$= 18.087 + 0.057 \text{ amu} + 2.034 \text{ amu}$$

$$= 20.18 \text{ amu}$$

2.5   Electron density is the probability that an electron will be found in a particular region of an atomic orbital.

2.7   a. Zr (zirconium)
       b. 22.99
       c. Cr (chromium)
       d. At (astatine)

2.9   a. helium, atomic number = 2, mass = 4.00 amu
       b. fluorine, atomic number = 9, mass = 19.00 amu
       c. manganese, atomic number = 25, mass = 54.94 amu

2.11   a. Total electrons = 11, valence electrons = 1
       b. Total electrons = 12, valence electrons = 2
       c. Total electrons = 16, valence electrons = 6
       d. Total electrons = 17, valence electrons = 7
       e. Total electrons = 18, valence electrons = 8

2.13   a. sulfur: $1s^2, 2s^2, 2p^6, 3s^2, 3p^4$
       b. calcium: $1s^2, 2s^2, 2p^6, 3s^2, 3p^6, 4s^2$

2.15   a. [Ne] $3s^2, 3p^4$
       a. [Ar] $4s^2$

2.17   a. Ca^{2+} and Ar are isoelectronic.
       b. Sr^{2+} and Kr are isoelectronic.
       c. S^{2-} and Ar are isoelectronic.
       d. Mg^{2+} and Ne are isoelectronic.
       e. P^{3-} and Ar are isoelectronic.

2.19   a. (smallest) F, N, Be (largest)
       b. (lowest) Be, N, F (highest)
       c. (lowest) Be, N, F, (highest)

End-of-Chapter Questions and Problems

2.21   a. Atomic number = 8, therefore 8 protons and 8 electrons.
          Mass number - atomic number = 16 - 8 = 8, therefore 8 neutrons.

        b. Atomic number = 15, therefore 15 protons and 15 electrons.
          Mass number - atomic number = 31 - 15 = 16, therefore 16 neutrons.

2.23

	Particle	Mass	Charge
a.	electron	5.4×10^{-4} amu	-1
b.	proton	1.00 amu	$+1$
c.	neutron	1.00 amu	0

2.25   a. An ion is a charged atom or group of atoms formed by the loss or gain of electrons.
        b. A loss of electrons by a neutral species results in a cation.
        c. A gain of electrons by a neutral species results in an anion.

2.27   From the periodic table, all isotopes of Rn have 86 protons. Isotopes differ in the number of neutrons.

2.29   a. 34
        b. 46

2.31   a. $^{1}_{1}H$ b. $^{14}_{6}C$

2.33

	Atomic Symbol	# Protons	# Neutrons	# Electrons	Charge
a.	$^{23}_{11}Na$	11	12	11	0
b.	$^{32}_{16}S^{2-}$	16	16	18	2−
c.	$^{16}_{8}O$	8	8	8	0
d.	$^{24}_{12}Mg^{2+}$	12	12	10	2+
e.	$^{39}_{19}K^{+}$	19	20	18	1+

2.35   a.   neutrons
        b.   protons
        c.   protons, neutrons
        d.   ion
        e.   neucleus, negative

2.37   The major postulates of Dalton's atomic theory include the following:

- All matter consists of tiny particles called atoms.
- Atoms cannot be created, divided, destroyed, or converted to any other type of atom.
- All atoms of a particular element have identical properties.
- Atoms of different elements have different properties.
- Atoms combine in simple whole-number ratios.
- Chemical change involves joining, separating, or rearranging atoms.

2.39   a. James Chadwick demonstrated the existence of the neutron in 1932. He accomplished this with a series of experiments using small nuclei as projectiles to study the nucleus.

        b. Louis DeBroglie theorized that electrons had wave-like as well as particle-like properties; this theory is known as *wave-particle duality*.

2.41   a. Dalton developed the Law of Multiple Proportions; determined the relative atomic weights of the elements known at that time; developed the first scientific atomic theory.

        b. Crookes developed the cathode ray tube and discovered "cathode rays;" characterized electron properties.

2.43   Our understanding of the nucleus is based on the gold foil experiment performed by Geiger and interpreted by Rutherford. In this experiment, Geiger bombarded a piece of gold foil with alpha particles, and observed that some alpha particles passed straight through the foil, others were deflected, and some simply bounced back. This led Rutherford to propose that the atom consisted of a small, dense nucleus (alpha particles bounced back), surrounded by a cloud of electrons (some alpha particles were deflected). The size of the nucleus is small when compared to the volume of the atom (alpha particles were able to pass through the foil).

2.45   Crookes used the cathode ray tube. He observed particles emitted by the cathode and traveling toward the anode. This ray was deflected by an electric field. Thomson measured the curvature of the ray influenced by the electric and magnetic fields. This measurement provided the mass-to-charge ratio of the negative particle. Thomson also gave the particle the name, electron.

2.47   A cathode ray is the negatively charged particle formed in a cathode ray tube. It was characterized as an electron, with a mass of nearly zero and a charge of -1.

2.49   radiowave   ↑
        microwave |
        infrared | Increasing
        visible | wavelength
        ultraviolet |
        x-ray |
        gamma ray |

2.51   Infrared radiation has greater energy than microwave radiation.

2.53   Spectroscopy is the measurement of intensity and energy of electromagnetic radiation.

2.55   According to Bohr, Planck, and others, electrons exist only in certain allowed regions, quantum levels, outside of the nucleus.

2.57   The most important points of Bohr's theory are:
   - Electrons are found in orbits at discrete distances from the nucleus.
   - The orbits are quantized - they are of discrete energies.
   - Electrons can only be found in these orbits, never in between (they are able to jump instantaneously from orbit to orbit).
   - Electrons can undergo transitions - if an electron absorbs energy, it will jump to a higher orbit; when the electron falls back to a lower orbit, it will release energy.

2.59   Bohr's atomic model was the first to successfully account for electronic properties of atoms, specifically, the interaction of atoms and light (spectroscopy). Although significantly modified, it still remains a useful model to predict bonding in simple systems.

2.61   a. sodium
        b. potassium
        c. magnesium

2.63   Group IA (or 1) is known collectively as the alkali metals and consists of lithium, sodium, potassium, rubidium, cesium, and francium.

2.65   Group VIIA (or 17) is known collectively as the halogens and consists of fluorine, chlorine, bromine, iodine, and astatine.

2.67   a. true
        b. true

2.69   a. The metals are: Na, Ni, Al.
        b. The representative metals are: Na, Al.
        c. The elements that tend to form positive ions are: Na, Ni, Al.
        d. The element that is inert is: Ar.

2.71   a. one
        b. one
        c. three
        d. seven
        e. zero (or eight)
        f. zero (or two)

2.73   A principal energy level is designated $n = 1, 2, 3$, and so forth. It is similar to Bohr's orbits in concept. A sublevel is a part of a principal energy level and is designated *s, p, d,* and *f.*

2.75   The *s* orbital represents the probability of finding an electron in a region of space surrounding the nucleus. A diagram of an *s* orbital is found in Figure 2.12 on p. 61.

2.77   Three *p* orbitals (p_x, p_y, p_z) can exist in a given principal energy level.

2.79   A 3*p* orbital is a higher energy orbital than a 2*p* orbital because it is a part of a higher energy principal energy level.

2.81 $2n^2 = 2(1)^2 = 2$ e⁻ for $n = 1$
$2n^2 = 2(2)^2 = 8$ e⁻ for $n = 2$
$2n^2 = 2(3)^2 = 18$ e⁻ for $n = 3$

2.83   a.   Al $1s^2, 2s^2, 2p^6, 3s^2, 3p^1$ (hence, a 3*p* orbital)
        b.   Na $1s^2, 2s^2, 2p^6, 3s^1$ (hence, a 3*s* orbital)
        c.   Sc $1s^2, 2s^2, 2p^6, 3s^2, 3p^6, 4s^2, 3d^1$ (hence, a 3*d* orbital)
        d.   Ca $1s^2, 2s^2, 2p^6, 3s^2, 3p^6, 4s^2$ (hence, a 4*s* orbital)
        e.   Fe $1s^2, 2s^2, 2p^6, 3s^2, 3p^6, 4s^2, 3d^6$ (hence, a 3*d* orbital)
        f.   Cl $1s^2, 2s^2, 2p^6, 3s^2, 3p^5$ (hence, a 3*p* orbital)

2.85   a.   Not possible; $n = 1$ level can have only *s*-level orbitals.
        b.   Possible; the electron configuration that is shown represents the carbon atom.
        c.   Not possible; $n = 2$ level can have only *s*- and *p*-orbitals.
        d.   Not possible; an *s*-orbital cannot contain three electrons.

2.87   a. Li^+ (Li loses 1 electron to attain outermost octet)
        b. O^{2-} (O gains 2 electrons to attain outermost octet)
        c. Ca^{2+} (Ca loses 2 electrons to attain outermost octet)
        d. Br^- (Br gains 1 electron to attain outermost octet)
        e. S^{2-} (S gains 2 electrons to attain outermost octet)
        f. Al^{3+} (Al loses 3 electrons to attain outermost octet)

2.89   a. O^{2-}, 10 e⁻; Ne, 10 e⁻; Isoelectronic
        b. S^{2-}, 18 e⁻; Cl^-, e⁻; Isoelectronic

2.91   a. Na^+; it has a noble gas electron configuration.
        b. S^{2-}; it has a noble gas electron configuration.
        c. Cl^-; it has a noble gas electron configuration.

2.93   a. $1s^2, 2s^2, 2p^6, 3s^2, 3p^6$
        b. $1s^2, 2s^2, 2p^6$

2.95   a.   (Smallest) F, O, N (Largest)
         b.   (Smallest) Li, K, Cs (Largest)
         c.   (Smallest) Cl, Br, I (Largest)

2.97   a.   (Smallest) O, N, F (Largest)
         b.   (Smallest) Cs, K, Li (Largest)
         c.   (Smallest) I, Br, Cl (Largest)

2.99   A positive ion is always smaller than its parent atom because the positive charge of the nucleus is shared among fewer electrons in the ion. As a result, each electron is pulled closer to the nucleus and the volume of the ion decreases.

2.101   The fluoride ion has a completed octet of electrons and an electron configuration resembling its nearest noble gas.

Chapter 3
Structure and Properties of
Ionic and Covalent Compounds
Solutions to the Odd-Numbered Problems

In-Chapter Questions and Problems

3.1   a. LiBr (one Li^+ and one Br^-)

        b. $CaBr_2$ (one Ca^{2+} and two Br^-)

        c. Ca_3N_2 (three Ca_2^+ and two N^{3-})

3.3   a. potassium cyanide

        b. magnesium sulfide

        c. magnesium acetate

3.5   a. $CaCO_3$

        b. $NaHCO_3$

        c. Cu_2SO_4

3.7   a. diboron trioxide

        b. nitrogen oxide

        c. iodine chloride

        d. phosphorus trichloride

3.9   a. P_2O_5

        b. SiO_2

3.11   a.   b.

3.13   a.   b.

3.15   a. Since carbon is less electronegative than oxygen, a likely skeletal structure would be:

$$\left[\begin{array}{c} O \\ \| \\ O-C-O-H \end{array} \right]^{-}$$

Counting valence electrons, we have:

1 carbon atom x 4 valence electrons =	4 electrons
3 oxygen atoms x 6 valence electrons =	18 electrons
1 hydrogen atom x 1 valence electron =	1 electron
1 negative charge on the ion =	1 electron
	For a total of **24 electrons**.

Distributing these electrons within the bicarbonate ion structure produces two resonance hybrids:

$$\left[\begin{array}{c} :\ddot{O}: \\ \| \\ :\ddot{O}-C-\ddot{O}-H \end{array} \right]^{-} \longleftrightarrow \left[\begin{array}{c} \overset{\bullet\bullet}{\ddot{O}:} \\ | \\ :\ddot{O}=C-\ddot{O}-H \end{array} \right]^{-}$$

   b. Since phosphorus is less electronegative than oxygen, the most reasonable skeletal structure is:

$$\left[\begin{array}{c} O \\ | \\ O-P-O \\ | \\ O \end{array} \right]^{3-}$$

Counting valence electrons, we have:

1 phosphorus atom x 5 electrons =	5 electrons
4 oxygen atoms x 6 electrons =	24 electrons
3 negative charges on the ion =	3 electrons
	For a total of **32 electrons**.

Distributing these electrons among the atoms of the phosphate ion structure produces the following Lewis structure:

$$\left[\begin{array}{c} :\ddot{O}: \\ \\ :\ddot{O}:P:\ddot{O}: \\ \\ :\ddot{O}: \end{array} \right]^{3-} \qquad \text{or} \qquad \left[\begin{array}{c} :\ddot{O}: \\ | \\ :\ddot{O}-P-\ddot{O}: \\ | \\ :\ddot{O}: \end{array} \right]^{3-}$$

3.17   a. The bonded nuclei are closer together when a double bond exists, in comparison to a single bond.

      b. The bond strength increases as the bond order increases. Therefore, a double bond is stronger than a single bond.

3.19   The Lewis structures of the two resonance forms of SeO_2 are:

$$\left[\; :\!\ddot{O}\!: \; Se :: \ddot{O}\!: \quad \longleftrightarrow \quad :\ddot{O}\!: :: Se : \ddot{O}\!: \; \right]$$

3.21   a.

$$H : \overset{\bullet\bullet}{\underset{\bullet\bullet}{P}} : H$$
$$H$$

Three groups and one lone pair of electrons surround the phosphorus atom; the structure is trigonal pyramidal (similar to the structure of ammonia).

      b.

$$H : \overset{H}{\underset{H}{\overset{\bullet\bullet}{\underset{\bullet\bullet}{Si}}}} : H$$

Four groups surround the silicon atom; the structure is tetrahedral (similar to the structure of methane).

3.23   a. $O - S$ Oxygen is more electronegative than sulfur; the bond is polar. The electrons are pulled toward the oxygen atom.

      b. $C \equiv N$ Nitrogen is more electronegative than carbon; the bond is polar. The electrons are pulled toward the nitrogen atom.

      c. $Cl - Cl$ There is no electronegativity difference between two identical atoms; the bond is nonpolar.

      d. $I - Cl$ Chlorine is more electronegative than iodine; the bond is polar. The electrons are pulled toward the chlorine atom.

3.25 a.

Cl
 \
  B—Cl
 /
Cl

Three groups surround the central atom forming 120° bond angles. Due to the symmetrical arrangement of the three B-Cl bonds, their polarities cancel and the molecule is nonpolar.

b.

H""N""H
     |
     H

Three groups and a lone pair of electrons surround the central atom. Due to the effect of the lone pair, the molecule is polar.

c. H–Cl

The H-Cl bond is polar due to the electronegativity difference between hydrogen and chlorine. Since H-Cl is the only bond in the molecule, the molecule is polar.

d.

    Cl
    |
Cl""Si""Cl
    |
    Cl

Four groups, all equivalent, surround the central atom. The structure is tetrahedral and the molecule is nonpolar.

3.27 a. H_2O is polar → higher melting and boiling point.
          C_2H_4 is nonpolar.

      b. CO is polar → higher melting and boiling point.
          CH_4 is nonpolar.

      c. NH_3 is polar → higher melting and boiling points.
          N_2 is nonpolar.

      d. Cl_2 is nonpolar.
          ICl is polar → higher melting and boiling points.

End-of-Chapter Questions and Problems

3.29 a. ionic
      b. covalent
      c. covalent
      d. covalent

3.31 a. covalent (2 nonmetals)
      b. covalent (2 nonmetals)
      c. covalent (2 nonmetals)
      d. ionic (metals and nonmetals)

3.33 a.

$$\text{Li}\cdot \ + \ \ddot{\underset{\cdot\cdot}{:\text{Br}}}\cdot \ \longrightarrow \ \text{Li}^+ \ + \ \left[\ddot{\underset{\cdot\cdot}{:\text{Br}}}:\right]^-$$

      b.

$$\cdot\text{Mg}\cdot \ + \ 2\ \ddot{\underset{\cdot\cdot}{:\text{Cl}}}\cdot \ \longrightarrow \ \text{Mg}^{2+} \ + \ 2\left[\ddot{\underset{\cdot\cdot}{:\text{Cl}}}:\right]^-$$

3.35 a.

$$\ddot{\underset{\cdot}{:\text{S}}}\cdot \ + \ 2\text{H}\cdot \ \longrightarrow \ \ddot{:\overset{\cdot\cdot}{\text{S}}}:\text{H}$$
$$\qquad\qquad\qquad\qquad\quad \text{H}$$

      b.

$$\cdot\ddot{\underset{\cdot}{\text{P}}}\cdot \ + \ 3\text{H}\cdot \ \longrightarrow \ \text{H}:\overset{\cdot\cdot}{\underset{\cdot\cdot}{\text{P}}}:\text{H}$$
$$\qquad\qquad\qquad\qquad\quad \text{H}$$

3.37 He has two valence electrons (electron configuration $1s^2$) and a complete N = 1 level. It has a stable electron configuration, with no tendency to gain or lose electrons, and satisfies the octet rule (2 e$^-$ for period 1). Hence, it is nonreactive.
      $\cdot\text{He}\cdot$

3.39 a. sodium ion
      b. copper(I) ion (or cuprous ion)
      c. magnesium ion
      d. iron(II) ion (or ferrous ion)
      e. iron(III) ion (or ferric ion)

3.41 a. sulfide ion
      b. chloride ion
      c. carbonate ion

3.43 a. K^+

      b. Br^-

3.45 a. $\text{SO}_4{}^{2-}$

      b. $\text{NO}_3{}^-$

3.47   a.   NaCl   (one 1+ cancels one 1-)
        b. $MgBr_2$ (one 2+ cancels two 1-)

3.49   a.   AgCN   (one 1+ cancels out 1-)
        b. NH_4Cl (one 1+ cancels out 1-)

3.51   a.   magnesium chloride
        b.   aluminum chloride

3.53   a.   nitrogen dioxide
        b.   sulfur trioxide

3.55   a. Al_2O_3
        b. Li_2S

3.57   a.   CO or CO_2
        b. SO_2 or SO_3

3.59   a. $NaNO_3$
        b. $Mg(NO_3)_2$

3.61   a. NH_4I
        b. $(NH_4)_2SO_4$

3.63   a.   copper(II) sulfide
        b.   copper(II) sulfate

3.65   a.   sodium hypochlorite
        b.   sodium chlorite

3.67   Ionic solid state compounds exist in regular, repeating, three-dimensional structures: the crystal lattice. The crystal lattice is made up of positive and negative ions. Solid state covalent compounds are made up of molecules, which may be arranged in a regular crystalline pattern or in an irregular (amorphous) structure.

3.69   The boiling points of ionic solids are generally much higher than those of covalent solids.

3.71   KCl would be expected to exist as a solid at room temperature; it is an ionic compound, and ionic compounds are characterized by high melting points.

3.73   Water will have a higher boiling point. Water is a polar molecule with strong intermolecular attractive forces, whereas carbon tetrachloride is a nonpolar molecule with weak intermolecular attractive forces. More energy, hence, a higher temperature is required to overcome the attractive forces among the water molecules.

3.75　a.　H·　　　　b.　He:　　　　c.　·C̈·　　　　d.　:N̈·

3.77　a.　Li$^+$　　　b.　Mg^{2+}　　　c.　[:C̈l̈:]$^-$　　　d.　[:P̈:]$^{3-}$

3.79　a.　NCl_3

:C̈l:N:C̈l:
　　:C̈l:

One nitrogen atom has 5 valence electrons, and three chlorine atoms contribute 21 valence electrons (3 x 7 valence electrons/atom of chlorine).

This produces a total of **26 valence electrons**.
This structure satisfies the octet rule for N and Cl.

b.　CH_3OH

　　H
　　¨　¨
H:C:O:H
　　¨　¨
　　H

One carbon atom has 4 valence electrons, one oxygen atom has 6 valence electrons, and four hydrogen atoms contribute 4 valence electrons (one for each hydrogen atom).

This produces a total of **14 valence electrons**.
This structure satisfies the octet rule for C and O.

c.　CS_2

:S::C::S:

One carbon atom contributes 4 valence electrons, and two sulfur atoms contribute 12 valence electrons (six from each sulfur atom) for a total of **16 valence electrons**.
This structure satisfies the octet rule for C and S.

3.81 a.   b.   c.

pyramidal,   tetrahedral around C,   linear
polar   angular around O   nonpolar
water soluble   polar   not water soluble
                            water soluble

3.83   Resonance can occur when more than one valid Lewis structure can be written for a
        molecule. Each individual structure that can be drawn is a resonance form. The true
        nature of the structure for the molecule is the resonance hybrid, which consists of the
        "average" of the resonance forms.

3.85

$$
\left[\; \underset{H}{\overset{H}{\text{H}}} \; \text{C} \; \text{C} \; \text{O} \; \right]^{-} \quad \longleftrightarrow \quad \left[\; \overset{H}{\text{H}} \; \text{C} \; \text{C} \; \text{O} \; \right]^{-}
$$

3.87

$$
\underset{\underset{H}{\overset{H}{|}}}{\text{H : C : C : O : H}}
$$

3.89

$$
\text{H : C : C : C : H}
$$

3.91   a. C and N   Polar covalent; N is more electronegative than C, resulting in an
                         unequal sharing of electrons in the bond.

       b. Si and P   Polar covalent; P is more electronegative than Si, resulting in the
                         unequal sharing of electrons in the bond.

       c. Na and Cl   Ionic; the bond is between a metal and a nonmetal.

d. Na and O   Ionic; the bond is between a metal and a nonmetal.

e. Ca and Br   Ionic; the bond is between a metal and a nonmetal.

3.93 a. C and N $[:C{\equiv}N:]^-$

b. Si and P (analogous to CN^-) $[:Si{\equiv}P:]^-$

c, d, and e are ionic compounds.

3.95 A molecule containing no polar bonds *must* be nonpolar. A molecule containing polar bonds may or may not itself be polar. It depends upon the number and arrangement of the bonds. For example:

- If a molecule contains only one bond, and that bond is polar, the molecule must be polar.
- If the molecule contains more than one polar bond, the molecule will be nonpolar if the arrangement of the bonds causes their effects to cancel. If not, the molecule will be polar.
- If lone pairs of electrons are present, their effect must be considered as well.
- If a molecule contains no polar bonds, it cannot be a polar molecule.

3.97 Polar compounds have strong intermolecular attractive forces. Higher temperatures are needed to overcome these forces and convert the solid to a liquid; hence, we predict higher melting points for polar compounds when compared to nonpolar compounds.

3.99 Yes. Many ionic compounds are water solube.

Chapter 4
Calculations and the Chemical Equation
Solutions to the Odd-Numbered Problems

In-Chapter Questions and Problems

4.1 $\dfrac{26.98 \text{ amu Al}}{1 \text{ atom Al}} \times \dfrac{1.661 \times 10^{-24} \text{ g Al}}{1 \text{ amu Al}} \times \dfrac{6.022 \times 10^{23} \text{ atoms Al}}{1 \text{ mol Al}} = 26.98 \dfrac{\text{g Al}}{\text{mol Al}}$

4.3 a. O atoms $= 2.50 \text{ mol O} \times \dfrac{6.022 \times 10^{23} \text{ atoms O}}{1 \text{ mol O}}$

O atoms $= 1.51 \times 10^{24}$ oxygen atoms

b. O atoms $= 2.50 \text{ mol O}_2 \times \dfrac{6.022 \times 10^{23} \text{ molecules O}_2}{1 \text{ mol O}_2} \times \dfrac{2 \text{ atoms O}}{1 \text{ molecule O}_2}$

O atoms $= 3.01 \times 10^{24}$ oxygen atoms

4.5 g He $= 3.50 \text{ mol He} \times \dfrac{4.00 \text{ g He}}{1 \text{ mol He}} = 14.0$ g He

4.7 a. $\underline{\begin{array}{lll} 3 \text{ atoms of hydrogen} \times & 1.01 \text{ amu/atom} = & 3.03 \text{ amu} \\ 1 \text{ atom of nitrogen} & \times\ 14.01 \text{ amu/atom} = & 14.01 \text{ amu} \end{array}}$
$ 17.04 \text{ amu}$

The mass of a single unit of NH_3 is 17.04 amu/formula unit. Therefore, the mass of 1 mole of formula units is 17.04 grams or 17.04 g/mol.

b. $\underline{\begin{array}{lll} 6 \text{ atoms of carbon} \times & 12.01 \text{ amu/atom} = & 72.06 \text{ amu} \\ 12 \text{ atoms of hydrogen} \times & 1.01 \text{ amu/atom} = & 12.12 \text{ amu} \\ 6 \text{ atoms of oxygen} & \times\ 16.00 \text{ amu/atoms} = & 96.00 \text{ amu} \end{array}}$
$ 180.18 \text{ amu}$

The mass of a single unit of $C_6H_{12}O_6$ is 180.18 amu/formula unit. Therefore, the mass of 1 mole of formula units is 180.18 grams or 180.18 g/mol.

c. $\underline{\begin{array}{lll} 1 \text{ atom of cobalt} & \times & 58.93 \text{ amu/atom} = & 58.93 \text{ amu} \\ 2 \text{ atoms of chlorine} \times & 35.45 \text{ amu/atom} = & 70.90 \text{ amu} \\ 12 \text{ atoms of hydrogen} \times\ 1.01 \text{ amu/atom} = & 12.12 \text{ amu} \\ 6 \text{ atoms of oxygen} \times & 16.00 \text{ amu/atom} = & 96.00 \text{ amu} \end{array}}$
$ 237.95 \text{ amu}$

The mass of a single unit of $CoCl_2 \cdot 6H_2O$ is 237.95 amu/formula unit. Therefore, the mass of 1 mole of formula units is 237.95 grams or 237.95 g/mol.

4.9   a.   DR
      b.   SR
      c.   DR
      d.   D

4.11   a. $KCl(aq) + AgNO_3(aq) \rightarrow KNO_3(aq) + AgCl(s)$

The solubility rules predict that silver chloride is insoluble; a precipitation reaction occurs.

      b. $CH_3COOK(aq) + AgNO_3(aq) \rightarrow$ no reaction

The solubility rules predict that both potential products, potassium nitrate (KNO_3) and silver acetate (CH_3COOAg), are soluble; no precipitation reaction occurs.

4.13   a. $4Fe(s) + 3O_2(g) \rightarrow 2Fe_2O_3(s)$

      b. $2C_6H_6(l) + 15O_2(g) \rightarrow 12CO_2(g) + 6H_2O(g)$

4.15   a. $5.00 \text{ mol } H_2 \times \dfrac{18.02 \text{ g } H_2O}{1 \text{ mol } H_2O} = 90.1 \text{ g } H_2O$

      b. $25.0 \text{ g LiCl} \times \dfrac{1 \text{ mol LiCl}}{42.39 \text{ g LiCl}} = 0.590 \text{ mol LiCl}$

4.17   a. $1 \text{ mol } C_2H_5OH \times \dfrac{3 \text{ mol } O_2}{1 \text{ mol } C_2H_5OH} = 3 \text{ mol } O_2$

      b. $1 \text{ mol } C_2H_5OH \times \dfrac{3 \text{ mol } O_2}{1 \text{ mol } C_2H_5OH} \times \dfrac{32 \text{ g } O_2}{1 \text{ mol } O_2} = 96.00 \text{ g } O_2$

4.19   Iron(III) oxide is Fe_2O_3

      a. $4Fe(s) + 3O_2(g) \rightarrow 2Fe_2O_3(s)$

      b. $5.00 \text{ g } Fe_2O_3 \times \dfrac{1 \text{ mol } Fe_2O_3}{159.7 \text{ g } Fe_2O_3} \times \dfrac{4 \text{ mol Fe}}{2 \text{ mol } Fe_2O_3} \times \dfrac{55.85 \text{ g Fe}}{1 \text{ mol Fe}} = 3.50 \text{ g Fe}$

4.21   a.   Step 1. Write down information about the reaction:

$$Sn(s) + 2 HF(aq) \rightarrow SnF_2(s) + H_2(g)$$

100.0 g   (excess)

Step 2. Convert the mass of Sn to moles of Sn:

$$100.0 \text{ g Sn x } \frac{1 \text{ mol Sn}}{118.69 \text{ g Sn}} = 0.8425 \text{ mol Sn}$$

Step 3. The reaction states that one mole of Sn will react to form one mole of SnF_2, so the mole ratio is 1:1. Use this conversion factor to calculate the mass of product:

$$0.8425 \text{ mol Sn x } \frac{1 \text{ mol SnF}_2}{1 \text{ mol Sn}} \text{ x } \frac{156.69 \text{ g SnF}_2}{1 \text{ mol SnF}_2} = 132.0 \text{ g SnF}_2$$

b. $$\% \text{ yield} = \frac{\text{actual yield}}{\text{theoretical yield}} \text{ x } 100 \%$$

$$\% \text{ yield} = \frac{5.00 \text{ g}}{132.00 \text{ g}} \text{ x } 100 \% = 3.79 \% \text{ yield}$$

End-of-Chapter Questions and Problems

4.23   Examples of other packaging units include a *ream* of paper (500 sheets of paper), a six-pack of soft drinks, a case of canned goods (24 cans), to name a few.

4.25   a. The average molar mass of Si (silicon) is 28.09 g.
        b. The average molar mass of Ag (silver) is 107.9 g.

4.27   The mass of Avogadro's number of argon atoms is the same as the average molar mass, 39.95 g.

4.29

$$2.00 \text{ mol Ne} \quad \text{x} \quad \frac{20.18 \text{ g Ne}}{1 \text{ mol Ne}} \quad = \quad 40.36 \text{ g Ne}$$

4.31 $$\frac{1.66 \text{ x } 10^{-24} \text{ g He}}{1 \text{ amu}} \text{ x } \frac{4.00 \text{ amu}}{1 \text{ atom He}} \text{ x } \frac{6.022 \text{ x } 10^{23} \text{ atoms He}}{1 \text{ mol He}} = 4.00 \frac{\text{g He}}{\text{mol He}}$$

4.33   a. $$20.0 \text{ g He x } \frac{1 \text{ mol He}}{4.00 \text{ g He}} = 5.00 \text{ mol He}$$

        b. $$0.040 \text{ kg Na x } \frac{10^3 \text{ g Na}}{1 \text{ kg Na}} \text{ x } \frac{1 \text{ mol Na}}{22.99 \text{ g Na}} = 1.7 \text{ mol Na}$$

c. $3.0 \text{ g Cl}_2 \times \dfrac{1 \text{ mol Cl}_2}{70.90 \text{ g Cl}_2} = 4.2 \times 10^{-2} \text{ mol Cl}_2$

4.35   The mass of 1.00 mol of silver is 107.87 g. Therefore, the mass of 15.0 mol of silver is:

$$15.0 \text{ mol Ag} \times \dfrac{107.87 \text{ g Ag}}{1 \text{ mol Ag}} = 1.62 \times 10^3 \text{ g Ag}$$

4.37   A molecule is a single unit made up of atoms joined by covalent bonds. An ion-pair is composed of positive and negatively charged ions joined by electrostatic attraction, the ionic bond. The ion pairs, unlike the molecule, do not form single units; the electrostatic charge is directed to other ions in a crystal lattice, as well.

4.39   a. $1 \text{ atom Na} \times \dfrac{22.99 \text{ amu Na}}{1 \text{ atom Na}} = 22.99 \text{ amu Na}$

$1 \text{ atom Cl} \times \dfrac{35.45 \text{ amu Cl}}{1 \text{ atom Cl}} = \underline{35.45 \text{ amu Cl}}$

58.44 amu NaCl

The mass of a single unit of NaCl is 58.44 amu/formula unit. Therefore, the mass of a mole of NaCl formula units is 58.44 g/mol.

b. $2 \text{ atoms Na} \times \dfrac{22.99 \text{ amu Na}}{1 \text{ atom Na}} = 45.98 \text{ amu Na}$

$1 \text{ atom S} \times \dfrac{32.06 \text{ amu S}}{1 \text{ atom S}} = 32.06 \text{ amu S}$

$4 \text{ atoms O} \times \dfrac{16.00 \text{ amu O}}{1 \text{ atom O}} = \underline{64.00 \text{ amu O}}$

$142.04 \text{ amu Na}_2\text{SO}_4$

The mass of a single unit of Na_2SO_4 is 142.04 amu/formula unit. Therefore, the mass of a mole of Na_2SO_4 formula units is 142.04 g/mol.

c. $3 \text{ atoms Fe} \times \dfrac{55.85 \text{ amu Fe}}{1 \text{ atom Fe}} = 167.55 \text{ amu Fe}$

$2 \text{ atoms P} \times \dfrac{33.97 \text{ amu P}}{1 \text{ atom P}} = 61.94 \text{ amu P}$

$8 \text{ atoms O} \times \dfrac{16.00 \text{ amu O}}{1 \text{ atom O}} = \underline{128.00 \text{ amu O}}$

$357.49 \text{ amu Fe}_3(\text{PO}_4)_2$

The mass of a single unit of $Fe_3(PO_4)_2$ is 357.49 amu/formula unit. Therefore, the mass of a mole of $Fe_3(PO_4)_2$ formula units is 357.49 g/mol.

4.41 $\dfrac{2 \text{ mol O}}{1 \text{ mol O}_2} \times \dfrac{16.00 \text{ g O}}{1 \text{ mol O}} = 32.00 \text{ g/mol O}_2$

4.43

1 atom Cu $\times \dfrac{63.55 \text{ amu Cu}}{1 \text{ atom Cu}} = 63.55 \text{ amu Cu}$

1 atom S $\times \dfrac{32.07 \text{ amu S}}{1 \text{ atom S}} = 32.07 \text{ amu S}$

9 atom O $\times \dfrac{16.00 \text{ amu O}}{1 \text{ atom O}} = 144.00 \text{ amu O}$

10 atom H $\times \dfrac{1.008 \text{ amu H}}{1 \text{ atom H}} = 10.08 \text{ amu H}$

$\overline{\hspace{3cm}}$

249.70 amu $CuSO_4 \bullet 5 \text{ } H_2O$

The molar mass of $CuSO_4 \bullet 5 \text{ } H_2O$ is 249.70 grams.

4.45 a. The formula weight of NaCl is 58.44 g/mol.

15.0 g NaCl $\times \dfrac{1 \text{ mol NaCl}}{58.44 \text{ g NaCl}} = 0.257 \text{ mol NaCl}$

b. The formula weight of Na_2SO_4 is 142.04 g/mol.

15.0 g Na_2O_4 $\times \dfrac{1 \text{ mol Na}_2SO_4}{142.04 \text{ g Na}_2SO_4} = 0.106 \text{ mol Na}_2SO_4$

4.47 a. The formula weight of H_2O is 18.02 g/mol.

1.000 mol H_2O $\times \dfrac{18.02 \text{ g H}_2O}{1 \text{ mol H}_2O} = 18.02 \text{ g H}_2O$

b. The formula weight of NaCl is 58.44 g/mol.

2.000 mol NaCl $\times \dfrac{58.44 \text{ g NaCl}}{1 \text{ mol NaCl}} = 116.9 \text{ g NaCl}$

4.49   a.   The formula weight of He is 4.00 g/mol.

$$10.0 \text{ mol He} \times \frac{4.00 \text{ g He}}{1 \text{ mol He}} = 40.0 \text{ g He}$$

b.   The formula weight of H_2 = 2.016 g/mol.

$$1.00 \times 10^2 \text{ mol } H_2 \times \frac{2.016 \text{ g } H_2}{1 \text{ mol } H_2} = 2.02 \times 10^2 \text{g H}$$

4.51   a.   The formula weight of Mg is 24.31 g/mol.

$$0.100 \text{ mol Mg} \times \frac{24.31 \text{ g Mg}}{1 \text{ mol Mg}} = 2.43 \text{ g Mg}$$

b.   The formula weight of $CaCO_3$ is 100.09 g/mol.

$$0.100 \text{ mol } CaCO_3 \times \frac{100.09 \text{ g } CaCO_3}{1 \text{ mol } CaCO_3} = 10.0 \text{ g } CaCO_3$$

4.53   a.   The formula weight of NaOH is 40.00 g/mol.

$$0.100 \text{ mol NaOH} \times \frac{40.00 \text{ g NaOH}}{1 \text{ mol NaOH}} = 4.00 \text{ g NaOH}$$

b.   The formula weight of H_2SO_4 is 98.08 g/mol.

$$0.100 \text{ mol } H_2SO_4 \times \frac{98.08 \text{ g } H_2SO_4}{1 \text{ mol } H_2SO_4} = 9.81 \text{ g } H_2SO_4$$

4.55   a.   The formula weight of KBr is 119.01 g/mol.

$$50.0 \text{ g KBr} \times \frac{119.01 \text{ g KBr}}{1 \text{ mol KBr}} = 0.420 \text{ mol KBr}$$

b.   The formula weight of $MgSO_4$ is 120.37 g/mol.

$$50.0 \text{g } MgSO_4 \times \frac{120.37 \text{ g } MgSO_4}{1 \text{ mol } MgSO_4} = 0.415 \text{ mol } MgSO_4$$

4.57 a. The formula weight of CS_2 is 76.13 g/mol.

$$50.0 \text{ g } CS_2 \times \frac{1 \text{ mol } CS_2}{76.13 \text{ g } CS_2} = 6.57 \times 10^{-1} \text{ mol } CS_2$$

b. The formula weight of $Al_2(CO_3)_3$ is 233.99 g/mol.

$$50.0 \text{ g } Al_2(CO_3)_3 \times \frac{1 \text{ mol } Al_2(CO_3)_3}{233.99 \text{ g } Al_2(CO_3)_3} = 2.14 \times 10^{-1} \text{ } Al_2(CO_3)_3$$

4.59 The ultimate basis for a correct chemical equation is the law of conservation of mass. No mass may be gained or lost in a chemical reaction, and the chemical equation must reflect this fact.

4.61 The subscript tells us the number of atoms or ions contained in one unit of the compound.

4.63 a. Heating an alkaline earth metal carbonate, for example:

$$MgCO_3 \text{ (s)} \xrightarrow{\Delta} MgO \text{ (s)} + CO_2 \text{ (g)}$$

b. The replacement of copper by zinc in copper sulfate,

$$Zn(s) + CuSO_4(aq) \rightarrow ZnSO_4(aq) + Cu(s)$$

4.65 Reaction of two soluble substances to form an insoluble product. For example:

$$2NaOH(aq) + FeCl_2(aq) \rightarrow Fe(OH)_2(s) + 2NaCl(aq)$$

4.67 The symbol Δ over the reaction arrow means that heat is necessary for the reaction to occur.

4.69 If we change the subscript, we change the identity of the compound.

4.71 A reactant is the starting material for a chemical reaction.

4.73 A product is the chemical species that results from a chemical reaction.

4.75 a. $2C_2H_6(g) + 7O_2(g) \rightarrow 4CO_2(g) + 6H_2O(g)$

b. $6K_2O(s) + P_4O_{10}(s) \rightarrow 4K_3PO_4(s)$

c. $MgBr_2(aq) + H_2SO_4(aq) \rightarrow 2HBr(g) + MgSO_4(aq)$

4.77 a. $Ca(s) + F_2(g) \rightarrow CaF_2(s)$

b. $2Mg(s) + O_2(g) \rightarrow 2MgO(s)$

c. $3H_2(g) + N_2(g) \rightarrow 2NH_3(g)$

4.79   a. $2C_4H_{10}(g) + 13O_2(g) \rightarrow 10H_2O(g) + 8CO_2(g)$

b. $Au_2S_3(s) + 3H_2(g) \rightarrow 2Au(s) + 3H_2S(g)$

c. $Al(OH)_3(s) + 3HCl(aq) \rightarrow AlCl_3(aq) + 3H_2O(l)$

d. $(NH_4)_2Cr_2O_7(s) \rightarrow Cr_2O_3(s) + N_2(g) + 4H_2O(g)$

e. $C_2H_5OH(l) + 3O_2(g) \rightarrow 2CO_2(g) + 3H_2O(g)$

4.81   a. $N_2(g) + 3H_2(g) \rightarrow 2NH_3(g)$

b. $HCl(aq) + NaOH(aq) \rightarrow NaCl(aq) + H_2O(l)$

4.83   a. $C_6H_{12}O_6(s) + 6O_2(g) \rightarrow 6H_2O(l) + 6CO_2(g)$

b. $Na_2CO_3(s) \xrightarrow{\Delta} Na_2O(s) + CO_2(g)$

4.85   The formula weight of B_2O_3 is 69.62 g/mol, and the formula weight of B_2H_6 is 27.67 g/mol.

$$20 \text{ g } B_2H_6 \times \frac{1 \text{ mol } B_2H_6}{27.67 \text{ g } B_2H_6} \times \frac{1 \text{ mol } B_2O_3}{1 \text{ mol } B_2H_6} \times \frac{69.62 \text{ g } B_2O_3}{1 \text{ mol } B_2O_3} = 50.3 \text{ g } B_2O_3$$

4.87   The formula weight of $CrCl_3$ is 158.35 g/mol, and the formula weight of Cr_2O_3 is 151.99 g/mol.

$$50.0 \text{ g } Cr_2O_3 \times \frac{1 \text{ mol } Cr_2O_3}{151.99 \text{ g } Cr_2O_3} \times \frac{2 \text{ mol } CrCl_3}{1 \text{ mol } Cr_2O_3} \times \frac{158.35 \text{ g } CrCl_3}{1 \text{ mol } CrCl_3} = 104 \text{ g } CrCl_3$$

4.89   a. $N_2(g) + 3H_2(g) \rightarrow 2NH_3(g)$

b. Three moles of H_2 will react with one mole of N_2, according to the coefficients in the balanced equation.

c. One mole of N_2 will produce two moles of the product NH_3, according to the coefficients in the balanced equation.

d. $140.0 \text{ g N}_2 \times \dfrac{1 \text{ mol N}_2}{28.02 \text{ g N}_2} \times \dfrac{3 \text{ mol H}_2}{1 \text{ mol N}_2} = 1.50 \text{ mol H}_2$

e. $1.50 \text{ mol H}_2 \times \dfrac{2 \text{ mol NH}_3}{3 \text{ mol H}_2} \times \dfrac{17.03 \text{ g NH}_3}{1 \text{ mol NH}_3} = 17.0 \text{ g NH}_3$

4.91 a.
$\begin{array}{lll}
5 \text{ C atoms} \times 12.01 \text{ amu/atom C} = & 60.05 \text{ amu} \\
11 \text{ H atoms} \times 1.008 \text{ amu/atom H} = & 11.09 \text{ amu} \\
1 \text{ N atom} \times 14.01 \text{ amu/atom N} = & 14.01 \text{ amu} \\
2 \text{ O atoms} \times 16.00 \text{ amu/atom O} = & 32.00 \text{ amu} \\
\underline{1 \text{ S atom} \times 32.06 \text{ amu/atom S} =} & \underline{32.06 \text{ amu}} \\
& 149.21 \text{ amu}
\end{array}$

The mass of a single unit of $C_5H_{11}NO_2S$ is 149.21 amu/formula unit. Therefore, the mass of a mole of $C_5H_{11}NO_2S$ formula units is 149.21 g /mol.

b. $1 \text{ mol C}_5\text{H}_{11}\text{NO}_2\text{S} \times \dfrac{2 \text{ mol O atoms}}{1 \text{ mol C}_5\text{H}_{11}\text{NO}_2\text{S}} \times \dfrac{6.02 \times 10^{23} \text{ O atoms}}{1 \text{ mol O atoms}}$

$= 1.20 \times 10^{24} \text{ O atoms}$

c. $1 \text{ mol C}_5\text{H}_{11}\text{NO}_2\text{S} \times \dfrac{2 \text{ mol O atoms}}{1 \text{ mol C}_5\text{H}_{11}\text{NO}_2\text{S}} \times \dfrac{16.00 \text{ g O}}{1 \text{ mol O atoms}} = 32.00 \text{ g O}$

d. $50.0 \text{ g C}_5\text{H}_{11}\text{NO}_2\text{S} \times \dfrac{1 \text{ mol C}_5\text{H}_{11}\text{NO}_2\text{S}}{149.21 \text{ g C}_5\text{H}_{11}\text{NO}_2\text{S}} \times \dfrac{2 \text{ mol O atoms}}{1 \text{ mol C}_5\text{H}_{11}\text{NO}_2\text{S}} \times$

$\dfrac{16.00 \text{ g O}}{1 \text{ mol O atoms}} = 10.7 \text{ g O}$

4.93 The formula weight of HgO is 216.59 g/mol.

$1.00 \times 10^2 \text{ g HgO} \times \dfrac{1 \text{ mol HgO}}{216.59 \text{ g HgO}} \times \dfrac{1 \text{ mol O}_2}{2 \text{ mol HgO}} \times \dfrac{32.00 \text{ g O}_2}{1 \text{ mol O}_2} = 7.39 \text{ g O}_2$

4.95 The balanced equation is:

$2C_2H_2(g) + 5O_2(g) \rightarrow 4CO_2(g) + 2H_2O(g)$

The formula weight of C_2H_2 is 26.04 g/mol.

$20.0 \text{ kg C}_2\text{H}_2 \times \dfrac{10^3 \text{ g C}_2\text{H}_2}{1 \text{ kg C}_2\text{H}_2} \times \dfrac{1 \text{ mol C}_2\text{H}_2}{26.04 \text{ g C}_2\text{H}_2} \times \dfrac{5 \text{ mol O}_2}{2 \text{ mol C}_2\text{H}_2} \times \dfrac{32.00 \text{ g O}_2}{1 \text{ mol O}_2}$

$$= 6.14 \times 10^4 \text{ g O}_2$$

4.97   Step 1. Write down information about the reaction:

$$C_{10}H_{20}(l) \ + \ H_2(g) \ \rightarrow \ C_{10}H_{22}(s)$$
(excess)   1.00 g

Step 2. Convert the mass of hydrogen to moles of hydrogen:

$$1.00 \text{ g H}_2 \times \frac{1 \text{ mol H}_2}{2.016 \text{ g H}_2} = 0.496 \text{ mol H}_2$$

Step 3. The reaction states that decene and hydrogen react in a 1:1 mole ratio. Use this conversion factor to calculate the mass of product:

$$0.496 \text{ mol H}_2 \times \frac{1 \text{ mol C}_{10}H_{22}}{1 \text{ mol H}_2} \times \frac{143 \text{ g C}_{10}H_{22}}{1 \text{ mol C}_{10}H_{22}} = 70.6 \text{ g C}_{10}H_{22}$$

4.99   Step 1. Write down information about the reaction:

$$N_2O_4(l) \ + \ 2 \text{ N}_2H_4(l) \ \rightarrow \ 3 \text{ N}_2(g) \ + \ 4 \text{ H}_2O(g)$$

1.00 kg   (excess)

Step 2. Convert the mass of N_2O_4 to moles of N_2O_4:

$$1.00 \text{ kg} \times \frac{10^3 \text{ g N}_2O_4}{1 \text{ kg N}_2O_4} \times \frac{1 \text{ mol N}_2O_4}{92.02 \text{ g N}_2O_4} = 10.9 \text{ mol N}_2O_4$$

Step 3. The reaction states that the ratio of moles of N_2O_4 to N_2 is 1:3. Use this conversion factor to calculate the mass of N_2:

$$10.9 \text{ mol N}_2O_4 \times \frac{2 \text{ mol N}_2}{1 \text{ mol N}_2O_4} \times \frac{28.02 \text{ g N}_2}{1 \text{ mol N}_2} = 9.13 \times 10^2 \text{ g N}_2$$

4.101   In problem 4.97, we found that the theoretical yield of decane was 70.6 g. Since only 65.4 g were obtained, the percent yield would be the fractional yield multiplied by 100%, or:

$$\frac{65.4 \text{ g}}{70.6 \text{ g}} \times 100 \% = 92.6 \%$$

4.103   In problem 4.99, we found that the theoretical yield of nitrogen was 9.13×10^2 g. Since only 75.0% was actually obtained, corresponding to a decimal fraction of 0.750,

$$9.13 \times 10^2 \text{ g N}_2 \times 0.750 = 6.85 \times 10^2 \text{ g N}_2$$

Chapter 5
States of Matter:
Gases, Liquids, and Solids
Solutions to the Odd-Numbered Problems

In-Chapter Questions and Problems

5.1 a. $725 \text{ mm Hg} \times \dfrac{1 \text{ atm}}{760 \text{ mm Hg}} = 0.954 \text{ atm}$

b. $29.0 \text{ cm Hg} \times \dfrac{10 \text{ mm Hg}}{1 \text{ cm Hg}} \times \dfrac{1 \text{ atm}}{760 \text{ mm Hg}} = 0.382 \text{ atm}$

c. $555 \text{ torr} \times \dfrac{1 \text{ atm}}{760 \text{ torr}} = 0.730 \text{ atm}$

5.3 a. $P_i V_i = P_f V_f$

$P_i = \dfrac{P_f V_f}{V_i}$

$P_i = \dfrac{(5.0 \text{ atm})(7.5 \text{ L})}{1.0 \text{ L}} = 37.5 \text{ atm} \approx 38 \text{ atm} \ (2 \text{ significant figures})$

b. $P_f = \dfrac{P_i V_i}{V_f}$

$P_f = \dfrac{(5.0 \text{ atm})(1.0 \text{ L})}{0.20 \text{ L}} = 25 \text{ atm}$

5.5 Initial temperature: $25°C + 273 = 298 \text{ K}$

a. $100°C + 273 = 373 \text{ K}$

$\dfrac{V_i}{T_i} = \dfrac{V_f}{T_f}$

$V_f = \dfrac{V_i T_f}{T_i}$

$$V_f = \frac{(3.00\ L)(373\ K)}{298\ K} = 3.76\ L$$

b. $°C = \dfrac{5}{9}(°F - 32) = \dfrac{5}{9}(150 - 32) = 66°C$

$K = °C + 273 = 66 + 273 = 339\ K$

$$\frac{V_i}{T_i} = \frac{V_f}{T_f}$$

$$V_f = \frac{V_i T_f}{T_i}$$

$$V_f = \frac{(3.00\ L)(339\ K)}{298\ K} = 3.41\ L$$

c. $\dfrac{V_i}{T_i} = \dfrac{V_f}{T_f}$

$$V_f = \frac{V_i T_f}{T_i}$$

$$V_f = \frac{(3.00\ L)(273\ K)}{298\ K} = 2.75\ L$$

5.7 $\dfrac{P_i V_i}{T_i} = \dfrac{P_f V_f}{T_f}$

$$P_f = = \frac{P_i V_i T_f}{V_f T_i}$$

$P_i = 760\ torr \times \dfrac{1\ atm}{760\ torr} = 1.00\ atm$

$V_i = 2.00\ L$

$T_f = T_i = 25.0°C + 273 = 298\ K$

$V_f = 10.0\ L$

$$P_f = \; = \; \frac{(1.00 \text{ atm})(2.00 \text{ L})(298 \text{ K})}{(10.0 \text{ L})(298 \text{ K})} = 0.200 \text{ atm}$$

5.9

$$\frac{V_i}{n_i} = \frac{V_f}{n_f}$$

$$n_f = \frac{n_i V_f}{V_i}$$

$$n_f = \frac{(1.00 \text{ mol H}_2)(100.0 \text{ L})}{22.4 \text{ L}} = 4.46 \text{ mol H}_2$$

5.11 $PV - nRT$

Solving for volume, V,

$$V = \frac{nRT}{P}$$

$$n = 10.0 \text{ g N}_2 \times \frac{1 \text{ mol N}_2}{28.00 \text{ g N}_2} = 0.357 \text{ mol N}_2$$

$$R = 0.0821 \; \frac{L \cdot atm}{K \cdot mol}$$

$$T = 30°C + 273 = 303 \text{ K}$$

$$P = 750 \text{ torr} \times \frac{1 \text{ atm}}{760 \text{ torr}} = 0.987 \text{ atm}$$

Substituting,

$$V = \frac{(0.357 \text{ mol N}_2)(0.0821 \text{ L} \cdot atm/K \cdot mol)(303 \text{ K})}{0.987 \text{ atm}}$$

$$V = 9.00 \text{ L}$$

5.13 $PV = nRT$

Solving for number of moles of N_2, n,

$$n = \frac{PV}{RT}$$

$$P = 1.00 \text{ atm} \qquad V = 5.00 \text{ L} \qquad R = 0.0821 \, \frac{\text{L} \cdot \text{atm}}{\text{K} \cdot \text{mol}} \qquad T = 273 \text{ K}$$

Substituting,

$$n = \frac{(1.00 \text{ atm})(5.00 \text{ L})}{(0.0821 \text{ L} \cdot \text{atm/K} \cdot \text{mol})(273 \text{ K})} = 0.223 \text{ mol N}_2$$

End-of-Chapter Questions and Problems

5.15   In all cases, gas particles are much farther apart than similar particles in the liquid or solid state. In most cases, particles in the liquid state are, on average, farther apart that those in the solid state. Water is the exception; liquid water's molecules are closer together than they are in the solid state.

5.17   Pressure is a force/unit area. Gas particles are in continuous, random motion. Collisions with the walls of the container result in a force (mass x acceleration) on the walls of the container. The sum of these collisional forces constitutes the pressure exerted by the gas.

5.19   Gases are easily compressed simply because there is a great deal of space between particles; they can be pushed closer together (compressed) because the space is available.

5.21   Gas particles are in continuous, random motion. They are free (minimal attractive forces between particles) to roam, up to the boundary of their container.

5.23   Gases exhibit more ideal behavior at low pressures. At low pressures, gas particles are more widely separated and therefore the attractive forces between particles are less. The ideal gas model assumes negligible attractive forces between gas particles.

5.25   The kinetic molecular theory states that the average kinetic energy of the gas particles increases as the temperature increases. Kinetic energy is proportional to $(\text{velocity})^2$. Therefore, as the temperature increases, the gas particle velocity increases and the rate of mixing increases as well.

5.27   The volume of the balloon is directly proportional to the pressure exerted on its inner walls. As the contents of the balloon cool, the average kinetic energy of the particles decreases, the pressure drops, and the balloon contracts.

5.29   Boyle's law states that the volume of a gas varies inversely with the gas pressure if the temperature and the number of moles of gas are held constant.

5.31   Volume will decrease according to Boyle's law. Volume is inversely proportional to the pressure exerted on the gas.

5.33   A volume of 5 L (ordinate) corresponds to a pressure of 1 atm (abscissa).

5.35   A volume of 2 L (ordinate) corresponds to a pressure of 2.5 atm (abscissa).

   $PV = k$

   $(2.5 \text{ atm}) (2 \text{ L}) = 5 \text{ L-atm} = k$

5.37 $P_i = 1.00 \text{ atm}$ $P_f = ? \text{ atm}$

   $V_i = 20.9 \text{ L}$ $V_f = 4.00 \text{ L}$

   $P_i V_i = P_f V_f$

   $P_f = \dfrac{P_i V_i}{V_f}$

   $P_f = \dfrac{(1.00 \text{ atm})(20.9 \text{ L})}{4.00 \text{ L}}$

   $P_f = 5.23 \text{ atm}$

5.39   Charles's law states that the volume of a gas varies directly with the absolute temperature if pressure and number of moles of gas are constant.

5.41   The Kelvin scale is the only scale that is directly proportional to molecular motion, and it is the motion that determines the physical properties of gases.

5.43   No. The volume is proportional to the temperature in K, not Celsius.

5.45 $V_i = 2.00 \text{ L}$ $V_f = ? \text{ L}$

   $T_i = 250°C$ $T_f = 500°C$

   $\dfrac{V_i}{T_i} = \dfrac{V_f}{T_f}$

   $V_f = \dfrac{V_i T_f}{T_i}$

   $V_f = \dfrac{(2.00 \text{ L})(500°C + 273 \text{ K})}{(250°C + 273 \text{ K})}$

   $V_f = 2.96 \text{ L}$

   The change in volume, $\Delta V = V_f - V_i$

$$\Delta V = 2.96 \text{ L} - 2.00 \text{ L}$$

$$\Delta V = 0.96 \text{ L}$$

5.47 $V_i = 1.25 \text{ L}$ $V_f = ? \text{ L}$

   $T_i = 20°C$ $T_f = 80°C$

$$\frac{V_i}{T_i} = \frac{V_f}{T_f}$$

$$V_f = \frac{V_i T_f}{T_i}$$

$$V_f = \frac{(1.25 \text{ L})(80°C + 273 \text{ K})}{(20°C + 273 \text{ K})}$$

$$V_f = \frac{(1.25 \text{ L})(353 \text{ K})}{(293 \text{ K})}$$

$V_f = 1.51 \text{ L}$

5.49   Examine each effect separately:

- Volume and temperature are <u>directly</u> proportional; increasing T <u>increases</u> V.
- Volume and pressure are <u>inversely</u> proportional; decreasing P <u>increases</u> V.

Therefore, both variables work together to <u>increase</u> the volume.

5.51 $$\frac{P_i V_i}{T_i} = \frac{P_f V_f}{T_f}$$

   $$P_f V_f T_i = P_i V_i T_f$$

   $$V_f = = \frac{P_i V_i T_f}{P_f T_i}$$

5.53 $P_i = 1.00 \text{ atm}$ $P_f = 125 \text{ atm}$

   $V_i = 2.25 \text{ L}$ $V_f = ? \text{ L}$

   $T_i = 16°C$ $T_f = 20°C$

Using the equation derived in question 6.29,

$$V_f = \frac{P_i V_i T_f}{P_f T_i}$$

and substituting:

$$V_f = \frac{(1.00 \text{ atm})(2.25 \text{ L})(20°C + 273 \text{ K})}{(125 \text{ atm})(16°C + 273 \text{ K})}$$

$$V_f = \frac{(1.00 \text{ atm})(2.25 \text{ L})(293 \text{ K})}{(125 \text{ atm})(289 \text{ K})}$$

$$V_f = 1.82 \times 10^{-2} \text{ L}$$

5.55   Avogadro's law states that equal volumes of any ideal gas contain the same number of moles if measured at constant temperature and pressure.

5.57 $n_i = 1.00 \text{ g He} \times \dfrac{1 \text{ mol He}}{4.00 \text{ g He}} = 0.25 \text{ mol He}$

$V_i = 1.00 \text{ L}$

$n_f = 6.00 \text{ g He} \times \dfrac{1 \text{ mol He}}{4.00 \text{ g He}} = 1.50 \text{ mol He}$

$V_f = ? \text{ L}$

$$\frac{V_i}{n_i} = \frac{V_f}{n_f}$$

$$V_f = \frac{V_i n_f}{n_i}$$

$$V_f = \frac{(1.00 \text{ L})(1.50 \text{ mol He})}{(0.25 \text{ mol He})} = 6.00 \text{ L}$$

5.59   No.   One mole of an ideal gas will occupy exactly 22.4 L; however, there is no completely ideal gas and careful measurement will show a different volume.

5.61   Standard temperature is 273K.

5.63 $PV = nRT$

P = 5.0 atm
V = 4.0 L
T = 32°C + 273 = 305 K
R = 0.0821 L x atm/K x mol

$$n = \frac{PV}{RT} = \frac{(5.0\,atm)(4.0\,L)}{(0.0821\,L\,x\,atm/K\,x\,mol)(305K)} = 0.799\,mol \approx 0.80\,mol$$

5.65 $PV = nRT$

$$n = 44.0\,g\,CO_2\,x\,\frac{1\,mol}{44.0\,g\,CO_2} = 1.00\,mol\,CO_2$$

T = 273 K
P = 1.00 atm
R = 0.0821 L x atm/K x mol

$$V = \frac{nRT}{P} = \frac{(1.00\,mol)(0.0821\,L\,x\,atm/K\,x\,mol)(273K)}{(1.00\,atm)} = 22.4\,L$$

5.67 $PV = nRT$

$$P = 725\,mmHg\,x\,\frac{1\,atm}{760\,mmHg} = 0.954\,atm$$

V = 7.55 L
T = 45°C + 273 = 318 K
R = 0.0821 L x atm/K x mol

$$n = \frac{PV}{RT} = \frac{(0/945\,atm)(7.55\,L)}{(0.0821\,L\,x\,atm/K\,x\,mol)(318K)} = 0.276\,mol$$

5.69 $P_i = 750$ torr $P_f = 1.00$ atm
V$_i$ = 65.0 mL   V$_f$ = ? L
T$_i$ = 22˚C   T$_f$ = 273 K

$$\frac{P_iV_i}{T_i} = \frac{P_fV_f}{T_f}$$

$$V_f = \frac{P_iV_iT_f}{P_fT_i}$$

$$V_f = \frac{\left(750 \text{ torr x } \dfrac{1 \text{ atm}}{760 \text{ torr}}\right)\left(650 \text{ mL x } \dfrac{1 \text{ L}}{10^3 \text{ mL}}\right)(273 \text{ K})}{(1.00 \text{ atm})(22°C + 273 \text{ K})}$$

$$V_f = \frac{(0.987 \text{ atm})(6.5 \times 10^{-2} \text{ L})(273 \text{ K})}{(1.00 \text{ atm})(295 \text{ K})}$$

$$V_f = 5.94 \times 10^{-2} \text{ L}$$

5.71   The molar volume of any gas at STP is 22.4 L, as shown by the following calculation:

$$PV = nRT$$

$$V = \frac{nRT}{P}$$

$n = 1.00 \text{ mol}$ $\qquad$ $R = 0.0821 \dfrac{L \cdot atm}{K \cdot mol}$ $\qquad$ $T = 273 \text{ K}$ $\qquad$ $P = 1.00 \text{ atm}$

$$V = \frac{(1.00 \text{ mol})(0.0821 \text{ L} \cdot \text{atm/K} \cdot \text{mol})(273 \text{ K})}{1.00 \text{ atm}}$$

$$V = 22.4 \text{ L}$$

5.73 $PV = nRT$

$$V = \frac{nRT}{P}$$

$n = 4.00 \text{ mol}$ $\qquad$ $R = 0.0821 \dfrac{L \cdot atm}{K \cdot mol}$ $\qquad$ $T = 27°C + 273 = 300 \text{ K}$

$$P = 8.25 \text{ torr x } \frac{1 \text{ atm}}{760 \text{ torr}} = 1.09 \times 10^{-2} \text{ atm}$$

$$V = \frac{(4.00 \text{ mol})(0.0821 \text{ L} \cdot \text{atm/K} \cdot \text{mol})(300 \text{ K})}{1.09 \times 10^{-2} \text{ atm}}$$

$$V = 9.08 \times 10^3 \text{ L}$$

5.75 $PV = nRT$

$$T = \frac{PV}{nR}$$

$$n = 1.75 \text{ g O}_2 \times \frac{1 \text{ mol O}_2}{32.0 \text{ g O}_2} = 5.47 \times 10^{-2} \text{ mol O}_2$$

$$T = \frac{(1.00 \text{ atm})(2.00 \text{ L})}{(5.47 \times 10^{-2} \text{ mol})(0.0821 \text{ L} \cdot \text{atm/K} \cdot \text{mol})}$$

$$T = 445 \text{ K}$$

$$T = 445 \text{ K} - 273 = 172°\text{C}$$

5.77 Dalton's law states that the total pressure of a mixture of gases is the sum of the partial pressures of the component gases.

5.79 $P_T = P_{N_2} + P_{F_2} + P_{He}$

$P_T = 0.40 \text{ atm} + 0.16 \text{ atm} + 0.18 \text{ atm}$

$P_T = 0.74 \text{ atm}$

5.81 Intermolecular forces in liquids are considerably stronger than intermolecular forces in gases. Particles are, on average, much closer together in liquids and the strength of attraction is inversely proportional to the distance of separation.

5.83 The vapor pressure of a liquid increases as the temperature of the liquid increases.

5.85 Evaporation is the conversion of a liquid to a gas at a temperature lower than the boiling point of the liquid. Condensation is the conversion of a gas to a liquid at a temperature lower than the boiling point of the liquid.

5.87 Viscosity is the resistance to flow caused by intermolecular attractive forces. Complex molecules may become entangled and not slide smoothly across one another.

5.89 All molecules exhibit London forces.

5.91 Only methanol exhibits hydrogen bonding. Methanol has an oxygen atom bonded to a hydrogen atom, a necessary condition for hydrogen bonding.

5.93 Solids are essentially incompressible because the average distance of separation among particles in the solid state is small. There is literally no space for the particles to crowd closer together.

5.95 a. ionic solids - high melting temperature, brittle

b. covalent solids - high melting temperature, hard

5.97   Beryllium. Metallic solids are good electrical conductors. Carbon forms covalent solids that are poor electrical conductors.

5.99   Mercury. Mercury is a liquid at room temperature, whereas chromium is a solid at room temperature. Liquids have higher vapor pressures than solids.

Chapter 6
Reactions and Solutions
Solutions to the Odd-Numbered Problems

In-Chapter Questions and Problems

6.1 A chemical analysis must be performed in order to determine the identity of all components, a qualitative analysis. If only one component is found, it is a pure substance; two or more components indicate a true solution.

6.3 After the container of soft drink is opened, CO_2 diffuses into the surrounding atmosphere; consequently, the partial pressure of CO_2 over the soft drink decreases and the equilibrium $[CO_2 \text{ (g)} \leftrightarrow CO_2 \text{ (aq)}]$ shifts to the left, lowering the concentration of CO_2 in the soft drink.

6.5 $$\% \text{ (W/V)} = \frac{\text{grams of solute}}{\text{milliliters of solution}} \times 100\%$$

$$\% \text{ (W/V)} = \frac{10.0 \text{ g NaCl}}{\left(0.0600 \text{ L} \times \dfrac{10^3 \text{ mL}}{1 \text{ L}} \right)} \times 100\%$$

$$\% \text{ (W/V)} = 16.7 \text{ \% NaCl}$$

6.7 $$\% \text{ (W/V)} = \frac{\text{grams of solute}}{\text{milliliters of solution}} \times 100\%$$

$$\% \text{ (W/V)} = \frac{15.0 \text{ g KCl}}{\left(0.200 \text{ L} \times \dfrac{10^3 \text{ mL}}{1 \text{ L}} \right)} \times 100\%$$

$$\% \text{ (W/V)} = 7.50 \text{ \% KCl}$$

6.9 $$\% \text{ (W/V)} = \frac{\text{grams of solute}}{\text{milliliters of solution}} \times 100\%$$

$$\% \text{ (W/V)} = \frac{20.0 \text{ g O}_2}{\left(78.0 \text{ L} \times \dfrac{10^3 \text{ mL}}{1 \text{ L}} \right)} \times 100\%$$

$$\% \text{ (W/V)} = 2.56 \times 10^{-2} \text{ \% oxygen}$$

6.11 $\% \, (W/W) = \dfrac{\text{grams solute}}{\text{grams solution}} \times 100\%$

$\% \, (W/W) = \dfrac{20.0 \text{ g O}_2}{(20.0 \text{ g O}_2 \; + \; 80.0 \text{ g N}_2)} \times 100\%$

$\% \, (W/W) = 20.0 \, \%$ oxygen

6.13 $\text{ppm} = \dfrac{\text{grams solute}}{\text{grams solution}} \times 10^6 \text{ppm} = \dfrac{20.0 \text{ g O}_2}{20.0 \text{ g O}_2 + 80.0 \text{ g N}_2} \times 10^6 \text{ppm} = 2.00 \times 10^5 \text{ ppm}$

6.15 $M_{HCl} = \dfrac{\text{mol HCl}}{L_{\text{solution}}}$

Solving for mol HCl,

$\text{mol HCl} = (M_{HCl})(L_{\text{solution}})$

$\text{mol HCl} = (0.250 \text{ M})(5.00 \times 10^2 \text{ mL} \times \dfrac{1 \text{ L}}{10^3 \text{ mL}})$

$\text{mol HCl} = 0.125 \text{ mol HCl}$

6.17 $(M_1)(V_1) = (M_2)(V_2)$

$M_1 = 12.0 \text{ M}$

$M_2 = 2.0 \text{ M}$

$V_2 = 1.0 \times 10^2 \text{ mL} \times \dfrac{1 \text{ L}}{10^3 \text{ mL}} = 1.0 \times 10^{-1} \text{ L}$

Solving for the initial volume of 12 M HCl, V_1,

$\dfrac{(M_2)(V_2)}{(M_1)} = \dfrac{(2.0 \text{ M})(1.0 \times 10^{-1} \text{ L})}{(12 \text{ M})}$

$V_1 = 1.7 \times 10^{-2} \text{ L}$ (or 17 mL) of 12 M HCl

To prepare the solution, dilute 1.7×10^{-2} L of 12 M HCl with sufficient water to produce 1.0×10^2 mL of total solution.

6.19 Pure water has the higher freezing point. The presence of a solute decreases the freeaing point.

6.21 $NH_4NO_3 \xrightarrow{H_2O} NH_4^+ + NO_3^-$

$$\frac{2 \text{ mol particles}}{1 \text{ mol } NH_4NO_3} \times \frac{5.0 \times 10^{-3} \text{ mol } NH_4NO_3}{L \text{ solution}} = 1.0 \times 10^{-2} \frac{\text{mol particles}}{L \text{ solution}}$$

and $1.0 \times 10^{-2} \dfrac{\text{mol particles}}{L \text{ solution}} = 1.0 \times 10^{-2}$ osmol

6.23 In Question 6.21, we found that the solution was 1.0×10^{-2} osmol; substituting this value for M (M represents the osmolarity) in the expression, $\pi = MRT$,

$\pi = MRT$

$$\pi = 1.0 \times 10^{-2} \frac{\text{mol particles}}{L \text{ solution}} \times \frac{0.0821 \text{ L} \cdot \text{atm}}{K \cdot \text{mol}} \times (25 + 273)K$$

$\pi = 0.24$ atm

6.25 The Lewis structures of these two compounds are:

Since oxygen is more electronegative than carbon, all carbon-oxygen bonds are polar. However, only carbon monoxide has a dipole moment. Hence, only carbon monoxide is polar. Carbon dioxide is linear and symmetrical. Consequently, carbon dioxide is nonpolar. Polar carbon monoxide is more soluble in water.

6.27

$$\frac{1.54 \text{ meq } Na^+}{1 \text{ L}} \times \frac{1 \text{ eq } Na^+}{10^3 \text{ meq } Na^+} \times \frac{1 \text{ mol charge}}{1 \text{ eq } Na^+} \times \frac{1 \text{ mol } Na^+}{1 \text{ mol charge}} = 1.54 \times 10^{-2} \frac{\text{mol } Na^+}{L}$$

End-of-Chapter Questions and Problems

6.29 $\% \text{ (W/V)} = \dfrac{\text{grams of solute}}{\text{milliliters of solution}} \times 100\%$

a. $\% \text{ (W/V)} = \dfrac{20.0 \text{ g NaCl}}{\left(1.00 \text{ L soln} \times \dfrac{10^3 \text{ mL soln}}{1 \text{ L soln}}\right)} \times 100\%$

% (W/V) = 2.00 % NaCl

b. $\% \text{ (W/V)} = \dfrac{33.0 \text{ g } C_6H_{12}O_6}{5.00 \times 10^2 \text{ mL soln}} \times 100\%$

$\% \text{ (W/V)} = 6.60 \% \text{ } C_6H_{12}O_6$

6.31 $\% \text{ (W/V)} = \dfrac{\text{grams of solute}}{\text{milliliters of solution}} \times 100\%$

a. $\% \text{ (W/V)} = \dfrac{50.0 \text{ g ethanol}}{\left(1.00 \text{ L soln} \times \dfrac{10^3 \text{ mL soln}}{1 \text{ L soln}}\right)} \times 100\%$

$\% \text{ (W/V)} = 5.00 \% \text{ ethanol}$

b. $\% \text{ (W/V)} = \dfrac{50.0 \text{ g ethanol}}{5.00 \times 10^2 \text{ mL soln}} \times 100\%$

$\% \text{ (W/V)} = 10.0 \% \text{ ethanol}$

6.33 $\% \text{ (W/W)} = \dfrac{\text{grams solute}}{\text{grams solution}} \times 100\%$

a. $\% \text{ (W/W)} = \dfrac{21.0 \text{ g NaCl}}{1.00 \times 10^2 \text{ g soln}} \times 100\%$

$\% \text{ (W/W)} = 21.0 \% \text{ NaCl}$

b. Use the density of the sodium chloride solution as a conversion factor to calculate the volume of the sodium chloride solution.

$$5.00 \times 10^2 \text{ mL soln} \times \dfrac{1.12 \text{ g soln}}{1 \text{ mL soln}} = 5.60 \times 10^2 \text{ g soln}$$

Then,
$\% \text{ (W/W)} = \dfrac{21.0 \text{ g NaCl}}{5.60 \times 10^2 \text{ g soln}} \times 100\%$

$\% \text{ (W/W)} = 3.75 \% \text{ NaCl}$

6.35 $\% \text{ (W/W)} = \dfrac{\text{grams solute}}{\text{grams solution}} \times 100\%$

Solve for g solute,

$$g \text{ solute} = \frac{\% (W/W)(g \text{ soln})}{100\%}$$

a. $g \text{ NaCl} = \dfrac{[0.900\% \ (W/W)](2.50 \times 10^{2} g \text{ soln})}{100\%}$

   $g \text{ NaCl} = 2.25 \text{ g NaCl}$

b. Assume that the density of the solution is 1.00 g/mL; then

   $g \text{ CH}_3\text{COONa} = \dfrac{[1.25\% \ (W/W)](2.50 \times 10^{2} g \text{ soln})}{100\%}$

   $g \text{ CH}_3\text{COONa} = 3.13 \text{ g NaC}_2\text{H}_3\text{O}_2$

6.37 $\%(W/V) = \dfrac{\text{grams of solute}}{\text{milliliters of solution}} \times 100\%$

   $\%(W/V) = \dfrac{14.6 \text{ g KNO}_3}{75.0 \text{ mL of solution}} \times 100\%$

   $\%(W/V) = 19.47\% \text{ KNO}_3 \approx 19.5\% \text{ KNO}_3$

6.39 $\%(W/V) = \dfrac{\text{grams of solute}}{\text{milliliters of solution}} \times 100\%$

   $\text{Grams of solute} = \dfrac{\%(W/V)(\text{milliliters of solution})}{100\%}$

   $\text{Grams of sugar} = \dfrac{1.00 \ \%(W/V)(100 \text{ mL of solution})}{100\%}$

   $\text{Grams of sugar} = 1.00 \text{ g sugar}$

6.41   First, ensure that both concentration units are the same; convert %(W/W) to ppm:
   $100\%(W/W) = 10^{6} \text{ ppm}$

   $0.04\%(W/W) \times \dfrac{10^{6} \text{ ppm}}{100\%(W/W)} = 4 \times 10^{2} \text{ ppm}$

   $4 \times 10^{2} > 50$ ppm; Therefore, the 0.04 %(W/W) solution is more concentrated.

6.43 $1.0 \text{ mg Cu}^{2+} \times \dfrac{1 \text{ g}}{10^3 \text{ mg}} = 1.0 \times 10^{-3} \text{ g Cu}^{2+}$

$0.50 \text{ kg} \times \dfrac{10^3 \text{ g}}{1 \text{ kg}} = 5.0 \times 10^2 \text{ g solution}$

$\text{ppt Cu}^{2+} = \dfrac{\text{g Cu}^{2+}}{\text{g solution}} \times 10^3 \text{ ppt} = \dfrac{1.0 \times 10^{-3} \text{ g Cu}^{2+}}{5.0 \times 10^2 \text{ g solution}} \times 10^3 \text{ ppt}$

$\text{ppt Cu}^{2+} = 2.0 \times 10^{-3} \text{ ppt}$

6.45   a. $M_{\text{NaCl}} = \dfrac{\text{mol NaCl}}{\text{L solution}}$

$M_{\text{NaCl}} = \dfrac{\left(20.0 \text{ g NaCl} \times \dfrac{1 \text{ mol NaCl}}{58.44 \text{ g NaCl}}\right)}{1.00 \text{ L solution}}$

$M_{\text{NaCl}} = 0.342 \text{ M NaCl}$

b. $M_{C_6H_{12}O_6} = \dfrac{\text{mol } C_6H_{12}O_6}{\text{L solution}}$

$M_{C_6H_{12}O_6} = \dfrac{\left(33.0 \text{ g } C_6H_{12}O_6 \times \dfrac{1 \text{ mol } C_6H_{12}O_6}{180.0 \text{ g } C_6H_{12}O_6}\right)}{\left(5.00 \times 10^2 \text{ mL solution} \times \dfrac{1 \text{ L soln}}{10^3 \text{ mL soln}}\right)}$

$M_{C_6H_{12}O_6} = 0.367 \text{ M } C_6H_{12}O_6$

6.47 $M = \dfrac{\text{mol solute}}{\text{L solution}}$

Solve for mol solute,   mol solute $= (M)(\text{L solution})$

a. mol NaCl $= (0.100 \text{ M})\left(2.50 \times 10^2 \text{ mL soln} \times \dfrac{1 \text{ L soln}}{10^3 \text{ mL soln}}\right)$

mol NaCl $= 2.50 \times 10^{-2} \text{ mol NaCl}$

and, 2.50×10^{-2} mol NaCl x $\dfrac{58.44 \text{ g NaCl}}{1 \text{ mol NaCl}}$ = 1.46 g NaCl

b. mol $C_6H_{12}O_6$ = (0.200 M)$\left(2.50 \times 10^2 \text{ mL soln x } \dfrac{1 \text{ L soln}}{10^3 \text{ mL soln}}\right)$

mol $C_6H_{12}O_6$ = 5.00×10^{-2} mol $C_6H_{12}O_6$

and, 5.00×10^{-2} mol $C_6H_{12}O_6$ x $\dfrac{180.0 \text{ g } C_6H_{12}O_6}{1 \text{ mol } C_6H_{12}O_6}$ = 9.00 g $C_6H_{12}O_6$

6.49 $M_{C_{12}H_{22}O_{11}} = \dfrac{\text{mol } C_{12}H_{22}O_{11}}{\text{L solution}}$

$M_{C_{12}H_{22}O_{11}} = \dfrac{\left(50.0 \text{ g } C_{12}H_{22}O_{11} \text{ x } \dfrac{1 \text{ mol } C_{12}H_{22}O_{11}}{342.3 \text{ g } C_{12}H_{22}O_{11}}\right)}{L}$

$M_{C_{12}H_{22}O_{11}}$ = 0.146 M $C_{12}H_{22}O_{11}$

6.51 $(M_1)(V_1) = (M_2)(V_2)$

M_1 = 1.00 M M_2 = 0.100 M

V_1 = ? V_2 = 0.500 L

Solve for V_1,

$V_1 = \dfrac{(M_2)(V_2)}{M_1}$

Substitute,

$V_1 = \dfrac{(0.100)(0.500)}{1.00}$

$V_1 = 5.00 \times 10^{-2}$ L

6.53 $(M_1)(V_1) = (M_2)(V_2)$

Solve for M_1,

$$M_1 = \frac{(M_2)(V_2)}{V_1}$$

Substitute,

$$M_1 = \frac{(2.00)(5.000 \times 10^{-1}\text{ L})}{5.00 \times 10^{-2}\text{ L}} = 20.0\text{ M}$$

6.55　$M = \dfrac{\text{mol}}{\text{L}} = \dfrac{2.25\text{ mol}}{2.50\text{ L}} = 0.900\text{ M}$

6.57　$M = \dfrac{\text{mol}}{\text{L}}$; mol $= (M)(L)$; mol $= (0.500\text{ M})(1.75\text{ L}) = 0.875\text{ mol}$

Then $(0.875\text{ mol}) \dfrac{(180\text{ g glucose})}{1\text{ mol glucose}} = 157.5\text{ g glucose} \approx 158\text{ g glucose}$

6.59　$(M_1)(V_1) = (M_2)(V_2)$;　$M_1 = \dfrac{(M_1)(V_1)}{V_2} = \dfrac{(0.500\text{ M})(50.0\text{ mL})}{(500.0\text{ mL})}$

$M = 0.500\text{ M}$

6.61　A colligative property is a solution property that depends on the concentration of solute particles rather than the identity of the particles.

6.63　Salt is an ionic substance that dissociates in water to produce positive and negative ions. These ions (or particles) lower the freezing point of water. If the concentration of salt particles is large, the freezing point may be depressed below the surrounding temperature, and the ice would melt.

6.65　Chemical properties depend on the identity of the substance, whereas colligative properties depend on concentration, not identity.

6.67　Raoult's law states that when a solute is added to a solvent, the vapor pressure of the solvent decreases in proportion to the concentration of the solute.

6.69　One mole of $CaCl_2$ produces three moles of particle in solution, whereas one mole of NaCl produces two moles of particles in solution. Therefore, a one molar $CaCl_2$ solution contains a greater number of particles than a one molar NaCl solution and will produce a greater freezing-point depression.

Overview of questions 6.71-6.75:

The *molar* concentration of both solutions is identical. Sodium chloride is an ionic compound that dissociates in water to produce two ions for each ion pair. Therefore, the molarity of particles is 2 x 0.50 M = 1.0 M. Sucrose, in contrast, is a covalent, nondissociating solute; the number of particles and molecules is identical. Consequently, the molarity of *particles* is 0.50 M. Armed with this information, we can now answer each question.

6.71   The NaCl solution would melt at a *lower* temperature since it has a larger number of particles per liter. So, the sucrose solution, with fewer particles per liter, would have the higher melting point. (Recall that solute particles depress the freezing point.)

6.73   The NaCl solution would have the lower vapor pressure because it has the larger number of particles per liter. The sucrose solution, with fewer particles per liter, would have the higher vapor pressure.

6.75   Using our definition of osmotic pressure: $\pi = MRT$

The concentration, M, must be represented as osmolarity.

$$M = M \ = \ \frac{0.50 \text{ mol NaCl}}{L} \ \text{x} \ \frac{2 \text{ mol particles}}{1 \text{ mol NaCl}} \ = \ \frac{1.0 \text{ mol particles}}{L}$$

Now, substituting in our osmotic pressure expression:

π = (1.0 mol particles/L) x (0.0821 Latm/Kmol particles) x 298 K = 24 atm

(Note that there are two significant figures in the answer; the initial data, the NaCl concentration, has only two significant figures.)

6.77   A $\rightarrow$ B (more dilute to more concentrated solution)

6.79   No net flow. (Both solutions have the same concentration of particles.)

6.81 $CaCl_2$ is 3 x 0.15 M = 0.45 M particles
NaCl is 2 x 0.15 M = 0.30 M particles; 0.45 M > 0.30 M; hypertonic

6.83   Glucose is 1 x 0.15 M = 0.15 M particles
NaCl is 2 x 0.15 M = 0.30 M particles; 0.15 M < 0.30 M; hypotonic

6.85   Water is often termed the "universal solvent" because it is a polar molecule and will dissolve, at least to some extent, most ionic and polar covalent compounds. The majority of our body mass is water and this water is an important part of the nutrient transport system due to its solvent properties. This is true in other animals and plants as well.

Because of its ability to hydrogen bond, water has a high boiling point and a low vapor pressure. Also, water is abundant and easily purified.

6.87   The shelf life is a function of the stability of the ammonia-water solution. The ammonia can react with the water to convert to the extremely soluble and stable ammonium ion Also, ammonia and water are polar molecules. Polar interactions, particularly hydrogen bonding, are strong and contribute to the long-term solution stability.

6.89   In the Lewis structure of water, there are two lone pairs of electrons on the oxygen atom. Remember that this causes the geometry, or shape, of the water molecule to be bent.

The partial (–) side of the water molecule attracts the positive sodium ion.

Several water molecules "hydrate" each sodium ion.

6.91   Polar; like dissolves like (H_2O is polar)

6.93

6.95   In dialysis, sodium ions move from a region of high concentration to a region of low concentration. If we wish to remove (transport) sodium ions from the blood, they can move to a region of lower concentration, the dialysis solution.

6.97   Elevated concentrations of sodium ions in the blood may cause confusion, stupor, or coma.

6.99   Elevated concentrations of sodium ions in the blood may occur whenever large amounts of water are lost. Diarrhea, diabetes, and certain high-protein diets are particularly problematic.

Chapter 7
Energy, Rate, and Equilibrium
Solutions to the Odd-Numbered Problems

In-Chapter Questions and Problems

7.1   a.   Exothermic. The reaction *produces* energy used to heat our homes.

      b.   Exothermic. $\Delta H°$ is *negative*, meaning that energy is *released* by the reaction.

      c.   Exothermic. 18.3 kcal of energy is shown as a *product* of the reaction.

7.3   He(g) has a greater entropy than Na(s). Gases have a greater degree of disorder than solids.

7.5 $\Delta G = \Delta H - T\Delta S$

substituting signs :
$\Delta G = (+) - T(-)$

ΔG must always be positive. A positive value for ΔG indicates a nonspontaneous process.

7.7 $Q = m_w \times \Delta T_w \times SH_w$

Solving for ΔT_w

$$\Delta T_w = \frac{Q}{m_w \times SH_w}$$

Substituting,

$$\Delta T_w = \frac{6.5 \times 10^2 \text{ cal}}{(50.0 \text{ g})(1.00 \text{ cal/g°C})} = 13°C$$

7.9 6.5×10^2 cal $\times \dfrac{4.18 \text{ J}}{1 \text{ cal}} = 2.7 \times 10^3$ J

7.11 $Q = m_w \times \Delta T_w \times SH_w$

$$Q = 1.00 \times 10^3 \text{ g} \times 3.0°C \times \frac{1.00 \text{ cal}}{\text{g} \cdot °C}$$

$Q = 3.0 \times 10^3$ cal in a 1.0 gram sample of candy

and,

$$3.0 \times 10^3 \text{ cal} \times \frac{1 \text{ nutritional Cal}}{10^3 \text{ cal}} = 3.0 \text{ nutritional Calories in a 1.0 gram sample of candy}$$

and,

$$\frac{3.0 \text{ nutritional Cal}}{1.0 \text{ g candy}} \times \frac{454 \text{ g candy}}{1 \text{ lb candy}} \times \frac{1 \text{ lb candy}}{16 \text{ oz candy}} \times \frac{2.5 \text{ oz}}{1 \text{ candy bar}} =$$

$$\frac{2.1 \times 10^2 \text{ nutritional Cal}}{\text{candy bar}}$$

7.13 Heat energy produced by the friction of striking the match provides the activation energy necessary for this combustion process.

7.15 If the enzyme catalyzed a process needed to sustain life, the substance interfering with that enzyme would be classified as a poison.

7.17 a. rate = $k[N_2]^n[O_2]^{n'}$ Note: n and n' must be experimentally determined.

    b. rate = $k[C_4H_6]^n$ Note: n must be experimentally determined.

7.19 At rush hour, approximately the same number of passengers enter and exit the train at any given stop. Throughout the trip, the number of passengers on the train may be essentially unchanged, but the identity of the individual passengers is continually changing.

7.21 Measure the concentrations of products and reactants at a series of times until no further concentration change is observed.

7.23 a. $K_{eq} = \dfrac{[N_2][O_2]^2}{[NO_2]^2}$

    b. $K_{eq} = [H_2]^2[O_2]$

7.25 A large value for the equilibrium constant favors product formation.

7.27

$$K_{eq} = \frac{[NH_3]^2}{[N_2][H_2]^3}$$

$$K_{eq} = \frac{[0.25]^2}{[0.11][1.91]^3}$$

$$K_{eq} = \frac{(6.25 \times 10^{-2})}{(0.11)(6.97)}$$

$$K_{eq} = 8.2 \times 10^{-2}$$

7.29   a.   A would decrease; the system would shift to remove excess B. B *reacts* with A to form products C and D.

b.   A would increase; addition of excess C shifts the equilibrium to the *left* producing both A and B.

c.   A would decrease; A would react with B to compensate for the loss of some D. The equilibrium shifts to the *right*.

d.   A would remain the same. The presence of a catalyst increases the rate of attainment of equilibrium, but does not affect the equilibrium position.

End-of-Chapter Questions and Problems

7.31   joule

7.33   An exothermic reaction is one in which energy is released during chemical change.

7.35   A fuel must release heat in the combustion (oxidation) process.   The release of heat during the process characterizes an exothermic reaction.

7.37   The temperature of the water (or solution) is measured in a calorimeter.   If the reaction being studied is exothermic, released energy heats the water and the temperature increases.   In an endothermic reaction, heat flows from the water to the reaction and the water temperature decreases.

7.39   Double-walled containers, used in calorimeters, provide a small airspace between the part of the calorimeter (inside wall) containing the sample solution and the outside wall, contacting the surroundings.   This makes heat transfer more difficult.

7.41   Free energy is the combined contribution of entropy and enthalpy for a chemical reaction.

7.43 The first law of thermodynamics, the law of conservation of energy, states that the energy of the universe is constant.

7.45 Enthalpy is a measure of heat energy.

7.47 $Q = m_w \times \Delta T_w \times SH_w$

$$Q = 2.00 \times 10^2 \text{ g } H_2O \times 6.00°C \times \frac{1.00 \text{ cal}}{\text{g } H_2O \text{ °C}}$$

$Q = 1.20 \times 10^3 \text{ cal}$

7.49 $1.20 \times 10^3 \text{ cal} \times \dfrac{4.18 \text{ J}}{1 \text{ cal}} = 5.02 \times 10^3 \text{ J}$

7.51 a. Entropy increases. Conversion of a solid to a liquid results in an increase in disorder of the substance. Solids retain their shape while liquids will flow and their shape is determined by their container.

b. Entropy increases. Conversion of a liquid to a gas results in an increase in disorder of the substance. Gas particles move randomly with very weak interactions between particles, much weaker than those interactions in the liquid state.

7.53 $\Delta G = \Delta H - T\Delta S$

substituting signs :
$\Delta G = (-) - T(+)$

ΔG must always be negative. A negative value for ΔG indicates a spontaneous process.

7.55 Isopropyl alcohol quickly evaporates (liquid $\rightarrow$ gas) after being applied to the skin. Conversion of a liquid to a gas requires heat energy. The heat energy is supplied by the skin. When this heat is lost, the skin temperature drops.

7.57 Decomposition of leaves and twigs to produce soil.

7.59 The activated complex is the arrangement of reactants in an unstable transition state as a chemical reaction proceeds. The activated complex must form in order to convert reactants to products.

7.61 The rate of a reaction is the change in concentration of a reactant or product per unit time. The rate constant is the proportionality constant that relates rate and concentration. The order is the exponent of each concentration term in the rate equation.

7.63   Increase; the rate is proportional to k_r.

7.65   A catalyst increases the rate of a reaction without itself undergoing change.

7.67   Assume a generalized exothermic reaction. In the catalyzed reaction, the energy released by the reaction is the same, but the activation energy is much lower. Equilibrium concentrations of products and reactants are the same in both the catalyzed and uncatalyzed reactions.

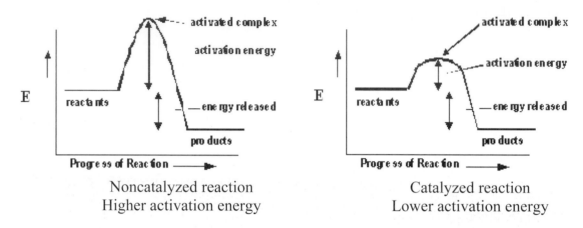

Noncatalyzed reaction
Higher activation energy

Catalyzed reaction
Lower activation energy

7.69   Enzymes are biological catalysts. The enzyme lysozyme catalyzes a process that results in the destruction of the cell walls of many harmful bacteria. This helps to prevent disease in organisms.

The breakdown of foods to produce material for construction and repair of body tissue, as well as energy, is catalyzed by a variety of enzymes. For example, amylase begins the hydrolysis of starch in the mouth.

7.71   An increase in concentration of reactants means that there are more molecules in a certain volume. The probability of collision is enhanced because they travel a shorter distance before meeting another molecule. The rate is proportional to the number of collisions per unit time.

7.73   Rate = k $[N_2O_4]$

7.75   A catalyst speeds up a chemical reaction by facilitating the formation of the activated complex, thus lowering the activation energy, the energy barrier for the reaction.

7.77 $K_{eq} = \dfrac{[products]}{[reactants]};$

A large K_{eq} indicates a large numerator; hence, products are favored.

7.79   Ice and water at $0^{\circ}C$.

7.81   LeChatelier's principle states that when a system at equilibrium is disturbed, the equilibrium shifts in the direction that minimizes the disturbance.

7.83   A dynamic equilibrium has fixed concentrations of all reactants and products; these concentrations do not change with time. However, the process is dynamic because products and reactants are continuously being formed and consumed. The concentrations do not change because the *rates* of production and consumption are equal.

7.85 $K_{eq} = \dfrac{[NO_2]^2}{[N_2O_4]}$

7.87   A physical equilibrium describes physical change; examples include the equilibrium between ice and water, or the equilibrium vapor pressure of a liquid.

        A chemical equilibrium describes chemical change; examples include the reactions shown in questions 7.93 and 7.94.

7.89   a. Equilibrium shifts to the left. Increasing the temperature increases the energy, a product of the reaction.

        b. No change; the number of moles of gaseous products and reactants are equal.

        c. No change; a catalyst has no effect on the equilibrium position of the reaction.

7.91   a. A slow reaction is an incomplete reaction. False; A slow reaction may go to completion, but take a longer period of time.

        b. The rate of forward and reverse reactions is never the same. False; The rate of forward and reverse reactions is equal in a dynamic equilibrium situation.

7.93   a. PCl_3 increases. Addition of product shifts the equilibrium to the left, favoring reactants.

        b. PCl_3 decreases. Added Cl_2 reacts with PCl_3 to produce products; the equilibrium shifts to the right.

        c. PCl_3 decreases. Removal of product shifts the equilibrium to the right, favoring the formation of more product.

        d. PCl_3 decreases. Decreasing the temperature removes heat from the system. Heat is a product; therefore the equilibrium shifts to the right.

        e. PCl_3 remains the same. Addition of a catalyst has no effect on the equilibrium position.

7.95   To determine the effect of pressure on equilibrium concentrations, focus on the number of moles of substances in the gaseous state. An increase in pressure would shift the equilibrium to the side of the reaction that has the least number of moles of gas. For *each* mole of *reactant* (in the gaseous state), *two* moles of *product* (in the gaseous state) are formed. Therefore, this reaction equilibrium would shift to the *left* upon an increase in pressure, and the concentration of H_2, a product, would *decrease*.

7.97 $$K_{eq} = \frac{[CO][H_2]}{[H_2O]}$$

7.99   False. The position of equilibrium is not affected by a catalyst, only the rate at which equilibrium is attained.

7.101   Removing the cap allows CO_2 to escape into the atmosphere. This corresponds to the removal of product (CO_2):

$$CO_2 \text{ (l)} \rightleftharpoons CO_2 \text{ (g)}$$

The equilibrium shifts to the right, dissolved CO_2 is lost, and the beverage goes "flat."

7.103 $$K_{eq} = \frac{[NH_3]^2}{[N_2][H_2]^3}$$

Chapter 8
Acids and Bases and Oxidation-Reduction
Solutions to the Odd-Numbered Problems

In-Chapter Questions and Problems

8.1   a. $HF(aq) + H_2O\ (l) \rightleftharpoons H_3O^+\ (aq) + F^-\ (aq)$

      b, $NH_3(aq) + H_2O\ (l) \rightleftharpoons NH_4^+(aq) + OH^-(aq)$

8.3   a.   HF and F^-; H_2O and H_3O^+

      b. NH_3 and NH_4^+ ; H_2O and OH^-

8.5   a. NH_4^+; based on Figure 8.2

      b. H_2SO_4; based on Figure 8.2

8.7 $[H_3O^+]\ [OH^-] = 1.0 \times 10^{-14}$

      $[OH^-] = \dfrac{1.0 \times 10^{-14}}{1.0 \times 10^{-3}}$

      $[OH^-] = 1.0 \times 10^{-11}$ M

8.9   Referring to the discussion of the decimal-based system, a solution of sodium hydroxide producing $[OH^-] = 1.0 \times 10^{-2}$ M corresponds to a pH of 12.00.

8.11   A pH of 8.50 is a noninteger. The calculation of the $[H_3O^+]$ is most easily accomplished with the aid of a calculator.

      $pH = -\log [H_3O^+]$

      and, $[H_3O^+] = 10^{-pH}$

      On the calculator:
         Enter 8.50
         Press "change sign" key
         Press 10^X key
         The result is $[H_3O^+] = 3.2 \times 10^{-9}$ M

8.13   The acid-base reaction is:

$$HCl(aq) + NaOH(aq) \rightarrow NaCl(aq) + H_2O(l)$$

$$(M_{acid})(V_{liters\ acid}) = (M_{base})(V_{liters\ base})$$

Solving for M_{base}, $M_{base} = \dfrac{(M_{acid})(V_{liters\ acid})}{(V_{liters\ base})}$

Substituting

$$M_{base} = \dfrac{(0.2000\ M)\left(20.00\ ml \times \dfrac{1\ L}{10^3\ mL}\right)}{\left(40.00\ ml \times \dfrac{1\ L}{10^3\ mL}\right)} = 0.1000\ M\ NaOH$$

8.15   The equilibrium reaction is:

$$CO_2 + H_2O \rightleftharpoons H_2CO_3 \rightleftharpoons H_3O^+ + HCO_3^-$$

An increase in the partial pressure of CO_2 is a stress on the left side of the equilibrium. The equilibrium will shift to the right in an effort to decrease the concentration of CO_2. This will cause the molar concentration of H_2CO_3 to increase.

8.17   The equilibrium reaction is:

$$CO_2 + H_2O \rightleftharpoons H_2CO_3 \rightleftharpoons H_3O^+ + HCO_3^-$$

In Question 8.15, the equilibrium shifts to the right. Therefore, the molar concentration of H_3O^+ should increase.

In Question 8.16, the equilibrium shifts to the left. Therefore, the molar concentration of H_3O^+ should decrease.

8.19   Propanoic acid is the acid; $[acid] = 2.00 \times 10^{-1}\ M$

Sodium propanoate is the conjugate base; $[conjugate\ base] = 2.00 \times 10^{-1}\ M$

The equilibrium is

$$C_2H_5COOH(aq) + H_2O(l) \rightleftharpoons H_3O^+(aq) + C_2H_5COO^-(aq)$$
$$\quad\quad acid \quad\quad\quad\quad\quad\quad\quad\quad\quad\quad\quad\quad conjugate\ base$$

and the hydronium ion concentration $= [H_3O^+] = \dfrac{[C_2H_5COOH]K_a}{[C_2H_5COO^-]}$

400

Substituting the values given in the problem

$$[H_3O^+] = \frac{[2.00 \times 10^{-1}](1.34 \times 10^{-5})}{[2.00 \times 10^{-1}]} = 1.34 \times 10^{-5}$$

and since

$$pH = -\log [H_3O^+]$$

$$pH = -\log 1.34 \times 10^{-5}$$

$$pH = 4.87$$

8.21   [acid] = 1.00 x 10⁻¹ M
[conjugate base] = 1.00 x 10⁻¹ M

$K_a = 1.8 \times 10^{-5}$
$pK_a = -\log K_a$
$pK_a = -\log(1.8 \times 10^{-5})$
$pK_a = 4.74$

Henderson-Hasselbalch equation: $pH = pK_a + \log\dfrac{[\text{conjugate base}]}{[\text{weak acid}]}$

Substituting,

$$pH = 4.74 + \log\frac{1.00 \times 10^{-1} \text{ M}}{1.00 \times 10^{-1} \text{ M}}$$

$$pH = 4.74 + \log(1)$$

$$pH = 4.74$$

Note that the log(1) = 0. So, the Henderson-Hasselbalch equation shows us that when the buffer concentrations of acid and conjugate base are equal, the pH will equal the pK_a.

8.23   [acid] = 2.00 x 10⁻¹ M
[conjugate base] = 2.00 x 10⁻¹ M

$K_a = 1.34 \times 10^{-5}$
$pK_a = -\log K_a$
$pK_a = -\log(1.34 \times 10^{-5})$
$pK_a = 4.87$

Henderson-Hasselbalch equation: $pH = pK_a + \log\frac{[\text{conjugate base}]}{[\text{weak acid}]}$

Substituting,

$pH = 4.76 + \log\dfrac{2.00 \times 10^{-1} \text{ M}}{2.00 \times 10^{-1} \text{ M}}$

$pH = 4.87$

8.25 $Ca \rightarrow Ca^{2+} + 2 \text{ e}^-$ (oxidation ½ reaction)
$\underline{S + 2 \text{ e}^- \rightarrow S^{2-}}$ (reduction ½ reaction)
$Ca + S \rightarrow CaS$ (complete reaction)

End-of-Chapter Questions and Problems

8.27   a.   An Arrhenius acid is a substance that dissociates, producing hydrogen ions.

b.   A Brønsted-Lowry acid is a substance that behaves as a proton donor.

8.29   The Brønsted-Lowry theory provides a broader view of acid-base theory than does the Arrhenius theory. Brønsted-Lowry emphasizes the role of the solvent in the dissociation process.

8.31   a. $HNO_2 \ (aq) + H_2O \ (l) \ \rightleftharpoons \ H_3O^+ \ (aq) + NO_2^- \ (aq)$

b. $HCN \ (aq) + H_2O \ (l) \ \rightleftharpoons \ H_3O^+ \ (aq) + CN^- \ (aq)$

8.33   a. HNO_2 and NO_2^-; H_2O and H_3O^+

b. HCN and CN^-; H_2O and H_3O^+

8.35   a.   Weak

b.   Weak

c.   Weak

8.37   a. CN^- and HCN; NH_3 and NH_4^+

b. CO_3^{2-} and HCO_3^-; Cl^- and HCl

8.39   Concentration refers to the quantity of acid or base contained in a specified volume of solvent. Strength refers to the degree of dissociation of the acid or base.

8.41   a. H_3O^+ is a Brønsted acid

b.   OH⁻ is a Brønsted base

c. H_2O can behave as both acid and base

8.43   a. H_2CO_3 is a Brønsted acid

        b. HCO_3^- can behave as both acid and base

        c. CO_3^{2-} is a Brønsted base

8.45 $CN^- + H^+ \rightarrow HCN$
       (base)   (conjugate acid)

8.47 $HI \rightarrow H^+ + I^-$
       (acid)   (conjugate base)

8.49 $[H_3O^+][OH^-] = 1.0 \times 10^{-14}$

Solving for $[H_3O^+]$, $[H_3O^+] = \dfrac{1 \times 10^{-14}}{[OH^-]}$

a. Substituting $[OH^-] = 1.0 \times 10^{-7}$ M

$$[H_3O^+] = \dfrac{1 \times 10^{-14}}{1.0 \times 10^{-7}} = 1.0 \times 10^{-7} \text{ M}$$

b. Substituting $[OH^-] = 1.0 \times 10^{-3}$ M

$$[H_3O^+] = \dfrac{1 \times 10^{-14}}{1.0 \times 10^{-3}} = 1.0 \times 10^{-11} \text{ M}$$

8.51   a. Neutral, $[H_3O^+] = 1.0 \times 10^{-7}$ M

        b. Basic, $[H_3O^+]$ is less than 1.0×10^{-7} M

8.53   a. $pH = -\log [H_3O^+]$

          $pH = -\log [1.0 \times 10^{-7}]$

          $pH = 7.00$

        b. $pH = -\log [H_3O^+]$

          Since $[H_3O^+][OH^-] = 1.0 \times 10^{-14}$

and, $[H_3O^+] = \dfrac{1 \times 10^{-14}}{[OH^-]}$

Substituting, $[H_3O^+] = \dfrac{1 \times 10^{-14}}{1 \times 10^{-9}}$

$[H_3O^+] = 1.0 \times 10^{-5} M$

and, $pH = -\log [1.0 \times 10^{-5}]$

$pH = 5.00$

8.55   a. $pH = -\log [H_3O^+]$

and, $[H_3O^+] = 10^{-pH}$

On the calculator:
Enter 1.00
    Press "change sign" key
    Press 10^x key
    The result is $[H_3O^+] = 1.0 \times 10^{-1} M$

and, $[H_3O^+][OH^-] = 1.0 \times 10^{-14}$

solving for $[OH^-]$

$$[OH^-] = \dfrac{1.0 \times 10^{-14}}{[H_3O^+]} = \dfrac{1.0 \times 10^{-14}}{1.0 \times 10^{-1}} = 1.0 \times 10^{-13} \ M$$

b. $pH = -\log [H_3O^+]$

and, $[H_3O^+] = 10^{-pH}$

On the calculator:
    Enter 9.00
    Press "change sign" key
    Press 10^x key
    The result is $[H_3O^+] = 1.0 \times 10^{-9} M$

and, $[H_3O^+][OH^-] = 1.0 \times 10^{-14}$

solving for $[OH^-]$

$$[OH^-] = \frac{1.0 \times 10^{-14}}{[H_3O^+]} = \frac{1.0 \times 10^{-14}}{1.0 \times 10^{-9}} = 1.0 \times 10^{-5} \text{ M}$$

8.57   a. $pH = -\log [H_3O^+]$

and, $[H_3O^+] = 10^{pH}$

On the calculator:
   Enter 1.30
   Press "change sign" key
   Press 10^x key
   The result is $[H_3O^+] = 5.0 \times 10^{-2}$ M

and, $[H_3O^+][OH^-] = 1.0 \times 10^{-14}$

solving for $[OH^-]$

$$[OH^-] = \frac{1.0 \times 10^{-14}}{[H_3O^+]} = \frac{1.0 \times 10^{-14}}{5.0 \times 10^{-2}} = 2.0 \times 10^{-13} \text{ M}$$

b. $pH = -\log [H_3O^+]$

and, $[H_3O^+] = 10^{pH}$

On the calculator:
   Enter 9.70
   Press "change sign" key
   Press 10^x key
   The result is $[H_3O^+] = 2.0 \times 10^{-10}$ M

and, $[H_3O^+][OH^-] = 1.0 \times 10^{-14}$

solving for $[OH^-]$

$$[OH^-] = \frac{1.0 \times 10^{-14}}{[H_3O^+]} = \frac{1.0 \times 10^{-14}}{2.0 \times 10^{-10}} = 5.0 \times 10^{-5} \text{ M}$$

8.59   A neutralization reaction is one in which an acid and a base react to produce water and a salt (a "neutral" solution).

8.61   a. $pH = -\log [H_3O^+]$

and, $[H_3O^+] = 10^{-pH}$

On the calculator:
    Enter 6.00
    Press "change sign" key
    Press 10^x key
    The result is $[H_3O^+] = 1.0 \times 10^{-6} M$

and, $[H_3O^+][OH^-] = 1.0 \times 10^{-14}$

solving for $[OH^-]$

$$[OH^-] = \frac{1.0 \times 10^{-14}}{[H_3O^+]} = \frac{1.0 \times 10^{-14}}{1.0 \times 10^{-6}} = 1.0 \times 10^{-8} \text{ M}$$

b. $pH = -\log [H_3O^+]$

and, $[H_3O^+] = 10^{-pH}$

On the calculator:
    Enter 5.20
    Press "change sign" key
    Press 10^x key
    The result is $[H_3O^+] = 6.3 \times 10^{-6} M$

and, $[H_3O^+][OH^-] = 1.0 \times 10^{-14}$

solving for $[OH^-]$

$$[OH^-] = \frac{1.0 \times 10^{-14}}{[H_3O^+]} = \frac{1.0 \times 10^{-14}}{6.3 \times 10^{-6}} = 1.6 \times 10^{-9} \text{ M}$$

c. $pH = -\log [H_3O^+]$

and, $[H_3O^+] = 10^{-pH}$

On the calculator:
    Enter 7.80
    Press "change sign" key
    Press 10^x key
    The result is $[H_3O^+] = 1.6 \times 10^{-8} M$

and, $[H_3O^+][OH^-] = 1.0 \times 10^{-14}$

solving for [OH⁻]

$$[OH^-] = \frac{1.0 \times 10^{-14}}{[H_3O^+]} = \frac{1.0 \times 10^{-14}}{1.6 \times 10^{-8}} = 6.3 \times 10^{-7} \text{ M}$$

8.63   One pH unit difference corresponds to a tenfold (10^1) difference in concentration of H_3O^+ or OH⁻.

  a.   4 - 2 = 2 pH units, therefore $(10^1)^2$ or 1×10^2

  b.   11 - 7 = 4 pH units, therefore $(10^1)^4$ or 1×10^4

  c.   12 - 2 = 10 pH units, therefore $(10^1)^{10}$ or 1×10^{10}

8.65 $pH = -\log [H_3O^+]$

and, $[H_3O^+] = 10^{-pH}$

  a.   pH = 5.0
       $[H_3O^+] = 10^{-5.0}$
       $[H_3O^+] = 1 \times 10^{-5}$

  b.   pH = 12.0
       $[H_3O^+] = 10^{-12.0}$
       $[H_3O^+] = 1 \times 10^{-12}$

  c.   pH = 5.5
       $[H_3O^+] = 10^{-5.5}$
       On the calculator:
           Enter 5.50
           Press "change sign" key
           Press 10^x key
           The result is $[H_3O^+] = 3.2 \times 10^{-6}$

8.67 $pH = -\log [H_3O^+]$

  a. $[H_3O^+] = 1.0 \times 10^{-6}$M
       $pH = -\log[1.0 \times 10^{-6}]$
       pH = 6.00

  b. $[H_3O^+] = 1.0 \times 10^{-8}$M
       $pH = -\log[1.0 \times 10^{-8}]$
       pH = 8.00

c. $[H_3O^+] = 5.6 \times 10^{-4} M$
$pH = -\log[5.6 \times 10^{-4}]$
$pH = 3.25$

8.69 $pH = -\log[H_3O^+]$

$pH = -\log 7.5 \times 10^{-4}$

$pH = 3.12$

8.71 $[H_3O^+][OH^-] = 1.0 \times 10^{-14}$

$$[H_3O^+] = \frac{1.0 \times 10^{-14}}{[OH^-]} = \frac{1.0 \times 10^{-14}}{[5.5 \times 10^{-4}]} = 1.8 \times 10^{-11}$$

$pH = -\log[H_3O^+]$
$pH = -\log 1.8 \times 10^{-11}$
$pH = 10.74$

8.73 a. NH_3 is a weak base.
NH$_4$Cl is a salt formed from NH_3.
Therefore, NH_3 and NH$_4$Cl can form a buffer solution.

b. HNO_3 is a strong acid; strong acids are not suitable for buffer preparation; they are completely dissociated.
Therefore, HNO_3 and KNO_3 cannot form a buffer solution.

8.75 a. A buffer solution contains components (a weak acid and its salt or a weak base and its salt) that enable the solution to resist large changes in pH when acids or bases are added.

b. Acidosis is a medical condition characterized by higher-than-normal levels of CO_2 in the blood and lower-than-normal blood pH.

8.77 a. Addition of strong acid is equivalent to adding H_3O^+. This is a stress on the right side of the equilibrium and the equilibrium will shift to the left. Consequently the [CH$_3$COOH] increases.

b. Water, in this case, is a solvent and does not appear in the equilibrium expression. Hence, it does not alter the position of the equilibrium.

8.79 $$K_a = \frac{[H_3O^+][\text{conjugate base}]}{[\text{acid}]}$$

$$[H_3O^+] = \frac{[acid]K_a}{[conjugate\ base]}$$

$$[H_3O^+] = \frac{[0.200]5.80 \times 10^{-7}}{[0.500]}$$

$$[H_3O^+] = 2.32 \times 10^{-7}\ M$$

8.81   The very fact that an acid has a K_a value means that it is a weak acid. The smaller the value of K_a, the weaker the acid.

8.83 $pH = pK_a + \log \dfrac{[conjugate\ base]}{[weak\ acid]}$; $pH = -\log 1.8 \times 10^{-5} + \log \dfrac{[1.0]}{[1.0]} = 4.74$

8.85

$pH = pK_a + \log \dfrac{[conjugate\ base]}{[weak\ acid]}$; $7.40 = -\log 4.5 \times 10^{-7} + \log \dfrac{[HCO_3^-]}{[H_2CO_3]}$

$7.40 = 6.35 + \log \dfrac{[HCO_3^-]}{[H_2CO_3]}$

$1.05 = \log \dfrac{[HCO_3^-]}{[H_2CO_3]}$

Taking the antilog of both sides,

$11.2 = \dfrac{[HCO_3^-]}{[H_2CO_3]}$

8.87   a.   Oxidation is the loss of electrons, loss of hydrogen atoms, or gain of oxygen atoms.

b.   An oxidizing agent removes electrons from another substance. In doing so, the oxidizing agent becomes reduced.

8.89   During an oxidation process in an oxidation-reduction reaction, the species oxidized *loses* electrons.

8.91   During an oxidation-reduction reaction, the species *oxidized* is the reducing agent.

8.93
$$Cl_2 \quad + \quad 2KI \quad \rightarrow \quad 2KCl \ + \ I_2$$

substance reduced   substance oxidized
oxidizing agent   reducing agent

8.95 $2KI \rightarrow 2K^+ + 2\ e^- + I_2$ (oxidation ½ reaction)
      $Cl_2 + 2\ e^- \rightarrow 2Cl^-$ (reduction ½ reaction)

8.97 An oxidation-reduction reaction must take place to produce electron flow in a voltaic cell.

8.99 Storage battery

Chapter 9
The Nucleus, Radioactivity, and Nuclear Medicine
Solutions to the Odd-Numbered Problems

In-Chapter Questions and Problems

9.1 Gamma radiation is very high-energy electromagnetic radiation. Other forms of electromagnetic radiation (in descending order of energy) are X-ray, ultraviolet, visible, infrared, microwave, and radiowave. [See Chapter 2, *An Environmental Perspective: Electromagnetic Radiation and Its Effects on Our Everyday Lives,* for more information.]

9.3 a. $^{85}_{36}\text{Kr} \rightarrow ^{85}_{37}\text{Rb} + ^{0}_{-1}\text{e}$

Always double-check your answer. Make sure that the sum of the atomic numbers is the same on both sides of the equation, and that the sum of the mass numbers are the same on both sides of the equation.
Atomic numbers: $36 = 37 + (-1)$
Mass numbers: $85 = 85 + 0$

b. $^{226}_{88}\text{Ra} \rightarrow ^{4}_{2}\text{He} + ^{222}_{86}\text{Rn}$

9.5 The half-life of sodium-24 is 15 hours. The number (n) of half-lives elapsed is:

$$n = 2.5 \text{ days} \times \frac{24 \text{ hours}}{1 \text{ day}} \times \frac{1 \text{ half - life}}{15 \text{ hours}} = 4 \text{ half - lives}$$

Then,
$$100 \text{ ng} \xrightarrow{\text{first half-life}} 50 \text{ ng}$$
$$\xrightarrow{\text{second half-life}} 25.0 \text{ ng}$$
$$\xrightarrow{\text{third half-life}} 12.5 \text{ ng}$$
$$\xrightarrow{\text{fourth half-life}} 6.3 \text{ ng}$$

Therefore, 6.3 ng of sodium-24 remain after 2.5 days.

9.7 The half-life of technetium-99 m is 6 hours. The number (n) of half-lives elapsed is:

$$n = 12 \text{ hours} \times \frac{1 \text{ half - life}}{6 \text{ hours}} = 2 \text{ half - lives}$$

Then, assume that the original amount is x g.

$$x \text{ g} \xrightarrow[\text{half-life}]{\text{first}} \frac{x}{2} \text{ g} \xrightarrow[\text{half-life}]{\text{second}} \frac{x}{4} \text{ g}$$

Therefore, 1/4 of the radioisotope remains after 2 half-lives, 12 hours.

9.9 Isotopes with short half-lives release their radiation rapidly. There is much more radiation per unit time observed with short half-life substances; hence, the signal is stronger and the sensitivity of the procedure is enhanced.

9.11 The rem takes into account the relative biological effect of the radiation in addition to the quantity of radiation. This provides a more meaningful estimate of potential radiation damage to human tissue.

9.13 Natural radioactivity is the spontaneous decay of a nucleus to produce high-energy particles or rays.

9.15 Alpha particles contain two protons and two neutrons; hence they have a charge of +2.

9.17 A beta particle is an electron with a –1 charge.

9.19 A positron has a positive charge and a beta particle has a negative charge.

9.21
- charge, $\alpha = +2$, $\beta = -1$
- mass, $\alpha = 4$ amu, $\beta = 0.000549$ amu
- velocity, $\alpha = 10\%$ of C, $\beta = 90\%$ of C

9.23 Chemical reactions involve joining, separating and rearranging atoms; valence electrons are critically involved. Nuclear reactions involve changes in nuclear composition only.

9.25 ^4_2He

9.27 $^{235}_{92}U$

9.29 Mass number – atomic number = number of neutrons
So ^1_1H, $1 - 1 = 0$ neutrons, ^2_1H, $2 - 1 = 1$ neutron, ^3_1H $3 - 1 = 2$ neutrons.

9.31 $^{15}_7N$

End-of-Chapter Questions and Problems

9.33 Alpha and beta particles are matter; gamma radiation is pure energy. Alpha particles are large and relatively slow moving. They are the least energetic and least penetrating. Gamma radiation moves at the speed of light, is highly energetic, and is most penetrating.

9.35 A helium atom has two electrons; an α particle has no electrons.

9.37 $^{60}_{27}Co \rightarrow ^{60}_{28}Ni + ^0_{-1}\beta + \gamma$

9.39 $\ _{11}^{23}\text{Na} + \ _1^2\text{H} \rightarrow \ _{11}^{24}\text{Na} + \ _1^1\text{H}$

9.41 $\ _{10}^{24}\text{Ne} \rightarrow \ _{-1}^0\text{e} + \ _{11}^{24}\text{Na}$

9.43 $\ _{55}^{140}\text{Cs} \rightarrow \ _{56}^{140}\text{Ba} + \ _{-1}^0\text{e}$

9.45 $\ _{83}^{209}\text{Bi} + \ _{24}^{54}\text{Cr} \rightarrow \ _{107}^{262}\text{Bh} + \ _0^1\text{n}$

9.47 $\ _{12}^{27}\text{Mg} \rightarrow \ _{13}^{27}\text{Al} + \ _{-1}^0\text{e}$

9.49 $\ _7^{12}\text{N} \rightarrow \ _6^{12}\text{C} + \ _{-1}^0\text{e}$

9.51 Natural radioactivity is a spontaneous process; artificial radioactivity is nonspontaneous and results from a nuclear reaction that produces an unstable nucleus.

9.53 • Nuclei for light atoms tend to be most stable if their neutron/proton ratio is close to 1.
• Nuclei with more than 84 protons tend to be unstable.
• Isotopes with a "magic number" of protons or neutrons (2, 8, 20, 50, 82, or 126 protons or neutrons) tend to be stable.
• Isotopes with even numbers of protons or neutrons tend to be more stable.

9.55 $\ _8^{20}\text{O}$; Oxygen – 20 has 20 – 8 = 12 neutrons, an n/p of 12/8, or 1.5. The n/p is probably too high for stability even though it does have a "magic number' of protons and an even number of protons and neutrons.

9.57 $\ _{24}^{48}\text{Cr}$; Chromium – 48 has 48 – 24 = 24 neutrons, an n/p of 24/24, or 1.0. It also has an even number of protons and neutrons. It would probably be stable.

9.59 The half-life of iodine-131 is 8.1 days. The number (n) of half-lives elapsed is:

$$n = 24 \text{ days x } \frac{1 \text{ half - life}}{8.1 \text{ days}} = 3.0 \text{ half-lives} \quad \text{(2 significant figures)}$$

Then,

$$3.2 \text{ mg} \xrightarrow[\text{half}-\text{life}]{\text{first}} 1.6 \text{ mg} \xrightarrow[\text{half}-\text{life}]{\text{second}} 0.80 \text{ mg} \xrightarrow[\text{half}-\text{life}]{\text{third}} 0.40 \text{ mg}$$

Therefore, 0.40 mg of iodine-131 remains after 24 days.

9.61 The half-life of iron-59 is 45 days. The number (n) of half-lives elapsed is:

$$n = 135 \text{ days x } \frac{1 \text{ half - life}}{45 \text{ days}} = 3.0 \text{ half-lives}$$

Then,

$$100 \text{ mg} \xrightarrow[\text{half-life}]{\text{first}} 50 \text{ mg} \xrightarrow[\text{half-life}]{\text{second}} 25 \text{ mg} \xrightarrow[\text{half-life}]{\text{third}} 13 \text{ mg}$$

Therefore, 13 mg of iron-59 remains after 135 days.

9.63 $$200 \text{ μg} \xrightarrow[\text{half-life}]{\text{first}} 100 \text{ μg} \xrightarrow[\text{half-life}]{\text{second}} 50 \text{ μg} \xrightarrow[\text{half-life}]{\text{third}} 25 \text{ μg}$$

So, 3 half-lives x $\dfrac{67 \text{ hours}}{1 \text{ half - life}}$ = 201 hours

9.65 Fission splits nuclei to produce energy.

9.67 a. The fission process involves the breaking down of large, unstable nuclei into smaller, more stable nuclei. This process releases some of the binding energy in the form of heat and/or light.

   b. The heat generated during the fission process could be used to generate steam, which is then used to drive a turbine to create electricity.

9.69 $^{3}_{1}\text{H} + ^{1}_{1}\text{H} \rightarrow ^{4}_{2}\text{He} + \text{energy}$

9.71 A "breeder" reactor creates the fuel that can be used by a conventional fission reactor during its fission process.

9.73 The reaction in a fission reactor that involves neutron production and causes subsequent reactions accompanied by the production of more neutrons in a continuing process.

9.75 High operating temperatures

9.77 Radiocarbon dating is a process used to determine the age of objects. The ratio of the masses of the stable isotope, carbon-12, and unstable isotope, carbon-14, is measured. Using this value and the half-life of carbon-14, the age of the coffin may be calculated.

9.79 $^{108}_{47}\text{Ag} + ^{4}_{2}\text{He} \rightarrow ^{112}_{49}\text{In}$

   $^{112}_{49}\text{In}$ is the intermediate isotope of indium.

9.81 a. Technetium-99m is used to study the heart (cardiac output, size, and shape), kidney (follow-up procedure for kidney transplant), and liver and spleen (size, shape, presence of tumors).

   b. Xenon-133 is used to locate regions of reduced ventilation and presence of tumors in the lung.

9.83 Radiation therapy provides sufficient energy to destroy molecules critical to the reproduction of cancer cells.

9.85 Background radiation, radiation from natural sources, is emitted by the sun as cosmic radiation, and from naturally radioactive isotopes found throughout our environment.

9.87 Level decreases. Radiation level is inversely proportional to the square of the distance from the source.

9.89 Positive effect. Potential damage is often directly proportional to the time of exposure.

9.91 Positive effect. The operator of the robotic device could be located far from the source (see Question 9.87), no physical contact with the source is necessary (see Question 9.88), and barriers of lead or other shielding can isolate the control and the robot (see Question 9.90).

9.93 Yes. Concrete has a higher density than wood and thus serves as a better radiation shield.

9.95 A film badge detects gamma radiation by darkening photographic film in proportion to the amount of radiation exposure over time. Badges are periodically collected and evaluated for their level of exposure. This mirrors the level of exposure of the personnel wearing the badges.

9.97 Relative biological effect is a measure of the damage to biological tissue caused by different forms of radiation.

9.99 a. The curie is the amount of radioactive material needed to produce 3.7×10^{10} atomic disintegrations per second.
b. The roentgen is the amount of radioactive material needed to produce 2×10^{9} ion pairs when passing through 1 cc of air at $0°C$.

Chapter 10
An Introduction to Organic Chemistry:
The Saturated Hydrocarbons
Solutions to the Odd-Numbered Problems

In-Chapter Questions and Problems

10.1   The student could test the solubility of the substance in water and in an organic solvent, such as hexane. Solubility in hexane would suggest an organic substance; whereas solubility in water would indicate an inorganic compound. The student could also determine the melting and boiling points of the substance. If the melting and boiling points are very high, an inorganic substance would be suspected.

10.3

Hexane   2-Methylpentane   3-Methylpentane

2,3-Dimethylbutane   2,2-Dimethylbutane

10.5

a.

$$2° \quad 2°$$
$$\begin{array}{cccc} & H & H & H & H \\ H-C-C-C-C-H \\ 1°H & H & H & H & 1° \end{array}$$

b.

$$\begin{array}{c} H \; 1° \\ H-C-H \\ 1°H \; 4° \quad H \; 1° \\ H-C-C-C-H \\ H \quad | \quad H \\ H-C-H \\ H \quad 1° \end{array}$$

c.

$$\begin{array}{c} 1°H \\ H-C-H \\ H \; H \; 3° \quad H \; H \; 1° \\ H-C-C-C-C-C-H \\ 1° \; H \; | \; 3° \; H \; H \; H \\ H-C-H \\ H \; 1° \end{array}$$

10.7   a. 2,3-Dimethylbutane
b. 2,2-Dimethylpentane
c. 2,2-Dimethylpropane
d. 1,2,3-Tribromopropane

10.9   a. The straight-chain isomers of molecular formula C_4H_9Br:

$$\begin{array}{cccc} H & H & H & H \\ H-C-C-C-C-Br \\ H & H & H & H \end{array} \qquad \begin{array}{cccc} H & H & Br & H \\ H-C-C-C-C-H \\ H & H & H & H \end{array}$$

1-Bromobutane   2-Bromobutane

b. The linear isomers of molecular formula $C_4H_8Br_2$:

$$\begin{array}{cccc} H & H & H & Br \\ H-C-C \; C-C-Br \\ H & H & H & H \end{array} \qquad \begin{array}{cccc} H & H & Br & Br \\ H-C-C-C-C-H \\ H & H & H & H \end{array} \qquad \begin{array}{cccc} H & Br & H & H \\ H-C-C-C-C-Br \\ H & H & H & H \end{array}$$

1,1-Dibromobutane   1,2-Dibromobutane   1,3-Dibromobutane

$$\begin{array}{cccc} Br & H & H & H \\ H-C-C-C-C-Br \\ H & H & H & H \end{array} \qquad \begin{array}{cccc} H & H & Br & H \\ H-C-C-C-C-H \\ H & H & Br & H \end{array} \qquad \begin{array}{cccc} H & Br & Br & H \\ H-C-C-C-C-Br \\ H & H & H & H \end{array}$$

1,4-Dibromobutane   2,2-Dibromobutane   2,3-Dibromobutane

10.11   a. *trans*-1-Bromo-2-ethylcyclobutane
        b. *trans*-1,2-Dimethylcyclopropane
        c. Propylcyclohexane

10.13 The following are the isomers of cyclopropane:

10.15   Three of the six axial hydrogen atoms of cyclohexane lie above the ring. The remaining
        three hydrogen atoms lie below the ring.

10.17   a. The combustion of cyclobutane:

$$\square + 6O_2 \longrightarrow 4CO_2 + 4H_2O + \text{heat energy}$$

        b. The monobromination of propane will produce two products, as shown in the
           following two equations:

$$CH_3CH_2CH_3 + Br_2 \xrightarrow{\text{Light or heat}} CH_3CH_2CH_2Br + HBr$$

$$CH_3CH_2CH_3 + Br_2 \xrightarrow{\text{Light or heat}} CH_3CHBrCH_3 + HBr$$

        c. The complete combustion of ethane:

$$2CH_3CH_3 + 7O_2 \rightarrow 4CO_2 + 6H_2O + \text{heat energy}$$

        d. The monochlorination of butane will produce two products, as shown in the following
           two equations:

$$CH_3CH_2CH_2CH_3 + Cl_2 \xrightarrow{\text{Light or heat}} CH_3CH_2CH_2CH_2Cl + HCl$$

$$CH_3CH_2CH_2CH_3 + Cl_2 \xrightarrow{\text{Light or heat}} CH_3CH_2CHClCH_3 + HCl$$

10.19   The products of the reaction in Problem 10.17b are 1-bromopropane and 2-bromopropane. The products of the reactions in Problem 10.17d are 1-chlorobutane, 2-chlorobutane.

End-of-Chapter Questions and Problems

10.21   The number of organic compounds is nearly limitless because carbon forms stable covalent bonds with other carbon atoms in a variety of different patterns. In addition, carbon can form stable bonds with other elements and functional groups, producing many families of organic compounds, including alcohols, aldehydes, ketones, esters, ethers, amines, and amides. Finally, carbon can form double or triple bonds with other carbon atoms to produce organic molecules with different properties.

10.23   The allotropes of carbon include graphite, diamond, and buckminsterfullerene.

10.25   Because ionic substances often form three-dimensional crystals made up of many positive and negative ions, they generally have much higher melting and boiling points than covalent compounds.

10.27   a. $LiCl > H_2O > CH_4$

  b. $NaCl > C_3H_8 > C_2H_6$

10.29   a. LiCl would be a solid; H_2O would be a liquid; and CH_4 would be a gas.

  b. NaCl would be a solid; both C_3H_8 and C_2H_6 would be gases.

10.31   a. Water-soluble inorganic compounds are good electrolytes because they dissociate into ions in water. The ions conduct an electrical charge.
  b. Inorganic compounds exhibit ionic bonding.
  c. Organic compounds have lower melting points.
  d. Inorganic compounds are more likely to be water soluble.
  e. Organic compounds are flammable.

10.33

a.
```
      H             H
 H——H      H-C-H
 H |        |   H
 |  |       H   |
H-C-C——C——C-C-H
 H H    H   H H
```

b.
```
 H  Br H  H
 |  |  |  |
H-C-C-C-C-H
 |  |  |  |
 H  H  Br H
```

10.35

a.

$CH_3CH_2CHCH_3$
       |
       CH_3

b.

           CH_3 CH_3
            | |
$CH_3CH_2CHCH_2CH_2CHCH_3$
            |
           CH_3

10.37

a.   b.   c.

10.39

a.   b.   c.

10.41   Structure b is not possible because there are five bonds to carbon-2. Structure d is not possible because there are five bonds to carbon-3. Structure e is not possible because there are five bonds to carbon-2 and carbon-3 and only three bonds to carbon-4. Structure f is not possible because there are five bonds to carbon-3.

10.43

a.
                 H
                 |
             H–C–H
        H | H H H
        | | | | |
    H–C–C–C–C–C–H
        | | | | |
        H H | H H
             H–C–H
                 |
                 H

b.
        H H H H H
        | | | | |
    H–C–C–C–C–C–H
        | | | | |
        H H H H H

420

10.45

a.

$$H-\overset{\overset{\displaystyle H}{|}}{\underset{\underset{\displaystyle H}{|}}{C}}-\overset{\overset{\displaystyle H}{|}}{\underset{\underset{\displaystyle H}{|}}{C}}-\overset{\overset{\displaystyle H-\overset{\overset{\displaystyle H}{|}}{\underset{\underset{\displaystyle H}{|}}{C}}-H}{|}}{\underset{\underset{\displaystyle H}{|}}{C}}-\overset{\overset{\displaystyle H}{|}}{\underset{\underset{\displaystyle H}{|}}{C}}-\overset{\overset{\displaystyle H}{|}}{\underset{\underset{\displaystyle H}{|}}{C}}-\overset{\overset{\displaystyle H-\overset{\overset{\displaystyle H}{|}}{\underset{\underset{\displaystyle H}{|}}{C}}-H}{|}}{\underset{\underset{\displaystyle H}{|}}{C}}-\overset{\overset{\displaystyle H}{|}}{\underset{\underset{\displaystyle H}{|}}{C}}-H$$

b.

$$H-\overset{\overset{\displaystyle H-\overset{\overset{\displaystyle H}{|}}{\underset{\underset{\displaystyle H}{|}}{C}}-H}{|}}{\underset{\underset{\displaystyle H-\overset{\overset{\displaystyle H}{|}}{\underset{\underset{\displaystyle H}{|}}{C}}-H}{|}}{C}}-\overset{\overset{\displaystyle H}{|}}{\underset{\underset{\displaystyle H}{|}}{C}}-\overset{\overset{\displaystyle H}{|}}{\underset{\underset{\displaystyle H}{|}}{C}}-H$$

10.47 An alcohol-Ethanol:

$$H-\overset{\overset{\displaystyle H}{|}}{\underset{\underset{\displaystyle H}{|}}{C}}-\overset{\overset{\displaystyle H}{|}}{\underset{\underset{\displaystyle H}{|}}{C}}-OH$$

An aldehyde-Ethanal:

$$H-\overset{\overset{\displaystyle H}{|}}{\underset{\underset{\displaystyle H}{|}}{C}}-\overset{\overset{\displaystyle O}{||}}{C}-H$$

A ketone-Propanone:

$$H-\overset{\overset{\displaystyle H}{|}}{\underset{\underset{\displaystyle H}{|}}{C}}-\overset{\overset{\displaystyle O}{||}}{C}-\overset{\overset{\displaystyle H}{|}}{\underset{\underset{\displaystyle H}{|}}{C}}-H$$

A carboxylic acid-Ethanoic acid:

$$H-\overset{\overset{\displaystyle H}{|}}{\underset{\underset{\displaystyle H}{|}}{C}}-\overset{\overset{\displaystyle O}{||}}{C}-OH$$

An amine - Ethanamine:

$$H-\overset{\overset{\displaystyle H}{|}}{\underset{\underset{\displaystyle H}{|}}{C}}-\overset{\overset{\displaystyle O}{||}}{C}-N\overset{\displaystyle H}{\underset{\displaystyle H}{}}$$

10.49 a. An alkane: C_nH_{2n+2} d. A cycloalkane: C_nH_{2n}
     b. An alkyne: C_nH_{2n-2} e. A cycloalkene: C_nH_{2n-2}
     c. An alkene: C_nH_{2n}

10.51 Alkanes have only carbon-to-carbon and carbon-to-hydrogen single bonds, as in the molecule ethane:

$$H-\overset{\overset{\displaystyle H}{|}}{\underset{\underset{\displaystyle H}{|}}{C}}-\overset{\overset{\displaystyle H}{|}}{\underset{\underset{\displaystyle H}{|}}{C}}-H$$

Alkenes have at least one carbon-to-carbon double bond, as in the molecule ethene:

$$\underset{H}{\overset{H}{\diagdown}}C=C\underset{H}{\overset{H}{\diagup}}$$

Alkynes have at least one carbon-to-carbon triple bond, as in the molecule ethyne:

$$H-C\equiv C-H$$

10.53 a. A carboxylic acid (Propanoic acid):

$$CH_3CH_2-\overset{\overset{\displaystyle O}{\|}}{C}-OH$$

b. An amine (Propanamine): $CH_3CH_2CH_2-NH_2$

c. An alcohol (1-Propanol): $CH_3CH_2CH_2-OH$

d. An ether (Diethyl ether): $CH_3CH_2-O-CH_2CH_3$

10.55

Aspirin
Acetylsalicylic acid

10.57 Hydrocarbons are nonpolar molecules, and hence are not soluble in water.

10.59 a. heptane > hexane > butane > ethane
b. $CH_3CH_2CH_2CH_2CH_2CH_2CH_2CH_2CH_3$ > $CH_3CH_2CH_2CH_2CH_3$ > $CH_3CH_2CH_3$

10.61 a. Heptane and hexane would be liquid at room temperature; butane and ethane would be gases.
b. $CH_3CH_2CH_2CH_2CH_2CH_2CH_2CH_2CH_3$ and $CH_3CH_2CH_2CH_2CH_3$ would be liquids at room temperature; $CH_3CH_2CH_3$ would be a gas.

10.63 Nonane: $CH_3CH_2CH_2CH_2CH_2CH_2CH_2CH_2CH_3$

Pentane: $CH_3CH_2CH_2CH_2CH_3$

Propane: $CH_3CH_2CH_3$

10.65 a. 2-Bromobutane:

$$\overset{\displaystyle Br}{\underset{\displaystyle |}{CH_3CHCH_2CH_3}}$$

b. 2-Chloro-2-methylpropane:

$$CH_3\overset{\overset{\displaystyle Cl}{|}}{\underset{\underset{\displaystyle CH_3}{|}}{-C-}}CH_3$$

c. 2,2-Dimethylhexane:

$$CH_3\overset{\overset{\displaystyle CH_3}{|}}{\underset{\underset{\displaystyle CH_3}{|}}{-C-}}CH_2CH_2CH_2CH_3$$

10.67

a. 2,2-Dibromobutane:

$$\begin{array}{ccccc} & H & Br & H & H \\ H- & C- & C- & C- & C-H \\ & H & Br & H & H \end{array}$$

b. 2-Iododecane:

$$\begin{array}{ccccccccccc} & H & I & H & H & H & H & H & H & H & H \\ H- & C- & C- & C- & C- & C- & C- & C- & C- & C- & C-H \\ & H & H & H & H & H & H & H & H & H & H \end{array}$$

c. 1,2-Dichloropentane:

$$\begin{array}{cccccc} & H & Cl & H & H & H \\ Cl- & C- & C- & C- & C- & C-H \\ & H & H & H & H & H \end{array}$$

d. 1-Bromo-2-methylpentane:

$$\begin{array}{ccccc} & & H & & \\ & & H-C-H & & \\ & H & | & H & H & H \\ H- & C- & C- & C- & C- & C-H \\ & Br & H & H & H & H \end{array}$$

10.69   a. 3-Methylpentane   c. 1-Bromoheptane
        b. 2,5-Dimethylhexane   d. 1-Chloro-3-methylbutane

10.71   a. 2-Chloropropane   d. 1-Chloro-2-methylpropane
        b. 2-Iodobutane   e. 2-Iodo-2-methylpropane
        c. 2,2-Dibromopropane

10.73   a. 2-Chlorohexane   c. 3-Chloropentane
        b. 1,4-Dibromobutane   d. 2-Methylheptane

10.75   a. The first pair of molecules are constitutional isomers: hexane and 2-methylpentane.
        b. The second pair of molecules are identical. Both are heptane.

10.77   a. Incorrect: 3-Methylhexane
        b. Incorrect: 2-Methylbutane
        c. Incorrect: 3-Methylheptane
        d. Correct

10.79

$$CH_3$$
$$|$$
a. $CH_3CHCH_2CHCH_3$ The name given in the problem is correct.
$$|$$
$$CH_3$$

$$CH_3$$
$$|$$
b. $CH_3CH_2CH_2CHCH_2CH_3$ The correct name is 3-methylhexane.

c. $I\!-\!CH_2CH_2CH_2CH_2CH_2\!-\!I$ The name given in the problem is correct.

d. $CH_3CH_2CH_2CH_2CH_2CHCH_2CH_2CH_3$ The correct name is 4-ethylnonane.
$$|$$
$$CH_2CH_3$$

     Br   Br
     | |
e. $CH_2CH_2CH_2CH_2CH_2CCH_2CH_3$ The name given in the problem is correct.
$$|$$
$$CH_3$$

10.81 Cycloalkanes are a family of molecules having carbon-to-carbon bonds in a ring structure.

10.83 The general formula for a cycloalkane is C_nH_{2n}.

10.85 a. Chlorocyclopropane
b. *cis*-1,2-Dichlorocyclopropane
c. *trans*-1,2-Dichlorocyclopropane
d. Bromocyclobutane

10.87 a. 1-Bromo-2-methylcyclobutane: b. Iodocyclopropane:

10.89 a. Incorrect: 1,2-Dibromocyclobutane
b. Incorrect: 1,2-Diethylcyclobutane
c. Correct
d. Incorrect: 1,2,3-Trichlorocyclohexane

10.91

a.

Br   Br

cis-1,3-Dibromocyclopentane

b.   CH₃

CH₃

trans-1,2-Dimethylcyclobutane

c.

Cl

Cl

cis-1,2-Dichlorocyclopropane

d.   CH₂CH₃

CH₂CH₃

trans-1,4-Diethylcyclohexane

10.93 a. *cis*-1,2-Dibromocyclopentane   c. *cis*-1,2-Dimethylcyclohexane
       b. *trans*-1,3-Dibromocyclopentane   d. *cis*-1,2-Dimethylcyclopropane

10.95 Conformational isomers are distinct isomeric structures that may be converted into one another by rotation about the bonds in the molecule.

10.97 In the chair conformation, the hydrogen atoms, and thus the electron pairs of C-H bonds, are farther from one another. As a result, there is less electron repulsion and the structure is more stable (more energetically favored). In the boat conformation, the electron pairs are more crowded. This causes greater electron repulsion, producing a less stable, less energetically favored conformation.

10.99 Because conformations are freely and rapidly interconverted, they cannot be separated from one another.

10.101 One conformation, termed the staggered conformation, is more stable than the other, termed the eclipsed conformation, because the electron pairs of the carbon-hydrogen bonds are farther from one another in the staggered conformation. This results in less electron repulsion and a more stable conformation. In the eclipsed conformation, the electrons are more crowded, resulting in greater electron repulsion and a less stable conformation.

10.103 a. $C_3H_8 + 5O_2 \rightarrow 4H_2O + 3CO_2$

       b. $C_7H_{16} + 11O_2 \rightarrow 8H_2O + 7CO_2$

       c. $C_9H_{20} + 14O_2 \rightarrow 10H_2O + 9CO_2$

       d. $2C_{10}H_{22} + 31O_2 \rightarrow 22H_2O + 20CO_2$

10.105 a. $8\,CO_2 + 10\,H_2O$

       b.

$$\underset{\text{2-Bromo-2-methylpropane}}{\overset{\displaystyle CH_3}{\underset{\displaystyle CH_3}{Br\!-\!\overset{|}{\underset{|}{C}}\!-\!CH_3}}} \quad + \quad \underset{\text{1-Bromo-2-methylpropane}}{\overset{\displaystyle CH_3}{\overset{|}{CH_3CHCH_2Br}}} + 2HBr$$

       c. $Cl_2 + $ light

10.107 The following molecules are all the constitutional isomers of C_6H_{14}:

CH₃

CH₃

$CH_3CH_2CH_2CH_2CH_2CH_3$ $CH_3CHCH_2CH_2CH_3$ $CH_3CH_2CHCH_2CH_3$

Hexane   2-Methylpentane   3-Methylpentane

CH₃

CH₃

$CH_3CHCHCH_3$ $CH_3CCH_2CH_3$

CH₃   CH₃

2,3-Dimethylbutane   2,2-Dimethylbutane

a. 2,3-Dimethylbutane produces only two monobrominated derivatives: 1-bromo-2,3-dimethylbutane and 2-bromo-2,3-dimethylbutane.

b. Hexane produces three monobrominated products: 1-bromohexane, 2-bromohexane, and 3-bromohexane. 2,2-Dimethylbutane also produces three monobrominated products: 1-bromo-2,2-dimethylbutane, 2-bromo-3,3-dimethylbutane, and 1-bromo-3,3-dimethylbutane.

c. 3-Methylpentane produces four monobrominated products: 1-bromo-3-methylpentane, 2-bromo-3-methylpentane, 3-bromo-3-methylpentane, and 1-bromo-2-ethylbutane.

10.109 The hydrocarbon is cyclooctane having a molecular formula of C_8H_{16}.

Chapter 11
The Unsaturated Hydrocarbons:
Alkenes, Alkynes, and Aromatics
Solutions to the Odd-Numbered Problems

In-Chapter Questions and Problems

11.1   a.

$$Br - \underset{\underset{H}{|}}{\overset{\overset{H}{|}}{C}} - \underset{\underset{H}{|}}{\overset{\overset{H}{|}}{C}} - C \equiv C - \underset{\underset{H}{|}}{\overset{\overset{H}{|}}{C}} - \underset{\underset{H}{|}}{\overset{\overset{H}{|}}{C}} - H$$

b.

$$H - \underset{\underset{H}{|}}{\overset{\overset{H}{|}}{C}} - C \equiv C - \underset{\underset{H}{|}}{\overset{\overset{H}{|}}{C}} - H$$

11.3

a. $Cl - C \equiv C - Cl$

b.

$$H - C \equiv C - \underset{\underset{H}{|}}{\overset{\overset{H}{|}}{C}} - \underset{\underset{H}{|}}{\overset{\overset{H}{|}}{C}} - \underset{\underset{H}{|}}{\overset{\overset{H}{|}}{C}} - \underset{\underset{H}{|}}{\overset{\overset{H}{|}}{C}} - \underset{\underset{H}{|}}{\overset{\overset{H}{|}}{C}} - \underset{\underset{H}{|}}{\overset{\overset{H}{|}}{C}} - \underset{\underset{H}{|}}{\overset{\overset{H}{|}}{C}} - I$$

11.5   a.

cis-3-Hexene   trans-3-Hexene

b.

cis-2,3-Dibromo-2-butene   *trans*-2,3-Dibromo-2-butene

11.7   Molecule c can exist as *cis*- and *trans*-isomers because there are two different groups on each of the carbon atoms attached by the double bond.

11.9   a.

b.

c.

11.11   The hydrogenation of the *cis*- and *trans*- isomers of 2-pentene would produce the same product, pentane.

11.13

a. $H_3C-C\equiv C-CH_3$ + 2 H$_2$ $\xrightarrow{\text{Ni}}$

        2-Butyne   Butane

b. $H_3C-C\equiv C-CH_2CH_3$ + 2 H$_2$ $\xrightarrow{\text{Ni}}$

        2-Pentyne   Pentane

11.15

a. $CH_3CH=CH_2$ + Br_2 $\longrightarrow$

$$\begin{array}{ccc} H & Br & H \\ | & | & | \\ H-C-C-CH \\ | & | & | \\ H & H & Br \end{array}$$

b. $CH_3CH=CHCH_3$ + Br_2 $\longrightarrow$

$$\begin{array}{cccc} H & H & Br & H \\ | & | & | & | \\ H-C-C-C-C-H \\ | & | & | & | \\ H & Br & H & H \end{array}$$

11.17

a. $CH_3C\equiv CCH_3$ + 2 Cl_2 $\longrightarrow$

$$\begin{array}{cccc} H & Cl & Cl & H \\ | & | & | & | \\ H-C-C-C-C-H \\ | & | & | & | \\ H & Cl & Cl & H \end{array}$$

b. $CH_3C\equiv CCH_2CH_3$ + 2 Cl_2 $\longrightarrow$

$$\begin{array}{ccccc} H & Cl & Cl & H & H \\ | & | & | & | & | \\ H-C-C-C-C-C-H \\ | & | & | & | & | \\ H & Cl & Cl & H & H \end{array}$$

11.19

a.

$$CH_3CH=CHCH_3 \quad + \quad H_2O \quad \xrightarrow{H^+} \quad \overset{\overset{\displaystyle OH}{|}}{CH_3CHCH_2CH_3}$$

b.

$$H_2C=CHCH_2CH_2\overset{\overset{\displaystyle CH_3}{|}}{CHCH_3} + H_2O \xrightarrow{H^+} \overset{\overset{\displaystyle OH}{|}}{CH_3CHCH_2CH_2}\overset{\overset{\displaystyle CH_3}{|}}{CHCH_3}$$
(major product)

$$H_2C=CHCH_2CH_2\overset{\overset{\displaystyle CH_3}{|}}{CHCH_3} + H_2O \xrightarrow{H^+} \overset{\overset{\displaystyle OH}{|}}{CH_2CH_2CH_2CH_2}\overset{\overset{\displaystyle CH_3}{/}}{CHCH_3}$$
(minor product)

c. These products will be formed in approximately equal amounts:

$$CH_3CH_2CH_2CH = CHCH_2CH_3 + H_2O \xrightarrow{\ H^+\ } \overset{\overset{\displaystyle OH}{\displaystyle |}}{CH_3CH_2CH_2CHCH_2CH_2CH_3}$$

$$CH_3CH_2CH_2CH = CHCH_2CH_3 + H_2O \xrightarrow{\ H^+\ } \overset{\overset{\displaystyle OH}{\displaystyle |}}{CH_3CH_2CH_2CH_2CHCH_2CH_3}$$

d.

$$\overset{\overset{\displaystyle Cl}{\displaystyle |}}{CH_3CHCH} = \overset{\overset{\displaystyle Cl}{\displaystyle \backslash}}{CHCHCH_3} + H_2O \xrightarrow{\ H^+\ } \overset{\overset{\displaystyle Cl\ \ OH\ \ \ Cl}{\displaystyle |\ \ \ |\ \ \ \ |}}{CH_3CHCHCH_2CHCH_3}$$
(only product)

11.21

a. $H_3C-C \equiv CH + H_2O \xrightarrow{\ H^+\ }$ (vinyl alcohol intermediate) $\longrightarrow$ (ketone product)

Or

$H_3C-C \equiv CH + H_2O \xrightarrow{\ H^+\ }$ (enol intermediate) $\longrightarrow$ (alcohol product)

b. $H_3C-C \equiv CCH_2CH_3 + H_2O \xrightarrow{\ H^+\ }$ (enol intermediate) $\longrightarrow$ (ketone product)

Or

$H_3C-C \equiv CCH_2CH_3 + H_2O \xrightarrow{\ H^+\ }$ (enol intermediate) $\longrightarrow$ (ketone product)

11.23 a. Reactant - *cis*-2-butene; only product—butane
b. Reactant - 1-butene; major product—2-butanol
c. Reactant - 2-butene; only product—2,3-dichlorobutane
d. Reactant - 1-pentene; major product—2-bromopentane

11.25 a.

b.

c.

d.

e.

f.

End-of-Chapter Questions and Problems

11.27   The longer the carbon chain of an alkene, the higher the boiling point.

11.29   The general formula for an alkane is C_nH_{2n+2}.
The general formula for an alkene is C_nH_{2n}.
The general formula for an alkyne is C_nH_{2n-2}.

11.31   Ethene is a planar molecule. All of the bond angles are 120°.

11.33   In alkanes, such as ethane, the four bonds around each carbon atom have tetrahedral geometry. The bond angles are 109.5°. In alkenes, such as ethene, each carbon is bonded by two single bonds and one double bond. The molecule is planar and each bond angle is approximately 120°.

11.35   Ethyne is a linear molecule. All of the bond angles are 180°.

11.37   In alkanes, such as ethane, the four bonds around each carbon atom have tetrahedral geometry. The bond angles are 109.5°. In alkenes, such as ethene, each carbon is bonded by two single bonds and one double bond. The molecule is planar and each bond angle is approximately 120°. In alkynes, such as ethyne, each carbon is bonded by one single bond and one triple bond. The molecular is linear and the bond angles are 180°.

11.39   a. 2-Pentyne > Propyne > Ethyne
b. 3-Decene > 2-Butene > Ethene

11.41   Identify the longest carbon chain containing the carbon-to-carbon double or triple bond. Replace the *–ane* suffix of the alkane name with *–ene* for an alkene or *-yne* for an alkyne. Number the chain to give the lowest number to the first of the two carbons involved in the double or triple bond. Determine the name and carbon number of each substituent group and place that information as a prefix in front of the name of the parent compound.

11.43   Geometric isomers of alkenes differ from one another in the placement of substituents attached to each of the carbon atoms of the double bond. Of the pair of geometric isomers, the *cis*-isomer is the one in which identical groups are on the same side of the double bond.

11.45 a.   2-Methyl-2-hexene:

$$H_3C \diagdown \atop H_3C \diagup C = C \diagup^{CH_2CH_2CH_3} \diagdown H$$

b.   *trans*-3-Heptene:

$$CH_3CH_2 \diagdown \atop H \diagup C = C \diagup^{H} \diagdown CH_2CH_2CH_3$$

c.   *cis*-1-Chloro-2-pentene:

$$ClCH_2 \diagdown \atop H \diagup C = C \diagup^{CH_2CH_3} \diagdown H$$

d.   *cis*-2-Chloro-2-methyl-3-heptene:

$$\begin{array}{c} CH_3 \\ | \\ H_3C - C - Cl \quad CH_2CH_2CH_3 \\ \diagdown \qquad \diagup \\ C = C \\ \diagup \qquad \diagdown \\ H \qquad\quad H \end{array}$$

e.   *trans*-5-Bromo-2,6-dimethyl-3-octene:

$$\begin{array}{c} CH_3 \\ | \\ H_3C - CH \qquad H \\ \diagdown \qquad \diagup \\ C = C \\ \diagup \qquad \diagdown \\ H \qquad CH - CH - CH_2CH_3 \\ \qquad\quad | \qquad | \\ \qquad\quad Br \quad CH_3 \end{array}$$

11.47 a.   3-Methyl-1-pentene
     b.   7-Bromo-1-heptene
     c.   5-Bromo-3-heptene
     d.   1-*t*-Butyl-4-methylcyclohexene

11.49 a. 1,3,5-Trifluoropentane:

$$\underset{|}{\overset{F}{\underset{CH_2}{|}}} - CH_2 - \underset{|}{\overset{F}{\underset{CH}{|}}} - CH_2 - \underset{|}{\overset{F}{\underset{CH_2}{|}}}$$

b. *cis*-2-Octene:

$$\underset{H_3C}{\overset{H}{\diagdown}} C = C \underset{CH_2CH_2CH_2CH_2CH_3}{\overset{H}{\diagup}}$$

c. Dipropylacetylene: $CH_3CH_2CH_2–C\equiv C–CH_2CH_2CH_3$

11.51 a. 2,3-Dibromobutane cannot exist as *cis* and *trans* isomers because it is an alkane.

b. 2-Heptene can exist as *cis* and *trans* isomers:

$$\underset{CH_3}{\overset{H}{\diagdown}} C = C \underset{CH_2CH_2CH_2CH_3}{\overset{H}{\diagup}} \qquad \underset{CH_3}{\overset{H}{\diagdown}} C = C \underset{H}{\overset{CH_2CH_2CH_2CH_3}{\diagup}}$$

   cis-2-Heptene   *trans*-2-Heptene

c. 2,3-Dibromo-2-butene can exist as *cis* and *trans* isomers:

$$\underset{CH_3}{\overset{Br}{\diagdown}} C = C \underset{CH_3}{\overset{Br}{\diagup}} \qquad \underset{CH_3}{\overset{Br}{\diagdown}} C = C \underset{Br}{\overset{CH_3}{\diagup}}$$

   cis-2,3-Dibromo-2-butene   *trans*-2,3-Dibromo-2-butene

d. Propene cannot exist as *cis* and *trans* isomers.

11.53 Alkenes b and c would not exhibit *cis-trans* isomerism.

11.55 Alkenes b and d can exist as both *cis*- and *trans*- isomers.

11.57 a. 1,5-Nonadiene
       b. 1,4,7-Nonatriene
       c. 2,5-Octadiene
       d. 4-Methyl-2,5-heptadiene

11.59

R₂C=CR₂ + H₂ —Pt, Pd, or Ni; heat or pressure→ R-C(H)(R)-C(R)(H)-R

$$\underset{R}{\overset{R}{\diagdown}}C=C\underset{R}{\overset{R}{\diagup}} \; + H_2 \;\; \xrightarrow[\text{heat or pressure}]{\text{Pt, Pd, or Ni}} \;\; R-\underset{R}{\overset{H}{C}}-\underset{H}{\overset{R}{C}}-R$$

11.61

$$\underset{R}{\overset{R}{\diagdown}}C=C\underset{R}{\overset{R}{\diagup}} \; + X_2 \;\; \longrightarrow \;\; R-\underset{R}{\overset{X}{C}}-\underset{X}{\overset{R}{C}}-R$$

11.63

$$\underset{R}{\overset{R}{\diagdown}}C=C\underset{R}{\overset{R}{\diagup}} \; + H_2O \;\; \xrightarrow{H^+} \;\; R-\underset{R}{\overset{H}{C}}-\underset{OH}{\overset{R}{C}}-R$$

11.65 The primary difference between complete hydrogenation of an alkene and an alkyne is that 2 moles of H_2 are required for the complete hydrogenation of an alkyne.

11.67 Addition of bromine (Br_2) to an alkene results in a color change from red to colorless. If equimolar quantities of Br_2 are added to hexene, the reaction mixture will change from red to colorless. This color change will not occur if cyclohexane is used.

11.69 a. H_2 d. $19\,O_2 \rightarrow 12\,CO_2 + 14\,H_2O$

b. H_2O e. Cl_2

c. HBr   f.

11.71

a.

$$H_3C-C\equiv C-CH_3 \; + 2H_2 \;\; \xrightarrow[\text{heat or pressure}]{\text{Pt, Pd, or Ni}} \;\; H_3C-\underset{H}{\overset{H}{C}}-\underset{H}{\overset{H}{C}}-CH_3$$

2-Butyne

436

b.

$$CH_3CH_2-C\equiv C-CH_3 \quad +2\ X_2 \longrightarrow$$

2-Pentyne

$$CH_3CH_2-\underset{\underset{X}{|}}{\overset{\overset{X}{|}}{C}}-\underset{\underset{X}{|}}{\overset{\overset{X}{|}}{C}}-CH_3$$

11.73

$$CH_2=CHCH_2CH_2CH_3 \qquad\qquad CH_3CH=CHCH_2CH_3$$

$$H_2C=\underset{\underset{\displaystyle CH_3}{|}}{C}-CH_2CH_3 \qquad H_2C=CH\underset{\overset{\displaystyle CH_3}{\backslash}}{C}HCH_3 \qquad CH_3-\underset{\underset{\displaystyle CH_3}{|}}{C}=CHCH_3$$

11.75 a.

$$CH_3\underset{\overset{\displaystyle Br}{|}}{C}HCH_2CH_3$$

b.

$$CH_3CH_2-\underset{\underset{\displaystyle I}{|}}{\overset{\overset{\displaystyle CH_3}{|}}{C}}-CH_2CH_2CH_3 \qquad\qquad CH_3\underset{\overset{\displaystyle I}{|}}{C}H\underset{\overset{\displaystyle CH_3}{|}}{C}HCH_2CH_2CH_3$$

(major product)   (minor product)

c.

11.77 A polymer is a macromolecule composed of repeating structural units called *monomers*.

11.79

Tetrafluoroethane   Teflon

11.81 The I.U.P.A.C. name for (a) is 2-pentene, for (b) is 3-bromo-1-propene, and for (c) is 2,3-dimethylcyclohexene.

a. These products will be formed in approximately equal amounts.

$$CH_3\overset{\displaystyle H}{\underset{\displaystyle H}{C}}=CCH_2CH_3 \ + \ H_2O \ \xrightarrow{H^+} \ CH_3\overset{\displaystyle OH}{C}HCH_2CH_2CH_3$$

$$CH_3\overset{\displaystyle H}{\underset{\displaystyle H}{C}}=CCH_2CH_3 \ + \ H_2O \ \xrightarrow{H^+} \ CH_3CH_2\overset{\displaystyle OH}{C}HCH_2CH_3$$

b.

$$H_2\overset{\displaystyle Br}{C}-\overset{\displaystyle H}{C}=CH_2 \ + \ H_2O \ \xrightarrow{H^+} \ H_2\overset{\displaystyle Br}{\underset{\displaystyle H}{C}}-\overset{\displaystyle H}{\underset{\displaystyle H}{C}}-\overset{\displaystyle H}{C}-OH \ + \ H_2\overset{\displaystyle Br}{\underset{\displaystyle H}{C}}-\overset{\displaystyle OH}{\underset{\displaystyle H}{C}}-\overset{\displaystyle H}{C}-$$

Minor Product   Major Produc

c. These products will be formed in approximately equal amounts.

+ H₂O $\xrightarrow{H^+}$

+ H₂O $\xrightarrow{H^+}$

438

11.83 a. (This is the minor product of this reaction.)

$$H_2C = CHCH_2\overset{\overset{\displaystyle CH_3}{|}}{C}HCH_3 \ + \ H_2O \ \xrightarrow{\ H^+\ } \ CH_2CH_2CH_2\overset{\overset{\displaystyle CH_3}{|}}{C}HCH_3$$
$$\hspace{10cm}\overset{\overset{\displaystyle OH}{|}}{}$$

b.

$$CH_3CH = CHCH_2CH_2CH_3 \ + \ HBr \ \longrightarrow \ CH_3CH_2\overset{\overset{\displaystyle Br}{|}}{C}HCH_2CH_2CH_3$$

OR

$$CH_3CH_2CH = CHCH_2CH_3 \ + \ HBr \ \longrightarrow \ CH_3CH_2\overset{\overset{\displaystyle Br}{|}}{C}HCH_2CH_2CH_3$$

c.

$+ \ HBr \ \longrightarrow$

d.

$+ \ H_2O \ \xrightarrow{\ H^+\ }$

11.85

a. $CH_2=CHCH_2CH=CHCH_3 \ + \ 2H_2 \ \xrightarrow[\text{heat}]{\text{Pt}} \ CH_3CH_2CH_2CH_2CH_2CH_3$

    1,4-Hexadiene   Hexane

b. $CH_3CH=CHCH=CHCH=CHCH_3 \ + \ 3H_2 \ \xrightarrow[\text{heat}]{\text{Ni}} \ CH_3CH_2CH_2CH_2CH_2CH_2CH_2CH_3$

    2,4,6-Octatriene   Octane

c.

$+ \ 2H_2 \ \xrightarrow[\text{pressure}]{\text{Pd}}$

    1,3-Cyclohexadiene   Cyclohexane

d.

1,3,5-Cyclooctatriene + 3H₂ → (Ni, heat) → Cyclooctane

11.87 The term *aromatic hydrocarbon* was first used as a term to describe the pleasant-smelling resins of tropical trees.

11.89 Resonance hybrids are molecules for which more than one valid Lewis structure can be written.

11.91 a. 2,4-Dibromotoluene:

b. 1,2,4-Triethylbenzene:

c. Isopropylbenzene:

d. 2-Bromo-5-chlorotoluene:

11.93 a.

b.

c.

d.

11.95 Kekulé proposed that single and double carbon-carbon bonds alternate around the benzene ring. To explain why benzene does not react like other unsaturated compounds, he proposed that the double and single bonds shift positions rapidly.

11.97 An addition reaction involves addition of a molecule to a double or triple bond in an unsaturated molecule. In a substitution reaction, one chemical group replaces another.

11.99

Benzene   Chlorine   Chlorobenzene

11.101

Pyrimidine

11.103

Purine

Chapter 12
Alcohols, Phenols, Thiols, and Ethers
Solutions to the Odd-Numbered Problems

In-Chapter Questions and Problems

12.1 a. 4-Methyl-1-pentanol
      b. 4-Methyl-2-hexanol
      c. 1,2,3-Propanetriol
      d. 4-Chloro-3-methyl-1-hexanol

12.3 a. primary alcohol
      b. secondary alcohol
      c. tertiary alcohol
      d. aromatic alcohol (phenol)
      e. secondary alcohol

12.5

a. $CH_3CH{=\!=}CH_2 + H_2O \xrightarrow{H^+} CH_3CH(OH)CH_3 + CH_3CH_2CH_2OH$

   major product · · · · · minor product

b. $CH_2{=\!=}CH_2 + H_2O \xrightarrow{H^+} CH_3CH_2OH$

c. $CH_3CH_2CH{=\!=}CHCH_2CH_3 + H_2O \xrightarrow{H^+} CH_3CH_2CH_2CH(OH)CH_2CH_3$

12.7 a. The major product is a secondary alcohol (2-propanol) and the minor product is a
         primary alcohol (1-propanol).
      b. The product, ethanol, is a primary alcohol.
      c. The product, 3-hexanol, is a secondary alcohol.

12.9 The following equation represents the reduction of the ketone, butanone. This reaction
      requires a catalyst.

$$CH_3CH_2\overset{\overset{\displaystyle O}{\|}}{C}CH_3 + H_2 \rightarrow CH_3CH_2CH(OH)CH_3$$

12.11 The following equation represents the reduction of the aldehyde, butanal. This reaction
       requires a catalyst. The product is 1-butanol.

$$CH3CH2CH2\overset{\overset{\displaystyle O}{\|}}{C}{-}H + H_2 \rightarrow CH_3CH_2CH_2CH_2OH$$

12.13 a. Ethanol
      b. 2-Propanol is the major product. 1-Propanol is the minor product.
      c. 2-Butanol

12.15 a. 2-Butanol is the major product. 1-Butanol is the minor product.
      b. 2-Methyl-2-propanol is the major product. 2-Methyl-1-propanol is the minor product.

12.17 a. $CH_3CH{=}CH_2$

      b. $CH_3CH{=}CHCH_3 \ + \ CH_3CH_2CH{=}CH_2$

12.19

  a.
$$\underset{\underset{CH_3}{|}}{\overset{\overset{CH_3}{|}}{CH_3CCH_2CH_2OH}} \rightarrow \underset{\underset{CH_3}{|}}{\overset{\overset{CH_3}{|}}{CH_3CCH_2}}\overset{\overset{O}{\|}}{C}{-}H$$

  b.
$$CH_3CH_2OH \ \rightarrow \ CH_3\overset{\overset{O}{\|}}{C}{-}H$$

12.21

  a.
$$\underset{}{\overset{\overset{OH}{|}}{CH_3CHCH_2CH_3}} \rightarrow CH_3\overset{\overset{O}{\|}}{C}CH_2CH_3$$

  b.
$$\overset{\overset{OH}{|}}{CH_3CHCH_2CH_2CH_3} \rightarrow CH_3\overset{\overset{O}{\|}}{C}CH_2CH_2CH_3$$

12.23 a. 1-Ethoxypropane
      b. 1-Methoxypropane

12.25 a. Ethyl propyl ether
      b. Methyl propyl ether

12.27

$$CH_3CH_2OH \ + \ CH_3CH_2OH \ \xrightarrow{\ H^+\ } \ CH_3CH_2{-}O{-}CH_2CH_3 \ + \ H_2O$$

            Ethanol   Diethyl ether   Water

12.29 The longer the hydrocarbon tail of an alcohol becomes, the less water soluble it will be.

12.31 a < d < c < b

12.33

    a. CH_3CH_2OH

    b. $CH_3CH_2CH_2CH_2OH$

    c. $CH_3\overset{\displaystyle |}{\underset{\displaystyle OH}{C}}HCH_3$

12.35 The I.U.P.A.C. rules for the nomenclature of alcohols require you to name the parent compound, that is the longest continuous carbon chain bonded to the –OH group. Replace the –*e* ending of the parent alkane with –*ol* of the alcohol. Number the parent chain so that the carbon bearing the hydroxyl group has the lowest possible number. Name and number all other substituents. If there are more than one hydroxyl groups, the –*ol* ending will be modified to reflect the number. If there are two –OH groups, the suffix –*diol* is used; if it has three –OH groups, the suffix –*triol* is used, etc.

12.37 a. 1-Heptanol
      b. 2-Propanol
      c. 2,2-Dimethylpropanol

12.39 a. 3-Hexanol:

$$H-\overset{\displaystyle H}{\underset{\displaystyle H}{C}}-\overset{\displaystyle H}{\underset{\displaystyle H}{C}}-\overset{\displaystyle OH}{\underset{\displaystyle H}{C}}-\overset{\displaystyle H}{\underset{\displaystyle H}{C}}-\overset{\displaystyle H}{\underset{\displaystyle H}{C}}-\overset{\displaystyle H}{\underset{\displaystyle H}{C}}-H$$

    b. 1,2,3-Pentanetriol:

$$H-\overset{\displaystyle OH}{\underset{\displaystyle H}{C}}-\overset{\displaystyle OH}{\underset{\displaystyle H}{C}}-\overset{\displaystyle OH}{\underset{\displaystyle H}{C}}-\overset{\displaystyle H}{\underset{\displaystyle H}{C}}-\overset{\displaystyle H}{\underset{\displaystyle H}{C}}-H$$

c. 2-Methyl-2-pentanol:

$$
\begin{array}{c}
\text{H}\text{OH}\text{H}\text{H}\text{H}\\
|||||\\
\text{H}-\text{C}-\text{C}-\text{C}-\text{C}-\text{C}-\text{H}\\
||||\\
\text{H}\text{H}\text{H}\text{H}\\
|\\
\text{H}-\text{C}-\text{H}\\
|\\
\text{H}
\end{array}
$$

12.41 a. Cyclopentanol
       b. Cyclooctanol
       c. 3-Methylcyclohexanol

12.43 a. Methyl alcohol
       b. Ethyl alcohol
       c. Ethylene glycol
       d. Propyl alcohol

12.45 a. 4-Methyl-2-hexanol:

$$
\begin{array}{c}
\text{OH}\text{CH}_3\\
||\\
\text{CH}_3\text{CHCH}_2\text{CHCH}_2\text{CH}_3
\end{array}
$$

       b. Isobutyl alcohol:

$$
\begin{array}{c}
\text{CH}_3\\
|\\
\text{H}_3\text{C}-\text{C}-\text{CH}_2\text{OH}\\
|\\
\text{H}
\end{array}
$$

       c. 1,5-Pentanediol:

$$
\begin{array}{c}
\text{OH}\text{OH}\\
||\\
\text{CH}_2\text{CH}_2\text{CH}_2\text{CH}_2\text{CH}_2
\end{array}
$$

       d. 2-Nonanol:

$$
\begin{array}{c}
\text{OH}\\
|\\
\text{CH}_3\text{CHCH}_2\text{CH}_2\text{CH}_2\text{CH}_2\text{CH}_2\text{CH}_2\text{CH}_3
\end{array}
$$

       e. 1,3,5-Cyclohexanetriol:

12.47 Denatured alcohol is 100% ethanol to which benzene or methanol is added. The additive makes the ethanol unfit to drink and prevents illegal use of pure ethanol.

12.49 Fermentation is the anaerobic degradation of sugar that involves no net oxidation. The alcohol fermentation, carried out by yeast, produces ethanol and carbon dioxide.

12.51 When the ethanol concentration in a fermentation reaches 12-12%, the yeast producing the ethanol are killed by it. To produce a liquor of higher alcohol concentration, the product of the original fermentation must be distilled.

12.53 The carbinol carbon is the one to which the hydroxyl group is bonded.

12.55 a. primary alcohol
       b. secondary alcohol
       c. tertiary alcohol
       d. tertiary alcohol
       e. tertiary alcohol

12.57 a. tertiary alcohol
       b. secondary alcohol
       c. primary alcohol
       d. tertiary alcohol

12.59

12.61

12.63

Secondary alcohol   Ketone

12.65 a. 2-Pentanol is the major product and 1-pentanol is the minor product.
b. 2-Pentanol and 3-pentanol
c. 3-Methyl-2-butanol is the major product and 3-methyl-1-butanol is the minor product.
d. 3,3-Dimethyl-2-butanol is the major product and 3,3-dimethyl-1-butanol is the minor product.

12.67 a. These products will be formed in approximately equal amounts.

$$CH_3CH=CHCH_2CH_2CH_3 + H_2O \xrightarrow{H^+} CH_3\overset{\overset{\displaystyle OH}{|}}{C}HCH_2CH_2CH_2CH_3$$

2-Hexene   2-Hexanol

or 3-Hexanol $CH_3CH_2\overset{\overset{\displaystyle OH}{|}}{C}HCH_2CH_2CH_3$

b.

Cyclopentene   Cyclopentanol

c.

$$CH_2=CHCH_2CH_2CH_2CH_2CH_2CH_3 + H_2O \xrightarrow{H^+} CH_3\overset{\overset{\displaystyle OH}{|}}{C}HCH_2CH_2CH_2CH_2CH_2CH_3$$

2-Octanol
(major product)

1-Octanol $\overset{\overset{\displaystyle OH}{|}}{C}H_2CH_2CH_2CH_2CH_2CH_2CH_2CH_3$
(minor product)

d.

1-Methylcyclohexanol
(major product)

or

2-Methylcyclohexanol
(minor product)

12.69 a. Butanone
b. N.R.
c. Cyclohexanone
d. N.R.

12.71 a. 3-Pentanone
b. Propanal (Upon further oxidation, propanoic acid would be formed.)
c. 4-Methyl-2-pentanone
d. N.R.
e. 3-Phenylpropanal (Upon further oxidation, 3-phenylpropanoic acid will be formed.)

12.73

$$CH_3CH_2OH \xrightarrow{\text{liver enzymes}} CH_3-\overset{\overset{\textstyle O}{\|}}{C}-H$$

Ethanol   Ethanal

The product, ethanal, is responsible for the symptoms of a hangover.

12.75 The reaction in which a water molecule is added to 1-butene is a hydration reaction.

$$CH_2=CHCH_2CH_3 + H_2O \xrightarrow{H^+} CH_3\overset{\overset{\textstyle OH}{|}}{C}HCH_2CH_3$$

2-Butanol is the major product of this reaction. 1-Butanol would also be formed as a minor product.

12.77

$$CH_3CH=CH_2 \xrightarrow{H_2O, \; H^+} \underset{\substack{| \\ CH_3CHCH_3}}{\overset{OH}{}} \xrightarrow{[O]} CH_3-\overset{\overset{\displaystyle O}{\parallel}}{C}-CH_3$$

| Propene | 2-Propanol | Propanone |
| (propylene) | (isopropanol) | (acetone) |

12.79 An oxidation product of cholesterol:

12.81 When discussing inorganic compounds, oxidation is a loss of electrons, whereas reduction is a gain of electrons.

12.83

$$CH_3CH_2CH_3 \quad < \quad CH_3CH_2CH_2OH \quad < \quad CH_3CH_2\overset{\overset{\displaystyle O}{\parallel}}{C}-H \quad < \quad CH_3CH_2\overset{\overset{\displaystyle O}{\parallel}}{C}-OH$$

12.85 Phenols arc compounds in which the hydroxyl group is attached to a benzene ring.

12.87 Picric acid: 2,4,6,-Trinitrotoluene:

Picric acid is water soluble because its polar hydroxyl group can form hydrogen bonds with water.

12.89 Hexachlorophene, hexylresorcinol, and *o*-phenylphenol are phenol compounds used as antiseptics or disinfectants.

12.91 Alcohols of molecular formula $C_4H_{10}O$

$$CH_3CH_2CH_2CH_2OH \qquad \underset{\displaystyle \overset{|}{CH_3}}{CH_3\overset{\displaystyle \overset{OH}{|}}{C}HCH_2CH_3} \qquad \underset{\displaystyle \overset{|}{CH_3}}{CH_3\overset{}{C}HCH_2OH} \qquad CH_3-\underset{\displaystyle \overset{|}{CH_3}}{\overset{\displaystyle \overset{OH}{|}}{C}}-CH_3$$

Ethers of molecular formula $C_4H_{10}O$

$$CH_3-O-CH_2CH_2CH_3 \qquad CH_3CH_2-O-CH_2CH_3 \qquad CH_3-O-\underset{\displaystyle \overset{|}{CH_3}}{CHCH_3}$$

12.93 Penthrane: 2,2-Dichloro-1,1-difluoro-1-methoxyethane
Enthrane: 2-Chloro-1-(difluoromethoxy)-1,1,2-trifluoroethane

12.95 a. $CH_3CH_2 - O - CH_2CH_3 \ + \ H_2O$

b. $CH_3CH_2 - O - CH_2CH_3 \ + \ CH_3 - O - CH_3$

$+ \ CH_3 - O - CH_2CH_3 \ + \ H_2O$

c.

$$CH_3 - O - CH_3 \ + \ CH_3 - O - \underset{\displaystyle \overset{|}{CH_3}}{\overset{\displaystyle \overset{CH_3}{|}}{C}HCH_3}$$

$$+ \ CH_3\underset{\displaystyle \overset{|}{CH_3}}{\overset{\displaystyle \overset{CH_3}{|}}{C}H} - O - \underset{\displaystyle \overset{|}{}}{CHCH_3} \ + \ H_2O$$

d.

$CH_2 - O - CH_2$ ⬠ $+ \ H_2O$

12.97 a. 2-Ethoxypentane
b. 2-Methoxybutane
c. 1-Ethoxybutane
d. Methoxycyclopentane

12.99 Cystine:

$$\overset{\displaystyle \text{COO}^-}{\underset{\displaystyle \text{N}^+\text{H}_3}{\text{H}-\text{C}}}-\text{CH}_2-\text{S}-\text{S}-\text{CH}_2-\overset{\displaystyle \text{COO}^-}{\underset{\displaystyle \text{N}^+\text{H}_3}{\text{C}-\text{H}}}$$

12.101 a. 1-Propanethiol
     b. 2-Butanethiol
     c. 2-Methyl-2-butanethiol
     d. 1,4-Cyclohexanedithiol

Chapter 13
Aldehydes and Ketones
Solutions to the Odd-Numbered Problems

In-Chapter Questions and Problems

13.1   a.

$$CH_3 — \overset{\overset{\displaystyle O}{\|}}{C} — CH_3$$

b.

$$CH_3\overset{\overset{\displaystyle OH}{|}}{C}HCH_2CHCH_3$$

13.3   a.

$$CH_3CH_2 — \overset{\overset{\displaystyle O}{\|}}{C} — OH$$

b.

$$CH3 — \overset{\overset{\displaystyle O}{\|}}{C} — OH$$

13.5   a. Butanal
       b. 2,4-Dimethylpentanal

13.7   a. I.U.P.A.C: 3,4-Dimethylpentanal
          Common: β,γ-Dimethylvaleraldehyde
       b. I.U.P.A.C: 2-Ethylpentanal
          Common: α-Ethylvaleraldehyde

13.9

     a. 3-Methylnonanal:

$$CH_3CH_2CH_2CH_2CH_2CH_2\overset{\overset{\displaystyle }{|}}{\underset{\underset{\displaystyle CH_3}{}}{C}}HCH_2\overset{\overset{\displaystyle O}{\|}}{C} — H$$

     b. β-Bromovaleraldehyde:

$$CH_3CH_2\underset{\underset{\displaystyle Br}{|}}{C}HCH_2\overset{\overset{\displaystyle O}{\|}}{C} — H$$

13.11   a. 3-Iodobutanone
        b. 4-Methyl-2-octanone

13.13 a. Methyl isopropyl ketone (I.U.P.A.C. name: 3-Methyl-2-butanone):

$$CH_3-\overset{\overset{\displaystyle O}{\|}}{C}-\underset{\underset{\displaystyle CH_3}{|}}{C}HCH_3$$

b. 4-Heptanone:

$$CH_3CH_2CH_2-\overset{\overset{\displaystyle O}{\|}}{C}-CH_2CH_2CH_3$$

13.15 Ethanal (acetaldehyde) is the aldehyde synthesized from ethanol in the liver.

$$CH_3-\overset{\overset{\displaystyle O}{\|}}{C}-H$$

13.17 The following equation represents the oxidation of 1-propanol to form propanal:

$$CH_3CH_2CH_2-OH \xrightarrow{H_2Cr_2O_7} CH_3CH_2-\overset{\overset{\displaystyle O}{\|}}{C}-H$$

Note that propanal may be further oxidized to form propanoic acid (a carboxylic acid).

13.19 The following equation represents the reaction between ethanal and Tollens' reagent:

$$CH_3\overset{\overset{\displaystyle O}{\|}}{C}-H + Ag(NH_3)_2^+ \longrightarrow CH_3\overset{\overset{\displaystyle O}{\|}}{C}-O^- + Ag^0$$

   Ethanal   Silver ammonia   Ethanoate   Silver
                complex   anion   metal

13.21 The following equation represents the hydrogenation of propanone:

$$CH_3-\overset{\overset{\displaystyle O}{\|}}{C}-CH_3 + H_2 \xrightarrow{Ni} CH_3-\underset{\underset{\displaystyle H}{|}}{\overset{\overset{\displaystyle OH}{|}}{C}}-CH_3$$

      Propanone

13.23 a. Reduction
       b. Reduction
       c. Reduction
       d. Oxidation
       e. Reduction

13.25 a. Hemiacetal
       b. Ketal
       c. Acetal
       d. Hemiketal

13.27

$$
\underset{\substack{\text{Propanal} \\ \text{Keto form}}}{
\begin{array}{c}
\text{H} \quad \text{H} \quad \text{O} \\
| \quad\; | \quad\; || \\
\text{H}-\text{C}-\text{C}-\text{C}-\text{H} \\
| \quad\; | \\
\text{H} \quad \text{H}
\end{array}
}
\;\rightleftharpoons\;
\underset{\substack{\text{Propanal} \\ \text{Enol form}}}{
\begin{array}{c}
\text{H} \qquad \text{OH} \\
| \qquad\; | \\
\text{H}-\text{C}-\text{C}=\text{C}-\text{H} \\
| \quad\; | \\
\text{H} \quad \text{H}
\end{array}
}
$$

13.29   The following equation represents an aldol condensation of two molecules of propanal:

$$
2\;\; CH_3CH_2-\overset{\overset{\displaystyle O}{||}}{C}-H \;\;\xrightarrow{OH^-}\;\; CH_3-CH_2-\overset{\overset{\displaystyle OH}{|}}{C}H-\overset{\underset{\displaystyle CH_3}{|}}{C}H-\overset{\overset{\displaystyle O}{||}}{C}-H
$$

The reaction below shows how each propanal molecule contributes to the product:

$$
\left(CH_3CH_2-\overset{\overset{\displaystyle O}{||}}{C}-H\right) + \boxed{CH_3CH_2-\overset{\overset{\displaystyle O}{||}}{C}-H} \;\;\xrightarrow{OH^-}
$$

$$
\left(CH_3-CH_2-\overset{\overset{\displaystyle OH}{|}}{C}H\right) \boxed{CH-\overset{\overset{\displaystyle O}{||}}{C}-H} \\ \underset{\displaystyle CH_3}{|}
$$

13.31   As the carbon chain length increases, the compounds become less polar and more hydrocarbonlike. As a result, their solubility in water decreases.

13.33   A good solvent should dissolve a wide range of compounds. Simple ketones are considered to be universal solvents because they have both a polar carbonyl group and nonpolar side chains. As a result, they dissolve organic compounds and are also miscible in water.

13.35

13.37   Alcohols have higher boiling points than aldehydes or ketones of comparable molecular weights because alcohol molecules can form intermolecular hydrogen bonds with one another. Aldehydes and ketones cannot form intermolecular hydrogen bonds.

13.39   To name an aldehyde using the I.U.P.A.C. nomenclature system, identify and name the longest carbon chain containing the carbonyl group. Replace the final $-e$ of the alkane name with $-al$. Number and name all substituents as usual. Remember that the carbonyl carbon is always carbon-1 and does not need to be numbered in the name of the compound.

13.41   The common names of aldehydes are derived from the same Latin roots as the corresponding carboxylic acids. For instance, methanal is formaldehyde; ethanal is acetaldehyde; propanal is propionaldehyde, etc.
        Substituted aldehydes are named as derivatives of the straight-chain parent compound. Greek letters are used to indicate the position of substituents. The carbon nearest the carbonyl group is the α-carbon, the next is the β-carbon, and so on.

13.43   a.   Methanal: b.   7,8-Dibromooctanal:

13.45 a. 3-Chloro-2-pentanone:

   b. Benzaldehyde:

14.47 a. 2-Butanone
   b. 2-Ethylhexanal

13.49 a. 3-Nitrobenzaldehyde
   b. 3,4-Dihydroxycyclopentanone

13.51 a. 3-Bromobutanal
   b. 2-Chloro-2-methyl-4-heptanone

13.53 a. 4,6-Dimethyl-3-heptanone
   b. 3,3-Dimethylcyclopentanone

13.55 a. Acetone   d.   Propionaldehyde
   b. Ethyl methyl ketone   e.   Methyl isopropyl ketone
   c. Acetaldehyde

13.57

a. 3-Hydroxybutanal   b. 2-Methylpentanal   c. 4-Bromohexanal

$$CH_3CHCH_2C-H$$
$$\quad\ |\qquad\quad OH$$

$$CH_3CH_2CH_2CHC-H$$
$$\qquad\qquad\ |\ CH_3$$

$$CH_3CH_2CHCH_2CH_2C-H$$
$$\qquad\ |\ Br$$

d. 3-Iodopentanal   e. 2-Hydroxy-3-methylheptanal

$$CH_3CH_2CHCH_2C-H$$
$$\qquad\quad |\ I$$

$$CH_3CH_2CH_2CH_2CHCHC-H$$
$$\qquad\qquad\qquad\quad CH_3\ |\quad OH$$

13.59 Acetone is a good solvent because it can dissolve a wide range of compounds. It has both a polar carbonyl group and nonpolar side chains. As a result, it dissolves organic compounds and is also miscible in water.

13.61 The oxidation of ethanol to ethanal occurs in the liver and is catalyzed by the enzyme alcohol dehydrogenase.

13.63 In organic molecules, oxidation may be recognized as a gain of oxygen or a loss of hydrogen. An aldehyde may be oxidized to form a carboxylic acid as in the following example in which ethanal is oxidized to produce ethanoic acid.

$$H_3C-\overset{\overset{\displaystyle O}{\|}}{C}-H \quad \xrightarrow{[O]} \quad H_3C-\overset{\overset{\displaystyle O}{\|}}{C}-OH$$

$\qquad$ Ethanal $\qquad\qquad\qquad$ Ethanoic acid

13.65 Addition reactions of aldehydes or ketones are those in which a second molecule is added to the double bond of the carbonyl group. An example is the addition of the alcohol ethanol to the aldehyde ethanal.

$$H_3C-\overset{\overset{\displaystyle O}{\|}}{C}-H \ + \ CH_3CH_2OH \ \xrightarrow{H^+} \ H_3C-\overset{\overset{\displaystyle OH}{|}}{\underset{\underset{\displaystyle H}{|}}{C}}-OCH_2CH_3$$

$\quad$ Ethanal $\qquad\qquad$ Ethanol $\qquad\qquad\qquad\qquad$ Hemiacetal

13.67

a. $\ CH_3CH_2\overset{\overset{\displaystyle OH}{|}}{C}HCH_3 \ \xrightarrow{[O]} \ CH_3CH_2\overset{\overset{\displaystyle O}{\|}}{C}CH_3$

$\qquad$ 2-Butanol $\qquad\qquad\qquad$ Butanone

b. $\ CH_3\overset{\overset{\displaystyle }{|}}{\underset{\underset{\displaystyle CH_3}{|}}{C}}HCH_2OH \ \xrightarrow{[O]} \ CH_3\overset{\overset{\displaystyle O}{\|}}{\underset{\underset{\displaystyle CH_3}{|}}{C}}HCH$

$\quad$ 2-Methyl-1-propanol $\qquad$ Methylpropanal

Note that methylpropanal can be further oxidized to methylpropanoic acid.

c.

Cyclopentanol   Cyclopentanone

13.69

Primary   Aldehyde   Carboxylic
alcohol   Acid

13.71 a. Reduction reaction

Ethanal   Ethanol

b. Reduction reaction

Cyclohexanone   Cyclohexanol

c. Oxidation reaction

2-Propanol   Propanone

13.73 Only (c) 3-methylbutanal and (f) acetaldehyde would give a positive Tollens' test.

13.75 a.

$$CH_3-\underset{\underset{O}{\parallel}}{C}-CH_3 \ + \ CH_3CH_2OH \ \longrightarrow \ CH_3-\underset{\underset{OCH_2CH_3}{|}}{\overset{\overset{OH}{|}}{C}}-CH_3$$

b.

$$CH_3-\underset{\underset{O}{\parallel}}{C}-H \ + \ CH_3CH_2OH \ \longrightarrow \ CH_3-\underset{\underset{OCH_2CH_3}{|}}{\overset{\overset{OH}{|}}{C}}-H$$

13.77 A hemiacetal is formed when an aldehyde reacts with one molecule of alcohol.

13.79 An acetal is formed when two molecules of alcohol react with an aldehyde.

13.81 a.

$$CH_3-\underset{\underset{O}{\parallel}}{C}-CH_3 \ + \ 2\,CH_3OH \longrightarrow \ CH_3-\underset{\underset{OCH_3}{|}}{\overset{\overset{OCH_3}{|}}{C}}-CH_3$$

b.

$$CH_3-\underset{\underset{O}{\parallel}}{C}-H \ + \ 2\,CH_3OH \longrightarrow \ CH_3-\underset{\underset{OCH_3}{|}}{\overset{\overset{OCH_3}{|}}{C}}-H$$

13.83

a. $H-\underset{\underset{O}{\parallel}}{C}-OH$

b. $H_3C-\underset{\underset{O}{\parallel}}{C}-OH$

13.85 a. methanal
       b. propanal

13.87   a. false   b. true   c. false   d. false

13.89

$$2 \quad \text{H}_3\text{C}-\overset{\overset{\displaystyle O}{\|}}{\text{C}}-\text{H} \quad \xrightarrow{\text{OH}^-} \quad \text{CH}_3\underset{\underset{\displaystyle \text{OH}}{|}}{\text{CH}}\text{CH}_2\overset{\overset{\displaystyle O}{\|}}{\text{C}}-\text{H}$$

13.91

$$\text{H}_3\text{C}-\overset{\overset{\displaystyle O}{\|}}{\text{C}}-\text{CH}_3 \qquad\qquad \text{H}_2\text{C}=\overset{\overset{\displaystyle \text{OH}}{|}}{\text{C}}-\text{CH}_3$$

Keto form of   Enol form of
Propanone   Propanone

13.93 a.

$$\text{CH}_3\text{CH}_2\text{CH}_2-\overset{\overset{\displaystyle \text{OH}}{|}}{\underset{\underset{\displaystyle \text{OCH}_2\text{CH}_3}{|}}{\text{C}}}-\text{CH}_3$$

b.

c.

13.95 (1) $2 \text{ CH}_3\text{CH}_2\text{OH}$

(2) $\text{KMnO}_4/\text{OH}^-$

(3) $\text{CH}_3\text{CH}=\text{CH}_2$

Chapter 14
Carboxylic Acids and Carboxylic Acid Derivatives
Solutions to the Odd-Numbered Problems

In-Chapter Questions and Problems

14.1   a. ketone
        b. ketone
        c. alkane

14.3   The carboxyl group consists of two very polar groups, the carbonyl group and the hydroxyl group. Thus, carboxylic acids are very polar, in addition to which, they can hydrogen bond to one another. Aldehydes are polar, as a result of the carbonyl group, but cannot hydrogen bond to one another. As a result, carboxylic acids have higher boiling points than aldehydes of the same carbon chain length.

14.5   a.   2,4-Dimethylpentanoic acid
        b.   2,4-Dichlorobutanoic acid

14.7   a.   2,3-Dihydroxybutanoic acid:

$$CH_3\overset{\overset{\displaystyle OH}{|}}{C}H\overset{\overset{\displaystyle}{|}}{\underset{\underset{\displaystyle OH}{|}}{C}}H-\overset{\overset{\displaystyle O}{||}}{C}-OH$$

        b.   2-Bromo-3-chloro-4-methylhexanoic acid:

$$CH_3CH_2\overset{\overset{\displaystyle}{|}}{\underset{\underset{\displaystyle CH_3}{|}}{C}}H\overset{\overset{\displaystyle Cl}{|}}{C}H\overset{\overset{\displaystyle}{|}}{\underset{\underset{\displaystyle Br}{|}}{C}}H-\overset{\overset{\displaystyle O}{||}}{C}-OH$$

14.9   a. α,γ-Dimethylvaleric acid
        b. α,γ-Dichlorobutyric acid

14.11 a. *o*-Toluic acid: b. 2,4,6-Tribromobenzoic acid:

c. 2,2,2-Triphenylethanoic acid:

14.13   a. Propanol would be oxidized to propanal, which would quickly be oxidized to propanoic acid:

$$CH_3CH_2\overset{\overset{\displaystyle O}{\|}}{C}—H \ \rightarrow \ CH_3CH_2\overset{\overset{\displaystyle O}{\|}}{C}—OH$$

  Propanal   Propanoic acid

b. Pentanal would be oxidized to pentanoic acid:

$$CH_3CH_2CH_2CH_2\overset{\overset{\displaystyle O}{\|}}{C}—OH$$

14.15 a.

$$CH_3-\overset{\overset{\displaystyle O}{\|}}{C}-H \longrightarrow CH_3-\overset{\overset{\displaystyle O}{\|}}{C}-OH$$

b.

$$HO-\overset{\overset{\displaystyle O}{\|}}{C}-CH_2-CH_2-\overset{\overset{\displaystyle O}{\|}}{C}-OH$$

14.17 a. Potassium propanoate   b. Barium butanoate

14.19 a. Propyl butanoate (propyl butyrate)   b. Ethyl butanoate (ethyl butyrate)

14.21 The following reaction between 1-butanol and ethanoic acid produces butyl ethanoate. It requires a trace of acid and heat. It is also reversible

$$CH_3CH_2CH_2CH_2OH + CH_3COOH \leftrightarrow CH_3\overset{\overset{\displaystyle O}{\|}}{C}-OCH_2CH_2CH_2CH_3.$$

b. The following reaction between ethanol and propanoic acid produces ethyl propanoate. It requires a trace of acid and heat. It is also reversible.

$$CH_3CH_2OH + CH_3CH_2COOH \leftrightarrow CH_3CH_2\overset{\overset{\displaystyle O}{\|}}{C}-OCH_2CH_3$$

14.23 a. CH_3COOH | $CH_3CH_2CH_2OH$
Ethanoic acid   Propanol

b. $CH_3CH_2CH_2CH_2CH_2COO^-K^+$ + $CH_3CH_2CH_2OH$
Potassium hexanoate   1-Propanol

c. $CH_3CH_2CH_2CH_2COO^-Na^+$ + CH_3OH
Sodium pentanoate   Methanol

d. $CH_3CH_2CH_2CH_2CH_2COOH$ + $CH_3CHCH_2CH_2CH_3$
                                                          |
                                                          OH
         Hexanoic acid   2-Pentanol

14.25 a.

$$CH_3CHC\!\!-\!\!OH \xrightarrow{\text{PCl}_3 \text{ or SOCl}_2} CH_3CHC\!\!-\!\!Cl$$

(with C=O and CH₃ substituent)

2-Methylpropanoic acid   2-Methylpropanoyl chloride

b.

$$CH_3CH_2CH_2CH_2CH_2C\!\!-\!\!OH \xrightarrow{\text{PCl}_3 \text{ or SOCl}_2} CH_3CH_2CH_2CH_2CH_2C\!\!-\!\!Cl$$

Hexanoic acid   Hexanoyl chloride

14.27 a.

$$H\!\!-\!\!C\!\!-\!\!OH \xrightarrow{\text{PCl}_3} H\!\!-\!\!C\!\!-\!\!Cl \;+\; \text{inorganic products}$$

Formic acid   Formyl chloride

b.

$$CH_3CH_2\!\!-\!\!C\!\!-\!\!OH \xrightarrow{\text{PCl}_3} CH_3CH_2\!\!-\!\!C\!\!-\!\!Cl \;+\; \text{inorganic products}$$

Propionic acid   Propionyl chloride

14.29 a.

$$CH_3CHCH_2C\!\!-\!\!Cl \xrightarrow[\text{ion}]{\substack{CH_3CHCH_2C\!\!-\!\!O^- \\ | \\ CH_3 \\ \text{3-Methylbutanoate}}} CH_3CHCH_2C\!\!-\!\!O\!\!-\!\!C\!\!-\!\!CH_2CHCH_3 \;+\; Cl^-$$

(with CH₃ substituents)

3-Methylbutanoyl chloride   3-Methylbutanoic anhydride

b.

End-of-Chapter Questions and Problems

14.31 Aldehydes are polar, as a result of the carbonyl group, but cannot hydrogen bond to one another. Alcohols are polar and can hydrogen bond as a result of the polar hydroxyl group. The carboxyl group of the carboxylic acids consists of both of these polar groups: the carbonyl group and the hydroxyl group. Thus, carboxylic acids are more polar than either aldehydes or alcohols, in addition to which, they can hydrogen bond to one another. As a result, carboxylic acids have higher boiling points than aldehydes or alcohols of the same carbon chain length.

14.33 a. 3-Hexanone
      b. 3-Hexanone
      c. Hexane

14.35

14.37 a. Heptanoic acid
      b. 1-Propanol
      c. Pentanoic acid
      d. Butanoic acid

14.39 The smaller carboxylic acids are water soluble. They have sharp, sour tastes and unpleasant aromas.

14.41 Citric acid is found naturally in citrus fruits. It is added to foods to give them a tart flavor or to act as a food preservative and antioxidant. Adipic acid imparts a tart flavor to soft drinks and is a preservative.

14.43 Determine the name of the parent compound, that is, the longest carbon chain containing the carboxyl group. Change the –*e* ending of the alkane name to –*oic* acid. Number the chain so that the carboxyl carbon is carbon-1. Name and number substituents in the usual way.

14.45 a. 2-Bromopentanoic acid:

$$\begin{array}{ccccc} H & H & H & H & O \\ | & | & | & | & \parallel \\ H-C-&C-&C-&C-&C-OH \\ | & | & | & | \\ H & H & H & Br \end{array}$$

b. 2-Bromo-3-methylbutanoic acid:

$$\begin{array}{c} H \\ | \\ H-C-H \\ \end{array}$$

$$\begin{array}{cccc} H & & H & O \\ | & | & | & \parallel \\ H-C-&C-&C-&C-OH \\ | & | & | \\ H & H & Br \end{array}$$

c. 2-Bromocyclohexanecarboxylic acid:

— COOH

Br

14.47 a. I.U.P.A.C. name: Methanoic acid
Common name: Formic acid
b. I.U.P.A.C. name: 3-Methylbutanoic acid
Common name: β-Methylbutyric acid
c. I.U.P.A.C. name: Cyclopentane carboxylic acid
Common name: Cyclovaleric carboxylic acid

14.49

Butanoic acid   Methylpropanoic acid

14.51 a. 4,4-Dimethylhexanoic acid:

$$CH_3CH_2\underset{\underset{\displaystyle CH_3}{|}}{\overset{\overset{\displaystyle CH_3}{|}}{C}}CH_2CH_2COOH$$

  b. 3-Bromo-4-methylpentanoic acid:

$$CH_3\underset{}{CH}\underset{\underset{\displaystyle Br}{|}}{CH}CH_2COOH$$

with CH_3 above the first CH.

  c. 2,3-Dinitrobenzoic acid:

  d. 3-Methylcyclohexanecarboxylic acid:

14.53 a. I.U.P.A.C. name: 2-Hydroxypropanoic acid
      Common name: α-Hydroxypropionic acid

  b. I.U.P.A.C. name: 3-Hydroxybutanoic acid
      Common name: β-Hydroxybutyric acid

c. I.U.P.A.C. name: 4,4-Dimethylpentanoic acid
   Common name: γ,γ-Dimethylvaleric acid

d. I.U.P.A.C. name: 3,3-Dichloropentanoic acid
   Common name: β,β-Dichlorovaleric acid

14.55 In organic molecules, oxidation may be recognized as a gain of oxygen or a loss of
       hydrogen. An aldehyde may be oxidized to form a carboxylic acid as in the following
       example in which ethanal is oxidized to produce ethanoic acid.

$$H_3C-\overset{\displaystyle O}{\overset{\|}{C}}-H \quad \xrightarrow{[O]} \quad H_3C-\overset{\displaystyle O}{\overset{\|}{C}}-OH$$

          Ethanal   Ethanoic acid

14.57 The following general equation represents the dissociation of a carboxylic acid.

$$R-\overset{\displaystyle O}{\overset{\|}{C}}-OH \quad \rightleftharpoons \quad R-\overset{\displaystyle O}{\overset{\|}{C}}-O^- \; + \; H^+$$

14.59 When a strong base is added to a carboxylic acid, neutralization occurs.

14.61 Soaps are made from water, a strong base, and natural fats or oils. The fats and oils are
       triesters of glycerol. In the presence of the strong base, the ester bonds are hydrolyzed
       and the salts of the long chain fatty acids are formed. The salts of fatty acids are soaps.

14.63 a. CH_3COOH

       b.

$$CH_3CH_2CH_2-\overset{\displaystyle O}{\overset{\|}{C}}-O-CH_3 \; + \; H_2O$$

       c. CH_3OH

14.65 a. The oxidation of 1-pentanol in the presence of an oxidizing agent yields pentanal.
       b. Continued oxidation of pentanal yields pentanoic acid.

14.67 Esters are mildly polar as a result of the polar carbonyl group within the structure.

14.69 Esters are formed from the reaction of a carboxylic acid with an alcohol. The name is derived by using the alkyl or aryl portion of the alcohol I.U.P.A.C. name as the first name. The –ic acid ending of the I.U.P.A.C. name of the carboxylic acid is replaced with –ate and follows the name of the alkyl or aryl group.

14.71 a. Methyl benzoate:

$$C_6H_5-\overset{\displaystyle O}{\overset{\displaystyle \|}{C}}-O-CH_3$$

b. Butyl decanoate:

$$CH_3CH_2CH_2CH_2CH_2CH_2CH_2CH_2CH_2-\overset{\displaystyle O}{\overset{\displaystyle \|}{C}}-O-CH_2CH_2CH_2CH_3$$

c. Methyl propionate:

$$CH_3CH_2-\overset{\displaystyle O}{\overset{\displaystyle \|}{C}}-O-CH_3$$

d. Ethyl propionate:

$$CH_3CH_2-\overset{\displaystyle O}{\overset{\displaystyle \|}{C}}-O-CH_2CH_3$$

14.73 a. Ethyl ethanoate
 b. Methyl propanoate
 c. Methyl 3-methylbutanoate
 d. Cyclopentyl benzoate

14.75 The following equation shows the general reaction for the preparation of an ester:

$$R-\overset{\displaystyle O}{\overset{\displaystyle \|}{C}}-OH \ + \ R\text{-}OH \ \underset{}{\overset{H^+,\ heat}{\rightleftharpoons}} \ R-\overset{\displaystyle O}{\overset{\displaystyle \|}{C}}-OR \ + \ H_2O$$

Carboxylic   Alcohol   Ester   Water
acid

14.77 The following equation shows the general reaction for the acid-catalyzed hydrolysis of an ester:

$$\underset{\text{Ester}}{R-\overset{\displaystyle O}{\overset{\|}{C}}-OR} \;+\; \underset{\text{Water}}{H_2O} \;\underset{\longrightarrow}{\overset{H^+,\;\text{heat}}{\rightleftharpoons}}\; \underset{\substack{\text{Carboxylic}\\\text{acid}}}{R-\overset{\displaystyle O}{\overset{\|}{C}}-OH} \;+\; \underset{\text{Alcohol}}{R\text{-}OH}$$

14.79 A hydrolysis reaction is the cleavage of any bond by the addition of a water molecule.

14.81

a. $CH_3CH_2CH_2-\overset{\displaystyle O}{\overset{\|}{C}}-O-CH_2CH_3$

b. $CH_3CH_2-\overset{\displaystyle O}{\overset{\|}{C}}-OH \;+\; CH_3CH_2OH$

c. $CH_3CH_2CH_2OH$

d. $CH_3CH_2\underset{\underset{\displaystyle Br}{|}}{CH}CH_2\overset{\displaystyle O}{\overset{\|}{C}}-O^- \;+\; CH_3CH_2OH$

14.83 Saponification is a reaction in which a soap is produced. More generally, it is the hydrolysis of an ester in the presence of a base. The following reaction shows the base-catalyzed hydrolysis of an ester:

$$CH_3(CH_2)_{14}-\overset{\displaystyle O}{\overset{\|}{C}}-O-CH_3 + NaOH \longrightarrow CH_3(CH_2)_{14}-\overset{\displaystyle O}{\overset{\|}{C}}-O^-\;Na^+ + CH_3OH$$

14.85

14.87 Compound A is

$$CH_3CH_2CH_2CH_2-\overset{\overset{\displaystyle O}{\|}}{C}-O-CH_3$$

Compound B is

$$CH_3CH_2CH_2CH_2-\overset{\overset{\displaystyle O}{\|}}{C}-OH$$

Compound C is CH_3OH

14.89 a.

$$CH_3CH_2-\overset{\overset{\displaystyle O}{\|}}{C}-OCH_2CH_2CH_3 \underset{\longrightarrow}{\overset{H^+,\ heat}{\longleftarrow}} CH_3CH_2-\overset{\overset{\displaystyle O}{\|}}{C}-OH\ +\ CH_3CH_2CH_2OH$$

Propyl propanoate   Propanoic   1-Propanol
                                   acid

b.

$$H-\overset{\overset{\displaystyle O}{\|}}{C}-OCH_2CH_2CH_2CH_3 \underset{\longrightarrow}{\overset{H^+,\ heat}{\longleftarrow}} H-\overset{\overset{\displaystyle O}{\|}}{C}-OH\ +\ CH_3CH_2CH_2CH_2OH$$

Butyl methanoate   Methanoic   1-Butanol
                                   acid

c.

$$H-\overset{\overset{\displaystyle O}{\|}}{C}-OCH_2CH_3 \underset{\longrightarrow}{\overset{H^+,\ heat}{\longleftarrow}} H-\overset{\overset{\displaystyle O}{\|}}{C}-OH\ +\ CH_3CH_2OH$$

Ethyl methanoate   Methanoic   Ethanol
                                   acid

d.

$$CH_3CH_2CH_2CH_2-\overset{\overset{\displaystyle O}{\|}}{C}-OCH_3 \underset{\longrightarrow}{\overset{H^+,\ heat}{\longleftarrow}} CH_3CH_2CH_2CH_2-\overset{\overset{\displaystyle O}{\|}}{C}-OH\ +\ CH_3OH$$

Methyl pentanoate   Pentanoic acid   Ethanol

14.91 a. PCl_3, PCl_5, or $SOCl_2$

b.

$$CH_3-\overset{\overset{\displaystyle O}{\|}}{C}-O^-$$

c.

14.93 a.   b.

$$2\ H_3C-\overset{\overset{\displaystyle O}{\|}}{C}-OH$$

14.95 a. Decanoic anhydride:

$$CH_3(CH_2)_8-\overset{\overset{\displaystyle O}{\|}}{C}-O-\overset{\overset{\displaystyle O}{\|}}{C}-(CH_2)_8CH_3$$

b. Acetic anhydride:

$$CH_3-\overset{\overset{\displaystyle O}{\|}}{C}-O-\overset{\overset{\displaystyle O}{\|}}{C}-CH_3$$

c. Valeric anhydride:

$$CH_3(CH_2)_3-\overset{\overset{\displaystyle O}{\|}}{C}-O-\overset{\overset{\displaystyle O}{\|}}{C}-(CH_2)_3CH_3$$

d. Benzoyl chloride:

14.97 Acid chlorides are noxious, irritating chemicals. They are slightly polar and have boiling points similar to comparable aldehydes or ketones. They cannot be dissolved in water because they react violently with it.

14.99 a.

$$CH_3CH_2OH \ + \ CH_3CH_2\overset{\overset{\displaystyle O}{\|}}{C}-O-\overset{\overset{\displaystyle O}{\|}}{C}-CH_2CH_3 \ \longrightarrow$$

$$CH_3CH_2\overset{\overset{\displaystyle O}{\|}}{C}-OCH_2CH_3 \ + \ CH_3CH_2\overset{\overset{\displaystyle O}{\|}}{C}-OH$$

b.

$$CH_3CH_2OH \ + \ CH_3-\overset{\overset{\displaystyle O}{\|}}{C}-O-\overset{\overset{\displaystyle O}{\|}}{C}-CH_3 \ \longrightarrow$$

$$CH_3-\overset{\overset{\displaystyle O}{\|}}{C}-O-CH_2CH_3 \ + \ CH_3-\overset{\overset{\displaystyle O}{\|}}{C}-OH$$

c.

$$CH_3CH_2OH \ + \ H-\overset{\overset{\displaystyle O}{\|}}{C}-O-\overset{\overset{\displaystyle O}{\|}}{C}-H \ \longrightarrow$$

$$II-\overset{\overset{\displaystyle O}{\|}}{C}-OCH_2CH_3 \ + \ H-\overset{\overset{\displaystyle O}{\|}}{C}-OH$$

14.101

a. Monoester

$$HO-\overset{\overset{\displaystyle O}{\|}}{\underset{\underset{\displaystyle OH}{|}}{P}}-O-CH_2CH_3$$

b. Diester

$$HO-\overset{\overset{\displaystyle O}{\|}}{\underset{\underset{\displaystyle OCH_2CH_3}{|}}{P}}-O-CH_2CH_3$$

c. Triester

$$CH_3CH_2O-\overset{\overset{\displaystyle O}{\|}}{\underset{\underset{\displaystyle OCH_2CH_3}{|}}{P}}-O-CH_2CH_3$$

14.103 ATP is the molecule used to store the energy released in metabolic reactions. The energy is stored in the phosphoanhydride bonds between two phosphoryl groups. The energy is released when the bond is hydrolyzed. A portion of the energy can be transferred to another molecule if the phosphoryl group is transferred from ATP to the other molecule.

14.105

$$\text{CH}_3\text{-}\overset{\displaystyle O}{\overset{\|}{C}}\text{-S-COENZYME A}$$

14.107　　　The structure of nitroglycerine:

$$
\begin{array}{c}
\quad\quad H \\
\quad\quad | \\
H-C-O-NO_2 \\
\quad\quad | \\
H-C-O-NO_2 \\
\quad\quad | \\
H-C-O-NO_2 \\
\quad\quad | \\
\quad\quad H
\end{array}
$$

Chapter 15
Amines and Amides
Solutions to the Odd-Numbered Problems

In-Chapter Questions and Problems

15.1   a. tertiary amine
        b. primary amine
        c. secondary amine

15.3

15.5   a. Methanol would have the higher boiling point than methylamine. The intermolecular hydrogen bonds between alcohol molecules will be stronger than those between two amines because oxygen is more electronegative than nitrogen.
        b. Water would have a higher boiling point than dimethylamine. The intermolecular hydrogen bonds between water molecules will be stronger than those between two amines because oxygen is more electronegative than nitrogen.
        c. Ethylamine will have a higher boiling point that methylamine because it has a higher molecular weight.
        d. Propylamine will have a higher boiling point than butane because propylamine molecules can form intermolecular hydrogen bonds, and the nonpolar butane cannot do so.

15.7   a. *N*-Methylaniline: b.   *N,N*-Dimethylaniline:

c. *N*-Ethylaniline:
d. *N*-Isopropylaniline:

```
                    H
                    |
      ⬡ — N              ⬡ — N — H
                    |                |
              CH₂CH₃          CH – CH₃
                                |
                              CH₃
```

15.9 a. 2-Propanamine:

```
      H   H   H
      |   |   |
  H — C — C — C — H
      |   |   |
      H   N   H
         / \
        H   H
```

b. 3-Octanamine:

```
              H     H
               \   /
      H   H    N    H   H   H   H   H
      |   |    |    |   |   |   |   |
  H — C — C — C — C — C — C — C — C — H
      |   |    |    |   |   |   |   |
      H   H    H    H   H   H   H   H
```

c. *N*-Ethyl-2-heptanamine:

```
      H   H   H   H   H   H   H
      |   |   |   |   |   |   |
  H — C — C — C — C — C — C — C — H
      |   |   |   |   |   |   |
      H   |   H   H   H   H   H
          |
        N — H
          |
      H — C — H
          |
      H — C — H
          |
          H
```

d. 2-Methyl-2-pentanamine:

```
        H     H
         \   /
      H   N   H   H   H
      |   |   |   |   |
  H — C — C — C — C — C — H
      |   |   |   |   |
      H   |   H   H   H
          |
      H — C — H
          |
          H
```

e. 4-Chloro-5-iodo-1-nonanamine:

$$H-\underset{\underset{H}{|}}{\overset{\overset{\displaystyle H\diagdown\;\diagup H}{N}}{C}}-\underset{\underset{H}{|}}{\overset{\overset{H}{|}}{C}}-\underset{\underset{H}{|}}{\overset{\overset{H}{|}}{C}}-\underset{\underset{Cl}{|}}{\overset{\overset{H}{|}}{C}}-\underset{\underset{I}{|}}{\overset{\overset{H}{|}}{C}}-\underset{\underset{H}{|}}{\overset{\overset{H}{|}}{C}}-\underset{\underset{H}{|}}{\overset{\overset{H}{|}}{C}}-\underset{\underset{H}{|}}{\overset{\overset{H}{|}}{C}}-\underset{\underset{H}{|}}{\overset{\overset{H}{|}}{C}}-H$$

f. *N,N*-Diethyl-1-pentanamine:

$$H-\overset{\overset{\displaystyle H}{|}}{C}-H$$
$$II-\overset{}{C}-H$$

$$H-\underset{\underset{H}{|}}{\overset{\overset{H}{|}}{C}}-\underset{\underset{H}{|}}{\overset{\overset{H}{|}}{C}}-N-\underset{\underset{H}{|}}{\overset{\overset{H}{|}}{C}}-\underset{\underset{H}{|}}{\overset{\overset{H}{|}}{C}}-\underset{\underset{H}{|}}{\overset{\overset{H}{|}}{C}}-\underset{\underset{H}{|}}{\overset{\overset{H}{|}}{C}}-\underset{\underset{H}{|}}{\overset{\overset{H}{|}}{C}}-H$$

15.11 a.

$-NH_3^+\;Br^-$

b.

$$CH_3CH_2-\underset{\underset{II}{|}}{\overset{\overset{H}{|}}{N^+}}-CH_3\;\;+\;OH^-$$

c. $CH_3-N^+H_3 + OH^-$

15.13 a. CH_3-NH_2

b.

$$CH_3-\underset{}{\overset{\overset{\displaystyle H}{|}}{N}}-CH_3$$

End-of-Chapter Questions and Problems

15.15 The nitrogen atom is more polar than the hydrogen atom in amines; thus, the N-H bond is polar and hydrogen bonding can occur between primary or secondary amine molecules. Thus, amines have a higher boiling point than alkanes, which are nonpolar. Because nitrogen is not as electronegative as oxygen, the N-H bond is not as polar as the O-H. As a result, intermolecular hydrogen bonds between primary and secondary amine molecules are not as strong as the hydrogen bonds between alcohol molecules. Thus, alcohols have a higher boiling point.

15.17   In systematic nomenclature, primary amines are named by determining the name of the parent compound, the longest continuous carbon chain containing the amine group. The *–e* ending of the alkane chain is replaced with *–amine*. Thus, an alkane becomes an alkanamine. The parent chain is then numbered to give the carbon bearing the amine group the lowest possible number. Finally, all substituents are named and numbered and added as prefixes to the "alkanamine" name.

15.19   Amphetamines elevate blood pressure and pulse rate. They also decrease the appetite.

15.21   a.   1-Butanamine would be more soluble in water because it has a polar amine group that can form hydrogen bonds with water molecules. Pentane is an alkane and therefore is nonpolar and water insoluble.
        b.   2-Pentanamine would be more soluble in water because it has a polar amine group that can form hydrogen bonds with water molecules. Cyclohexane is an alkane and therefore is nonpolar and water insoluble.

15.23   Triethylamine molecules cannot form hydrogen bonds with one another, but 1-hexanamine molecules are able to do so. As a result of the greater intermolecular attraction between 1-hexanamine molecules, it has a higher boiling point.

15.25   a.   2-Butanamine   c.   Cyclopentanamine

        b.   3-Hexanamine   d.   2-Methyl-2-propanamine

15.27   a.   Diethylamine: $CH_3CH_2-NH-CH_2CH_3$

        b.   Butylamine: $CH_3CH_2CH_2CH_2NH_2$

        c.   3-Decanamine:

$$\overset{\displaystyle NH_2}{\underset{\displaystyle |}{CH_3CH_2CHCH_2CH_2CH_2CH_2CH_2CH_2CH_3}}$$

        d.   3-Bromo-2-pentanamine:

$$\overset{\displaystyle NH_2}{\underset{\displaystyle |}{CH_3CHCHCH_2CH_3}}$$
$$\underset{\displaystyle Br}{\overset{\displaystyle |}{}}$$

e. Triphenylamine:

15.29 a. 2-Pentanamine:

$$\underset{\displaystyle CH_3CHCH_2CH_2CH_3}{\overset{\displaystyle NH_2}{|}}$$

b. 2-Bromo-1-butanamine:

$$\underset{\underset{\displaystyle Br}{|}}{\overset{\displaystyle NH_2}{\underset{\displaystyle CH_2CHCH_2CH_3}{|}}}$$

c. Ethylisopropylamine:

$$\underset{\displaystyle CH_3CH_2-NH-CHCH_3}{\overset{\displaystyle CH_3}{|}}$$

d. Cyclopentanamine:

— NH$_2$

15.31 Structural formulas for the eight isomeric amines with the formula $C_4H_{11}N$:

$CH_3CH_2CH_2CH_2NH_2$
1-Butanamine
(Primary amine)

$$\underset{\displaystyle CH_3CHCH_2CH_2CH_3}{\overset{\displaystyle NH_2}{|}}$$
2-Butanamine
(Primary amine)

$$\underset{\displaystyle CH_3CHCH_2NH_2}{\overset{\displaystyle CH_3}{|}}$$
2-Methyl-1-propanamine
(Primary amine)

$$CH_3-\underset{\underset{NH_2}{|}}{\overset{\overset{CH_3}{|}}{C}}-CH_3$$

2-Methyl-2-Propanamine
(Primary amine)

$$CH_3CH_2-\underset{\overset{|}{CH_3}}{\overset{\overset{CH_3}{|}}{N}}-CH_3$$

N,N-Dimethylethanamine
(Tertiary amine)

$$CH_3CH_2-NH-CH_2CH_3$$

N-Ethylethanamine
(Secondary amine)

$$\underset{CH_3CHCH_3}{\overset{\overset{NH-CH_3}{|}}{}}$$

N-Methyl-2-propanamine
(Secondary amine)

$$CH_3CH_2CH_2-NH-CH_3$$

N-Methyl-1-propanamine
(Secondary amine)

15.33 a. Cyclohexanamine is a primary amine.
      b. Dibutylamine is a secondary amine
      c. 2-Methyl-2-heptanamine is a primary amine.
      d. Tripentylamine is a tertiary amine.

15.35 a.

      b.

      c.

      d.

15.37 a. H_2O
      b. HBr
      c. $CH_3CH_2CH_2-N^+H_3$

d.

$$CH_3CH_2 - \overset{\overset{\displaystyle H}{\overset{\displaystyle |}{}}}{\underset{\underset{\displaystyle H}{\displaystyle |}}{N^+}} - CH_2CH_3 \quad Cl^-$$

15.39 Lower molecular weight amines are soluble in water because the N-H bond is polar and can form hydrogen bonds with water molecules. As the size of the organic substituents becomes larger, the entire molecule becomes more hydrocarbonlike and, thus, more hydrophobic overall.

15.41 Drugs containing amine groups are generally administered as ammonium salts because the salt is more soluble in water and, hence, in body fluids.

15.43 Putrescine (1,4-Butanediamine): $H_2N-CH_2-CH_2-CH_2-CH_2-NH_2$

Cadaverine (1,5-Pentanediamine): $H_2N-CH_2-CH_2-CH_2-CH_2-CH_2-NH_2$

15.45 a.

Pyridine

Indole

b. The pyridine ring is found in vitamin B_6, which is a water-soluble vitamin required for the synthesis and degradation of amino acids.

The indole ring is found in lysergic acid diethylamide, which is a hallucinogenic drug. It is also found in strychnine, which has been used as a rat poison.

15.47 Morphine has been used as a pain reliever or analgesic. Codeine is used as an analgesic and cough suppressant. Cocaine has been used as an anesthetic for the sinuses and eyes. Quinine is used to treat malaria. Vitamin B_6 is a water-soluble vitamin required by the body.

15.49 Amides have very high boiling points because the amide group consists of two very polar functional groups, the carbonyl group and the amino group. Strong intermolecular hydrogen bonding between the N-H bond of one amide and the C=O group of a second amide results in very high boiling points.

15.51 The I.U.P.A.C. names of amides are derived from the I.U.P.A.C. names of the carboxylic acids from which they are derived. The *–oic acid* ending of the carboxylic acid is replaced with the *–amide* ending.

15.53 Barbiturates are often called "downers" because they act as sedatives. They are sometimes used as anticonvulsants for epileptics and people suffering from other disorders that manifest as neurosis, anxiety, or tension.

15.55 a. I.U.P.A.C. name: Propanamide
Common name: Propionamide

b. I.U.P.A.C. name: Pentanamide
Common name: Valeramide

c. I.U.P.A.C. name: *N,N*-Dimethylethanamide
Common name: *N,N*-Dimethylacetamide

15.57 a. Ethanamide:

$$\overset{\overset{\displaystyle O}{\|}}{H_3C - C - NH_2}$$

b. *N*-Methylpropanamide:

$$CH_3CH_2 - \overset{\overset{\displaystyle O}{\|}}{C} - \overset{\overset{\displaystyle H}{|}}{N} - CH_3$$

c. *N,N*-Diethylbenzamide:

$$\text{C}_6\text{H}_5 - \overset{\overset{\displaystyle O}{\|}}{C} - \overset{}{\underset{\underset{\displaystyle CH_2CH_3}{|}}{N}} - CH_2CH_3$$

d. 3-Bromo-4-methylhexanamide:

$$CH_3CH_2\overset{}{\underset{\underset{\displaystyle Br}{|}}{CH}}\overset{\overset{\displaystyle CH_3}{|}}{CH}CH_2 - \overset{\overset{\displaystyle O}{\|}}{C} - NH_2$$

e. *N,N*-Dimethylacetamide:

$$CH_3 - \overset{\overset{\displaystyle O}{\|}}{C} - \overset{}{\underset{\underset{\displaystyle CH_3}{|}}{N}} - CH_3$$

15.59 *N,N*-Diethyl-*m*-toluamide:

$$CH_3$$

$$\overset{\text{O}}{\underset{\|}{C}}-\overset{CH_2CH_3}{\underset{|}{N}}-CH_2CH_3$$

Hydrolysis of this compound would release the carboxylic acid *m*-toluic acid and the amine *N*-ethylethanamine (diethylamine).

15.61 Amides are not proton acceptors (bases) because the highly electronegative carbonyl oxygen has a strong attraction for the nitrogen lone pair of electrons. As a result, they cannot "hold" a proton. They are not proton donors because there are no hydrogen atoms bonded to the oxygen.

15.63 Lidocaine hydrochloride:

Amide group

$$NH-\overset{\text{O}}{\underset{\|}{C}}-CH_2-\overset{+}{NH}-CH_2CH_3 \qquad Cl^-$$

$$H_3C \qquad CH_3 \qquad CH_2CH_3$$

15.65 Penicillin BT:

Carboxyl group

Amide group

Amide group

$$COOH$$

$$CH_3$$

$$N$$

$$CH_3$$

$$S$$

$$CH_3(CH_2)_3SCH_2-\overset{\text{O}}{\underset{\|}{C}}-\overset{N}{\underset{H}{|}}$$

15.67 a.

$$CH_3-\overset{\text{O}}{\underset{\|}{C}}-NH-CH_3 + H_3O^+ \longrightarrow CH_3COOH + CH_3NH_3{}^+$$

N-Methylethanamide   Ethanoic acid   Methanamine

b.

$$CH_3CH_2CH_2 - \overset{\overset{\displaystyle O}{\|}}{C} - NH - CH_3 \ + \ H_3O^+ \longrightarrow$$

N-Methylbutanamide

$$CH_3CH_2CH_2COOH \ + \ CH_3NH_3^+$$

Butanoic acid   Methanamine

c.

$$\underset{\underset{\displaystyle CH_3CHCH_2}{|}}{\overset{\overset{\displaystyle CH_3}{|}}{}} - \overset{\overset{\displaystyle O}{\|}}{C} - NH - CH_2CH_3 \ + \ H_3O^+ \longrightarrow$$

N-Ethyl-3-methylbutanamide

$$\underset{CH_3CHCH_2COOH}{\overset{\overset{\displaystyle CH_3}{|}}{}} \ + \ CH_3NH_3^+$$

3-Methylbutanoic acid   Methanamine

15.69   a.

$$CH_3CH_2 - \overset{\overset{\displaystyle O}{\|}}{C} - O - \overset{\overset{\displaystyle O}{\|}}{C} - CH_2CH_3$$

b.

$$CH_3CH_2 - \overset{\overset{\displaystyle O}{\|}}{C} - NH_2 \ + \ NH_4^+Cl^-$$

c.

$$CH_3CH_2CH_2 - \overset{\overset{\displaystyle O}{\|}}{C} - Cl \ + \ 2 \ CH_3CH_2NH_2$$

15.71   The following is the general structure of an amino acid:

$$H_2N-\overset{\displaystyle\overset{H}{|}}{\underset{\displaystyle\underset{R}{|}}{C}}-\overset{\displaystyle\overset{O}{||}}{C}-OH$$

15.73 Within the cell, the actual structures would have a protonated amino group and a carboxylate group.

Glycine:

$$H_2N-\overset{\displaystyle\overset{H}{|}}{\underset{\displaystyle\underset{H}{|}}{C}}-\overset{\displaystyle\overset{O}{||}}{C}-OH$$

Alanine:

$$H_2N-\overset{\displaystyle\overset{H}{|}}{\underset{\displaystyle\underset{CH_3}{|}}{C}}-\overset{\displaystyle\overset{O}{||}}{C}-OH$$

15.75 The structure of alanine is shown below. An asterisk denotes the chiral carbon.

$$H_2N-\overset{\displaystyle\overset{H}{|}}{\underset{\displaystyle\underset{CH_3}{|}}{C^*}}-\overset{\displaystyle\overset{O}{||}}{C}-OH$$

15.77 In an acyl group transfer reaction, the acyl group of an acid chloride is transferred from the Cl of the acid chloride to the N of an amine or ammonia. The product is an amide.

15.79 A neurotransmitter is a chemical that carries messages or signals from a nerve to a target cell.

15.81 a. Symptoms that result from a deficiency of dopamine are: tremors, monotonous speech, loss of memory and problem-solving ability, and loss of motor function.

b. Parkinson's disease is the name of this condition.

c. Symptoms that result from an excess of dopamine are schizophrenia and intense satiety sensations.

15.83 In proper amounts, dopamine causes a pleasant, satisfied feeling. This feeling becomes intense as the amount of dopamine increases. Several drugs, including cocaine, heroin, amphetamines, alcohol, and nicotine increase the levels of dopamine. It is thought that the intense satiety response this brings about may contribute to addiction to these substances.

15.85 Epinephrine is a component of the flight or fight response. It stimulates glycogen breakdown to provide the body with glucose to supply the needed energy for this stress response.

15.87 The amino acid tryptophan is the starting material from which serotonin is made.

15.89 The perception of pain, thermoregulation, and sleep are physiological responses that are affected by serotonin.

15.91 Histamine promotes the itchy skin rash associated with poison ivy and insect bites, the respiratory symptoms characteristic of hay fever, and the secretion of stomach acid.

15.93 Inhibitory neurotransmitters

15.95 When acetylcholine is released from a nerve cell, it binds to receptors on the surface of muscle cells. This binding stimulates the muscle cell to contract. To stop the contraction, the acetylcholine is then broken down to choline and acetate ion. This is catalyzed by the enzyme acetylcholinesterase.

15.97 Organophosphates inactivate acetylcholinesterase by binding covalently to it. Since acetylcholine is not broken down, nerve transmission continues, resulting in muscle spasm. Pyridine aldoxime methiodide (PAM) is an antidote to organophosphate poisoning because it displaces the organophosphate, thereby allowing acetylcholinesterase to function.

Chapter 16
Carbohydrates
Solutions to the Odd-Numbered Problems

In-Chapter Questions and Problems

16.1 It is currently recommended that 45 - 65% of the calories in the diet should be carbohydrates. Of that amount, no more than 10% should be simple sugars.

16.3 An aldose is a sugar with an aldehyde functional group. A ketose is a sugar with a ketone functional group.

16.5 a. ketose
b. aldose
c. ketose
d. aldose
e. ketose
f. aldose

16.7

16.9 a. D-
b. L-
c. D-
d. D-
e. D-
f. L-

16.11

```
        CHO
   H ——— OH
   H ——— OH
   H ——— OH
       CH₂OH
```

D-Ribose

16.13

```
        CHO
  HO ——— H
  HO ——— H
  HO ——— H
       CH₂OH
```

L-Ribose

16.15

α-D-Galactose   β-D-Galactose

16.17 α-Amylase and β-amylase are digestive enzymes that break down the starch
       amylose. α-Amylase cleaves glycosidic bonds of the amylose chain at random,
       producing shorter polysaccharide chains. β-Amylase sequentially cleaves maltose
       (a disaccharide of glucose) from the reducing end of the polysaccharide chain.

End-of-Chapter Questions and Problems

16.19 A monosaccharide is the simplest sugar and consists of a single saccharide unit.
       A monosaccharide cannot be broken down into a simpler molecule by hydrolysis.
       A disaccharide is made up of two monosaccharides joined covalently by a
       glycosidic bond. Hydrolysis of a disaccharide yields two monosaccharides.

16.21 The molecular formula for a simple sugar is $(CH_2O)_n$. Typically n is an integer
       from 3 to 7.

16.23 Mashed potato flakes, rice, and corn starch would contain amylose and amylopectin, both of which are polysaccharides. A candy bar contains sucrose, a disaccharide. Orange juice contains fructose, a monosaccharide. It may also contain sucrose if the label indicates that sugar has been added.

16.25 Four kilocalories of energy are released for each gram of carbohydrate "burned" or oxidized.

16.27

16.29 An *aldose* is a sugar that contains an aldehyde (carbonyl) group.

16.31 A tetrose is a sugar with a four-carbon backbone.

16.33 A ketopentose is a sugar with a five-carbon backbone and containing a ketone (carbonyl) group.

16.35 a. β-D-Glucose is a hemiacetal.
       b. β-D-Fructose is a hemiketal.
       c. α-D-Galactose is a hemiacetal.

16.37 The two aldotrioses of molecular formula C_3H_6O:

16.39 *Stereoisomers* are pairs of molecules having the same structural formula and bonding pattern but differing in the arrangement of the atoms in space.

16.41 A chiral carbon is a carbon atom that is bonded to four different groups.

16.43 A polarimeter is an instrument that is used to measure the optical activity of molecules. It has a monochromatic light source. Light waves from the source are directed through a polarizer. The light that emerges is plane polarized. This passes through the sample and into the analyzer, which will measure the degree to which the plane of light has been rotated by the sample. If the plane of light is not altered by the sample, the compound is optically inactive. However, if the plane of light is rotated either clockwise or counterclockwise, the sample is optically active.

16.45 A Fischer Projection is a two-dimensional drawing of a molecule that shows a chiral carbon at the intersection of two lines. Horizontal lines at the intersection represent bonds projecting out of the page and vertical lines represent bonds that project into the page.

16.47 Dextrose is a common name used for D-glucose.

16.49 D- and L-Glyceraldehyde are a pair of enantiomers; that is, they are nonsuperimposable mirror images of one another. In D-glyceraldehyde, the hydroxyl group on the chiral carbon farthest from the aldehyde group (C-2) is on the right of the structure. In L-glyceraldehyde, the hydroxyl group on C-2 is on the left of the structure.

16.51 a.   b.   c.

16.53 Anomers are isomers that differ in the arrangement of bonds around the hemiacetal carbon.

16.55 A hemiacetal is a member of the family of organic compounds formed in the reaction of one molecule of alcohol with an aldehyde. They have the following general structure:

$$
\begin{array}{c}
\text{OH} \\
\text{R} \!-\!\!\!\!-\!\!\!\!-\!\!\!\! \text{OR} \\
\text{H}
\end{array}
$$

16.57 The reaction between an aldehyde and an alcohol yields a hemiacetal. Thus, when the aldehyde portion of a glucose molecule reacts with the C-5 hydroxyl group, the product is an intramolecular hemiacetal.

16.59 When the carbonyl group at C-1 of D-glucose reacts with the C-5 hydroxyl group, a new chiral carbon is created (C-1). In the α-isomer of the cyclic sugar, the C-1 hydroxyl group is below the ring and in the β-isomer the C-1 hydroxyl group is above the ring.

16.61 β-Maltose and α-lactose would give positive Benedict's tests. Glycogen would give only a weak reaction because it is a long polymer and thus there are fewer reducing ends for a given mass of the carbohydrate.

16.63 Enantiomers are stereoisomers that are nonsuperimposable mirror images of one another. For instance:

$$
\begin{array}{cc}
\begin{array}{c}
\text{O} \\
\| \\
\text{C}-\text{H} \\
| \\
\text{H}-\text{C}-\text{OH} \\
| \\
\text{CH}_2\text{OH}
\end{array}
&
\begin{array}{c}
\text{O} \\
\| \\
\text{C}-\text{H} \\
| \\
\text{HO}-\text{C}-\text{H} \\
| \\
\text{CH}_2\text{OH}
\end{array}
\end{array}
$$

D-Glyceraldehyde   L-Glyceraldehyde

16.65 The linear structure of an aldehyde sugar forms a cyclic structure by formation of an intramolecular hemiacetal. The carbonyl group of the monosaccharide, in this case an aldose, reacts with a hydroxyl group on one of the other carbon atoms. The product is a cyclic intramolecular hemiacetal.

16.67 A ketal is the product formed in the reaction of two molecules of alcohol with a ketone. They have the following general structure:

$$
\begin{array}{c}
\text{OR} \\
\text{R} \!-\!\!\!\!-\!\!\!\!-\!\!\!\! \text{OR} \\
\text{R}
\end{array}
$$

16.69 A glycosidic bond is the bond formed between the hydroxyl group of the C-1 carbon of one sugar and a hydroxyl group of another sugar.

16.71 The structure of β-Maltose is:

16.73 Milk is the major source of lactose.

16.75 Since galactose is one of the two monosaccharides making up lactose (milk sugar), eliminating milk and milk products from the diet allows the patient to avoid most of the ill effects of galactosemia.

16.77 Lactose intolerance is the inability to produce the enzyme lactase that hydrolyzes the milk sugar lactose into its component monosaccharides, glucose and galactose. As a result of the undigested lactose in the intestine, uncomfortable symptoms, including abdominal cramps and diarrhea, occur.

16.79 A polymer is a very large molecule formed by the combination of many small molecules called monomers.

16.81 The major storage form of sugar in plants is starch.

16.83 Both amylose and cellulose are linear polymers of glucose units. However, the glucose units of amylose are joined by $\alpha(1 \rightarrow 4)$ glycosidic bonds, and those of cellulose are bonded together by $\beta(1 \rightarrow 4)$ glycosidic bonds.

16.85 The major physiological purpose of glycogen is to serve as a storage molecule for glucose. This represents an energy reservoir for the body. Glycogen synthesis and degradation in the liver are involved in regulation of blood glucose levels.

16.87 α-Amylase and β-amylase are produced in the salivary glands and in the pancreas.

Chapter 17
Lipids and Their Functions in Biochemical Systems
Solutions for the Odd-Numbered Problems

In-Chapter Questions and Problems

17.1   a.   Oleic acid: $CH_3(CH_2)_7CH=CH(CH_2)_7COOH$

       b.   Lauric acid: $CH_3(CH_2)_{10}COOH$

       c.   Linoleic acid: $CH_3(CH_2)_4CH-CH\ CH_2\ CH-CH(CH_2)_7COOH$

       d.   Stearic acid: $CH_3(CH_2)_{16}COOH$

17.3

$$CH_3(CH_2)_{10}COOH + CH_3CH_2OH \xrightarrow{H+,\ heat} CH_3(CH_2)_{\overline{10}}\overset{\displaystyle O}{\overset{\|}{C}}-OCH_2CH_3 + H_2O$$

Lauric acid   Ethanol   Ethyl dodecanoate
Dodecanoic acid

17.5

$$CH_3CH_2\overset{\displaystyle O}{\overset{\|}{C}}-OCH_2CH_2CH_2CH_3 + H_2O \xrightarrow{H+,\ heat} CH_3CH_2CH_2CH_2OH + CH_3CH_2COOH$$

Butyl propanoate   1-Butanol   Propanoic acid

17.7

$$CH_3CH_2\overset{\displaystyle O}{\overset{\|}{C}}-OCH_2CH_2CH_2CH_3 + KOH \longrightarrow CH_3CH_2COO^-K^+ + CH_3CH_2CH_2CH_2OH$$

Butyl propanoate   Potassium propanoate   1-Butanol

17.9

$$CH_3(CH_2)_5CH==CH(CH_2)_7COOH + H_2 \xrightarrow{Ni} CH_3(CH_2)_{14}COOH$$

cis-9-hexadecenoic acid   Hexadecanoic acid

17.11

$$\begin{array}{c} \text{H} \\ | \\ \text{H}-\text{C}-\text{OH} \\ | \\ \text{H}-\text{C}-\text{OH} \quad + 2 \;\; CH_3(CH_2)_{16}COOH \longrightarrow \\ | \\ \text{H}-\text{C}-\text{OH} \\ | \\ \text{H} \end{array} \qquad \begin{array}{c} \text{H} \quad\quad\quad \text{O} \\ | \quad\quad\quad\;\; \| \\ \text{H}-\text{C}-\text{O}-\text{C}-(CH_2)_{16}CH_3 \\ | \\ \text{H}-\text{C}-\text{O}-\text{C}-(CH_2)_{16}CH_3 \\ | \quad\quad\quad\;\; \| \\ \text{H}-\text{C}-\text{OH} \quad \text{O} \\ | \\ \text{H} \end{array}$$

17.13 a. Mono, di, and triglycerides of oleic acid:

$$CH_3(CH_2)_7CH=CH(CH_2)_7-\overset{\overset{\displaystyle O}{\|}}{C}-O-\underset{\underset{\displaystyle \quad\quad}{}}{CH_2}$$
$$CH-OH$$
$$CH_2-OH$$

$$CH_3(CH_2)_7CH=CH(CH_2)_7-\overset{\overset{\displaystyle O}{\|}}{C}-O-CH_2$$
$$CH_3(CH_2)_7CH=CH(CH_2)_7-\overset{\overset{\displaystyle O}{\|}}{C}-O-CH$$
$$CH_2-OH$$

$$CH_3(CH_2)_7CH=CH(CH_2)_7-\overset{\overset{\displaystyle O}{\|}}{C}-O-CH_2$$
$$CH_3(CH_2)_7CH=CH(CH_2)_7-\overset{\overset{\displaystyle O}{\|}}{C}-O-CH$$
$$CH_3(CH_2)_7CH=CH(CH_2)_7-\overset{\overset{\displaystyle O}{\|}}{C}-O-CH_2$$

b. Mono, di, and triglycerides of capric acid:

$$CH_3(CH_2)_8 - \overset{\overset{\textstyle O}{\|}}{C} - O - CH_2$$
$$| \\ CH - OH$$
$$| \\ CH_2 - OH$$

$$CH_3(CH_2)_8 - \overset{\overset{\textstyle O}{\|}}{C} - O - CH_2$$
$$CH_3(CH_2)_8 - \overset{\overset{\textstyle O}{\|}}{C} - O - CH$$
$$| \\ CH_2 - OH$$

$$CH_3(CH_2)_8 - \overset{\overset{\textstyle O}{\|}}{C} - O - CH_2$$
$$CH_3(CH_2)_8 - \overset{\overset{\textstyle O}{\|}}{C} - O - CH$$
$$CH_3(CH_2)_8 - \overset{\overset{\textstyle O}{\|}}{C} - O - CH_2$$

c. Mono, di, and triglycerides of palmitic acid:

$$CH_3(CH_2)_{14} - \overset{\overset{\textstyle O}{\|}}{C} - O - CH_2$$
$$| \\ CH - OH$$
$$| \\ CH_2 - OH$$

$$CH_3(CH_2)_{14} - \overset{\overset{\textstyle O}{\|}}{C} - O - CH_2$$
$$CH_3(CH_2)_{14} - \overset{\overset{\textstyle O}{\|}}{C} - O - CH$$
$$| \\ CH_2 - OH$$

$$CH_3(CH_2)_{14} - \overset{\displaystyle O}{\overset{\|}{C}} - O - CH_2$$

$$CH_3(CH_2)_{14} - \overset{\displaystyle O}{\overset{\|}{C}} - O - CH$$

$$CH_3(CH_2)_{14} - \overset{\displaystyle O}{\overset{\|}{C}} - O - CH_2$$

d.   Mono, di, and triglycerides of lauric acid:

$$CH_3(CH_2)_{10} - \overset{\displaystyle O}{\overset{\|}{C}} - O - CH_2$$
$$CH - OH$$
$$CH_2 - OH$$

$$CH_3(CH_2)_{10} - \overset{\displaystyle O}{\overset{\|}{C}} - O - CH_2$$
$$CH_3(CH_2)_{10} - \overset{\displaystyle O}{\overset{\|}{C}} - O - CH$$
$$CH_2 - OH$$

$$CH_3(CH_2)_{10} - \overset{\displaystyle O}{\overset{\|}{C}} - O - CH_2$$
$$CH_3(CH_2)_{10} - \overset{\displaystyle O}{\overset{\|}{C}} - O - CH$$
$$CH_3(CH_2)_{10} - \overset{\displaystyle O}{\overset{\|}{C}} - O - CH_2$$

17.15   Structure of the steroid nucleus:

17.17   Cholesterol is carried in low density lipoprotein (LDL) particles in the plasma. These bind to specific LDL receptors within cell membranes. This binding stimulates receptor-mediated endocytosis, the invagination of the cell membrane that draws the LDL particles into the cytoplasm of the cell. The process encases the LDL within a vesicle or endosome in the cytoplasm.

17.19 Membrane transport resembles enzyme catalysis because both processes exhibit a high degree of specificity. The specificity is the result of the precise fit of the substrate into a site on the enzyme molecule or the precise fit of the molecule or ion being transported into a site on the permease protein.

End-of-Chapter Questions and Problems

17.21 The four main groups of lipids are fatty acids, glycerides, nonglyceride lipids, and complex lipids.

17.23 Lipid-soluble vitamins are transported into cells of the small intestine in association with dietary fat molecules. Thus, a diet low in fat reduces the amount of vitamins A, D, E, and K that enters the body.

17.25 A saturated fatty acid is one in which the hydrocarbon tail has only carbon-to-carbon single bonds. Thus, each carbon atom is bonded to the maximum number of hydrogen atoms. An unsaturated fatty acid has at least one carbon-to-carbon double bond.

17.27 As the length of the hydrocarbon chains of fatty acids increases, the melting points increase.

17.29 The melting points of fatty acids increase as the length of the hydrocarbon chains increase. This is because the intermolecular attractive forces, including van der Waals forces, increase as the length of the hydrocarbon chain increases.

17.31 a. Decanoic acid:

$$CH_3(CH_2)_8 - \overset{\overset{\text{O}}{\|}}{C} - OH$$

b. Stearic acid:

$$CH_3(CH_2)_{16} - \overset{\overset{\text{O}}{\|}}{C} - OH$$

17.33 a. I.U.P.A.C. name: Hexadecanoic acid
    Common name: Palmitic acid
 b. I.U.P.A.C. name: Dodecanoic acid
    Common name: Lauric acid

17.35 a. Esterification of glycerol with three molecules of myristic acid:

$$
\begin{array}{l}
\text{CH}_2\text{OH} \\
| \\
\text{CHOH} \quad + \quad 3 \ \ \text{CH}_3(\text{CH}_2)_{12} - \overset{\displaystyle \overset{O}{\|}}{C} - \text{OH} \quad \longrightarrow \\
| \\
\text{CH}_2\text{OH}
\end{array}
$$

$$
\begin{array}{l}
\text{CH}_3(\text{CH}_2)_{12} - \overset{\displaystyle \overset{O}{\|}}{C} - O - \text{CH}_2 \\
\text{CH}_3(\text{CH}_2)_{12} - \overset{\displaystyle \overset{O}{\|}}{C} - O - \text{CH} \qquad + \ \ 3 \ \text{H}_2\text{O} \\
\text{CH}_3(\text{CH}_2)_{12} - \overset{\displaystyle \overset{O}{\|}}{C} - O - \text{CH}_2
\end{array}
$$

b. Acid hydrolysis of tristearoyl glycerate:

$$
\begin{array}{l}
\text{CH}_3(\text{CH}_2)_{16} - \overset{\displaystyle \overset{O}{\|}}{C} - O - \text{CH}_2 \\
\text{CH}_3(\text{CH}_2)_{16} - \overset{\displaystyle \overset{O}{\|}}{C} - O - \text{CH} \qquad + \ \ 3 \ \text{H}_2\text{O} \quad \longrightarrow \\
\text{CH}_3(\text{CH}_2)_{16} - \overset{\displaystyle \overset{O}{\|}}{C} - O - \text{CH}_2
\end{array}
$$

$$
3 \ \ \text{CH}_3(\text{CH}_2)_{16} - \overset{\displaystyle \overset{O}{\|}}{C} - \text{OH} \quad + \quad
\begin{array}{l}
\text{CH}_2\text{OH} \\
| \\
\text{CHOH} \\
| \\
\text{CH}_2\text{OH}
\end{array}
$$

c. Reaction of decanoic acid with KOH:

$$
\text{CH}_3(\text{CH}_2)_8 - \overset{\displaystyle \overset{O}{\|}}{C} - \text{OH} \ + \ \text{KOH} \quad \longrightarrow \quad \text{CH}_3(\text{CH}_2)_8 - \overset{\displaystyle \overset{O}{\|}}{C} - O^- K^+ \ + \ \text{H}_2\text{O}
$$

d. Hydrogenation of linoleic acid:

$$CH_3(CH_2)_4CH=CHCH_2CH=CH(CH_2)_7-\overset{\displaystyle O}{\overset{\displaystyle \|}{C}}-OH$$

$$+\ 2\ H_2$$

$$\Big\downarrow Ni$$

$$CH_3(CH_2)_{16}-\overset{\displaystyle O}{\overset{\displaystyle \|}{C}}-OH$$

17.37 An essential fatty acid is one that the body cannot synthesize and thus must be supplied in the diet. The essential fatty acid linoleic acid is required for the synthesis of arachidonic acid, a precursor for the synthesis of the prostaglandins, a group of hormonelike molecules.

17.39 Aspirin effectively decreases the inflammatory response by inhibiting the synthesis of all prostaglandins. Aspirin works by inhibiting cyclooxygenase, the first enzyme in prostaglandin biosynthesis. This inhibition results from the transfer of an acetyl group from aspirin to the enzyme. Since cyclooxygenase is found in all cells, synthesis of all prostaglandins is inhibited.

17.41 Prostaglandins stimulate smooth muscle contraction, especially uterine contractions during labor. They enhance fever and swelling associated with the inflammatory response. Some prostaglandins cause bronchial dilation. Others inhibit secretion of acid into the stomach and stimulate the secretion of a mucous layer that protects the stomach lining.

17.43 A glyceride is a lipid ester that contains the glycerol molecule and from 1 to 3 fatty acids.

17.45 An emulsifying agent is a molecule that aids in the suspension of triglycerides in water. They are amphipathic molecules, such as lecithin, that serve as bridges holding together the highly polar water molecules and the nonpolar triglycerides.

17.47 A triglyceride with three saturated fatty acid tails would be a solid at room temperature. The long, straight fatty acid tails would stack with one another because of strong intermolecular and intramolecular attractions.

17.49

$$CH_3(CH_2)_{14} - \overset{\overset{\displaystyle O}{\|}}{C} - O - CH_2$$

$$CH_2(CH_2)_6 - \overset{\overset{\displaystyle O}{\|}}{C} - O - CH$$

H   C=C   CH₂(CH₂)₆

CH₃(CH₂)₄CH₂   H

CH₃(CH₂)₄CH₂   CH₂(CH₂)₆

C=C

H   H

$$CH_2(CH_2)_6 - \overset{\overset{\displaystyle O}{\|}}{C} - O - CH_2$$

17.51

$$CH_3(CH_2)_8 - \overset{\overset{\displaystyle O}{\|}}{C} - O - CH_2$$

$$CH_3(CH_2)_{10} - \overset{\overset{\displaystyle O}{\|}}{C} - O - CH$$

$$CH_2 - O - \overset{\overset{\displaystyle O}{\|}}{\underset{\underset{\displaystyle O^-}{|}}{P}} - O^-$$

17.53 Triglycerides consist of three fatty acids esterified to the three hydroxyl groups of glycerol. In phospholipids, there are only two fatty acids esterified to glycerol. A phosphoryl group is esterified (phosphoester linkage) to the third hydroxyl group.

17.55 A sphingolipid is one that is not derived from glycerol, but rather from sphingosine, a long-chain, nitrogen-containing (amino) alcohol. Like phospholipids, sphingolipids are amphipathic.

17.57 A glycosphingolipid or glycolipid is one that is built on a ceramide backbone structure. Ceramide is a fatty acid derivative of sphingosine.

17.59 Sphingomyelins are important structural lipid components of nerve cell membranes. They are found in the myelin sheath that surrounds and insulates cells of the central nervous system.

17.61 Cholesterol is readily soluble in the hydrophobic region of biological membranes. It is involved in regulating the fluidity of the membrane.

17.63 Progesterone is the most important hormone associated with pregnancy. It is needed for the successful initiation and completion of the pregnancy. It prepares the lining of the uterus to accept the fertilized egg, facilitates development of the fetus, and suppresses ovulation during pregnancy. Testosterone is needed for development of male secondary sexual characteristics. Estrone is required for proper development of female secondary sexual characteristics.

17.65 Cortisone is used to treat rheumatoid arthritis, asthma, gastrointestinal disorders, and many skin conditions.

17.67 Myricyl palmitate (beeswax) is made up of the fatty acid palmitic acid and the alcohol myricyl alcohol:

$$CH_3(CH_2)_{28}CH_2OH$$

17.69 Isoprenoids are a large, diverse collection of lipids that are synthesized from the isoprene unit:

$$\underset{\displaystyle CH_2=C-CH=CH_2}{\overset{\displaystyle CH_3}{\vert}}$$

17.71 Some biologically important terpenes include the steroids and bile salts, lipid-soluble vitamins, certain plant hormones, and chlorophyll.

17.73 The four major types of plasma lipoproteins are chylomicrons, high density lipoproteins, low density lipoproteins, and very low density lipoproteins.

17.75 The terms *good* and *bad* cholesterol refer to two classes of lipoprotein complexes. The high density lipoproproteins, or HDL, are considered to be good cholesterol because a correlation has been made between elevated levels of HDL and a reduced incidence of atherosclerosis. Low density lipoproteins, or LDL, are considered to be bad cholesterol because evidence suggests that high levels of LDL is associated with increased risk of atherosclerosis.

17.77 Atherosclerosis results when cholesterol and other substances coat the arteries causing a narrowing of the passageways. As the passageways become narrower, greater pressure is required to provide adequate blood flow. This results in higher blood pressure (hypertension).

17.79 If the LDL receptor is defective, it cannot function to remove cholesterol-bearing LDL particles from the blood. The excess cholesterol, along with other substances, will accumulate along the walls of the arteries causing atherosclerosis.

17.81 The basic structure of a biological membrane is a bilayer of phospholipid molecules arranged so that the hydrophobic hydrocarbon tails are packed in the center and the hydrophilic head groups are exposed on the inner and outer surfaces.

17.83 A peripheral membrane protein is bound to only one surface of the membrane, either inside or outside the cell.

17.85 Cholesterol is freely soluble in the hydrophobic layer of a biological membrane. It moderates the fluidity of the membrane by disrupting the stacking of the fatty acid tails of membrane phospholipids.

17.87 Specific membrane proteins on human and mouse cells were labeled with red and green fluorescent dyes, respectively. The cells were fused into single-celled hybrids and were observed using a microscope with an ultraviolet light source. The ultraviolet light caused the dyes to fluoresce. Initially, the dyes were localized in regions of the membrane representing the original human or mouse cell. Within an hour, the proteins were evenly distributed throughout the membrane of the fused cell.

17.89 If the fatty acyl tails of membrane phospholipids are converted from saturated to unsaturated, the fluidity of the membrane will increase. Each carbon-to-carbon double bond that is added will introduce a "kink" into the fatty acyl tail. As a result, the tails cannot pack together as they would if they were saturated. The result is increased fluidity.

17.91 Both simple diffusion and facilitated diffusion are means of passive transport. Simple diffusion involves the net movement of a solute directly across a membrane from a region of higher concentration to a region of lower concentration. Facilitated diffusion also involves movement of a substance from a region of high concentration to an area of low concentration. However, facilitated diffusion requires a channel protein or permease through which the solute must pass.

17.93 Both active transport and facilitated diffusion require a protein channel or permease through which solutes pass into or out of the cell. Active transport requires an energy input to transport molecules or ions against the gradient (from an area of lower concentration to an area of higher concentration). Facilitated diffusion is a means of passive transport in which molecules or ions pass from regions of higher concentration to regions of lower concentration through a permease protein. No energy is expended by the cell in facilitated diffusion.

17.95 An antiport transport mechanism is one in which one molecule or ion is transported into the cell while a different molecule or ion is transported out of the cell.

17.97 Each permease or channel protein has a binding site that has a shape and charge distribution that is complementary to the molecule or ion that it can bind and transport across the cell membrane. Any molecule or ion that does not fit into this binding site will not be transported by the permease.

17.99 One ATP is hydrolyzed to transport 3 Na^+ out of the cell and 2 K^+ into the cell.

Chapter 18
Protein Structure and Function
Solutions to the Odd-Numbered Problems

In-Chapter Questions and Problems

18.1   a.   Glycine (gly)

b.   Proline (pro)

c.   Threonine (thr)

$$H_3{}^+N - \underset{\underset{H}{|}}{\overset{\overset{H}{|}}{C}} - \overset{\overset{O}{||}}{C} - O^-$$

$$H - \overset{+}{\underset{\underset{\diagdown\diagup}{|}}{N}} - \underset{\underset{H}{|}}{\overset{\overset{H}{|}}{C}} - \overset{\overset{O}{||}}{C} - O^-$$

$$H_3{}^+N - \underset{\underset{CH_3}{\underset{|}{CH - OH}}}{\overset{\overset{H}{|}}{C}} - \overset{\overset{O}{||}}{C} - O^-$$

d.   Aspartate (asp)

e.   Lysine (lys)

$$H_3{}^+N - \underset{\underset{\underset{O\diagup\diagdown O^-}{C}}{\underset{|}{CH_2}}}{\overset{\overset{H}{|}}{C}} - \overset{\overset{O}{||}}{C} - O^-$$

$$H_3{}^+N - \underset{\underset{\underset{\underset{+NH_3}{CII_2}}{\underset{|}{CH_2}}}{\underset{\underset{|}{CH_2}}{\underset{|}{CH_2}}}}{\overset{\overset{H}{|}}{C}} - \overset{\overset{O}{||}}{C} - O^-$$

18.3   a.   Alanyl-phenylalanine:

$$H_3{}^+N - \underset{\underset{CH_3}{|}}{\overset{\overset{H}{|}}{C}} - \overset{\overset{O}{||}}{C} - \underset{\underset{H}{|}}{N} - \underset{\underset{CH_2}{|}}{\overset{\overset{H}{|}}{C}} - \overset{\overset{O}{||}}{C} - O^-$$

b. Lysyl-alanine:

$$H_3{}^+N - \overset{\overset{\displaystyle H}{|}}{\underset{\underset{\displaystyle CH_2}{|}}{C}} - \overset{\overset{\displaystyle O}{\|}}{C} - \overset{\overset{\displaystyle H}{|}}{\underset{\underset{\displaystyle H}{|}}{N}} - \overset{\overset{\displaystyle H}{|}}{\underset{\underset{\displaystyle CH_3}{|}}{C}} - \overset{\overset{\displaystyle O}{\|}}{C} - O^-$$

CH₂ — CH₂ — CH₂ — CH₂ — ⁺NH₃ (chain on first residue)

c. Phenylalanyl-tyrosyl-leucine:

18.5   The primary structure of a protein is the amino acid sequence of the protein chain. Regular, repeating folding of the peptide chain caused by hydrogen bonding between the amide nitrogens and carbonyl oxygens of the peptide bond is the secondary structure of a protein. The two most common types of secondary structure are the α-helix and the β-pleated sheet. Tertiary structure is the further folding of the regions of α-helix and β-pleated sheet into a compact, globular structure. Formation and maintenance of the tertiary structure results from weak attractions between amino acid R groups. The binding of two or more peptides to produce a functional protein defines the quaternary structure.

18.7   Oxygen is efficiently transferred from hemoglobin to myoglobin in the muscle because myoglobin has a greater affinity for oxygen.

18.9   High temperature disrupts the hydrogen bonds and other weak interactions that maintain protein structure.

18.11 Vegetables vary in amino acid composition. No single vegetable can provide all of the amino acid requirements of the body. By eating a variety of different vegetables, all the amino acid requirements of the human body can be met.

18.13 An enzyme is a protein that serves as a biological catalyst, speeding up biological reactions.

18.15 A transport protein is a protein that transports materials across the cell membrane or throughout the body.

18.17 Enzymes speed up reactions that might take days or weeks to occur on their own. They also catalyze reactions that might require very high temperatures or harsh conditions if carried out in the laboratory. In the body, these reactions occur quickly under physiological conditions.

18.19 Transferrin is a transport protein that carries iron from the liver to the bone marrow, where it is used to produce the heme group for hemoglobin and myoglobin. Hemoglobin transports oxygen in the blood.

18.21 Egg albumin is a nutrient protein that serves as a source of protein for the developing chick. Casein is the nutrient storage protein in milk, providing protein, a source of amino acids, for mammals.

18.23 The general structure of an L-α-amino acid:

$$H_3{}^+N - \underset{\underset{R}{|}}{\overset{\overset{COO^-}{|}}{C}} - H$$

18.25 A zwitterion is a neutral molecule with equal numbers of positive and negative charges. Under physiological conditions, amino acids are zwitterions.

18.27 A chiral carbon is one that has four different atoms or groups of atoms attached to it.

18.29 Interactions between the R groups of the amino acids in a polypeptide chain are important for the formation and maintenance of the tertiary and quaternary structures of proteins.

18.31 The following are the structures of the nine amino acids with hydrophobic R groups:

$$H_3{}^+N - \underset{\underset{H}{|}}{\overset{\overset{H}{|}}{C}} - \overset{\overset{O}{||}}{C} - O^- \qquad H_3{}^+N - \underset{\underset{CH_3}{|}}{\overset{\overset{H}{|}}{C}} - \overset{\overset{O}{||}}{C} - O^-$$

Glycine   Alanine

$$H_3{}^+N - \underset{\underset{\underset{H_3C \quad CH_3}{/ \ \backslash}}{CH}}{\overset{\overset{H}{|}}{C}} - \overset{\overset{O}{||}}{C} - O^-$$

Valine

Leucine

$$H_3{}^+N-\overset{\displaystyle H}{\underset{\displaystyle \underset{\displaystyle CH_3}{\overset{\displaystyle |}{\underset{\displaystyle |}{CHCH_3}}}}{\overset{|}{C}}}-\overset{\displaystyle O}{\overset{||}{C}}-O^-$$

Isoleucine

Phenylalanine

Proline

Tryptophan

Methionine

18.33 A peptide bond is an amide bond between two amino acids in a peptide chain.

18.35 Linus Pauling and his colleagues carried out X-ray diffraction studies of protein. Interpretation of the the pattern formed when X-rays were diffracted by a crystal of pure protein led Pauling to conclude that peptide bonds are both planar (flat) and rigid and that the N-C bonds are shorter that expected. In other words, they deduced that the peptide bond has a partially double bond character because it exhibits resonance. There is no free rotation about the amide bond because the carbonyl group of the amide bond has a strong attraction for the amide nitrogen lone pair of electrons. This can best be described using a resonance model:

The partially double bonded character of the resonance structure restricts free rotation.

18.37 a. His-trp-cys:

b. Gly-leu-ser:

c. Arg-ile-val:

18.39 The primary structure of a protein is the sequence of amino acids bonded to one another by peptide bonds.

18.41 The primary structure of a protein determines its three-dimensional shape because the location of R groups along the protein chain is determined by the primary structure. The interactions among the R groups, based on their location in the chain, will govern how the protein folds. This, in turn, dictates its three-dimensional structure and biological function.

18.43 The genetic information in the DNA dictates the order in which amino acids will be added to the protein chain. The order of the amino acids is the primary structure of the protein.

18.45 The secondary structure of a protein is the folding of the primary structure into an α-helix or β-pleated sheet.

18.47 a. α-Helix
       b. β-Pleated sheet

18.49 A fibrous protein is one that is composed of peptides arranged in long sheets or fibers.

18.51 A parallel β-pleated sheet is one in which the hydrogen-bonded peptide chains have their amino termini aligned head to head.

18.53 The tertiary structure of a protein is the globular, three-dimensional structure of a protein that results from folding the regions of secondary structure.

18.55

$$
\begin{array}{c}
\text{H} \\
| \\
\text{H}_3{}^+\text{N} - \text{C} - \text{COO}^- \\
| \\
\text{CH}_2 \\
| \\
\text{S} \\
| \\
\text{S} \\
| \\
\text{CH}_2 \\
| \\
\text{H}_3{}^+\text{N} - \text{C} - \text{COO}^- \\
| \\
\text{H}
\end{array}
$$

18.57 The tertiary structure is a level of folding of a protein chain that has already undergone secondary folding. The regions of α-helix and β-pleated sheet are folded into a globular structure.

18.59 Quaternary protein structure is the aggregation of two or more folded peptide chains to produce a functional protein.

18.61 A glycoprotein is a protein with sugars as prosthetic groups. Glycoproteins are often receptors on the cell surface.

18.63 Hydrogen bonding maintains the secondary structure of a protein and contributes to the stability of the tertiary and quaternary levels of structure.

18.65 The partially double bonded character of the resonance structure restricts free rotation.

18.67 The code for the primary structure of a protein is carried in the genetic information (DNA).

18.69 The function of hemoglobin is to carry oxygen from the lungs to oxygen-demanding tissues throughout the body.

18.71 Hemoglobin is a protein composed of four subunits—two α-globin and two β-globin subunits. Each subunit holds a heme group, which in turn carries an Fe^{3+} ion.

18.73 The function of the heme group in hemoglobin and myoglobin is to bind to molecular oxygen.

18.75 Because carbon monoxide binds tightly to the heme groups of hemoglobin, it is not easily removed or replaced by oxygen. As a result, the effects of oxygen deprivation (suffocation) occur.

18.77 When sickle cell hemoglobin (HbS) is deoxygenated, the amino acid valine fits into a hydrophobic pocket on the surface of another HbS molecule. Many such sickle cell hemoglobin molecules polymerize into long rods that cause the red blood cell to sickle. In normal hemoglobin, glutamic acid is found in place of the valine. This negatively charged amino acid will not "fit" into the hydrophobic pocket.

18.79 When individuals have one copy of the sickle cell gene and one copy of the normal gene, they are said to carry the *sickle cell trait*. These individuals will not suffer serious side effects, but may pass the trait to their offspring. Individuals with two copies of the sickle cell globin gene exhibit all the symptoms of the disease and are said to have *sickle cell anemia*.

18.81 *Denaturation* is the process by which the organized structure of a protein is disrupted, resulting in a completely disorganized, nonfunctional form of the protein.

18.83 Heat is an effective means of sterilization because it destroys the proteins of microbial life-forms, including fungi, bacteria, and viruses.

18.85 Even relatively small fluctuations in blood pH can be life threatening. It is likely that these small changes would alter the normal charges on the proteins and modify their interactions. These changes can render a protein incapable of carrying out its functions.

18.87 Proteins become polycations at low pH because the additional protons will protonate the carboxylate groups. As these negative charges are neutralized, the charge on the proteins will be contributed only by the protonated amino groups ($-N^+H_3$).

18.89 The low pH of the yogurt denatures the proteins of microbial contaminants, inhibiting their growth.

18.91 An essential amino acid is one that must be provided in the diet because it cannot be synthesized in the body.

18.93 A complete protein is one that contains all of the essential and nonessential amino acids.

18.95 Chymotrypsin catalyzes the hydrolysis of peptide bonds on the carbonyl side of aromatic amino acids.

Site of chymotrypsin-catalyzed hydrolysis

Phenylalanyl alanine

Phenylalanine

Alanine

18.97   In a vegetarian diet, vegetables are the only source of dietary protein. Because individual vegetable sources do not provide all the needed amino acids, vegetables must be mixed to provide all essential and nonessential amino acids required for biosynthesis.

18.99   Synthesis of digestive enzymes must be carefully controlled because the active enzyme would digest, and thus destroy, the cell that produces it.

Chapter 19
Enzymes
Solutions to the Odd-Numbered Problems

In-Chapter Questions and Problems

19.1 a. transferase   d. oxidoreductase
      b. transferase   e. hydrolase
      c. isomerase

19.3   a. Pyruvate kinase catalyzes the transfer of a phosphoryl group from
        phosphoenolpyruvate to adenosine diphosphate.

Phosphoenolpyruvate   Pyruvate

b. Alanine transaminase catalyzes the transfer of an amino group from alanine to α-
   ketoglutarate, producing pyruvate and glutamate.

Alanine

+

α-Ketoglutarate   Glutamate

c. Triose phosphate isomerase catalyzes the isomerization of the ketone dihydroxyacetone phosphate to the aldehyde glyceraldehyde-3-phosphate.

Dihydroxyacetone phosphate   Glyceraldehyde-3-phosphate

d. Pyruvate dehydrogenase catalyzes the oxidation and decarboxylation of pyruvate, producing acetyl coenzyme A and CO_2.

Pyruvate   Coenzyme A   Acetyl coenzyme A

19.5   a.   sucrose
       b.   pyruvate
       c.   succinate

19.7   The induced-fit model assumes that the enzyme is flexible. Both the enzyme and the substrate are able to change shape to form the enzyme-substrate complex. The lock-and-key model assumes that the enzyme is inflexible (the lock) and the substrate (the key) fits into a specific rigid site (the active site) on the enzyme to form the enzyme-substrate complex.

19.9   An enzyme might distort a bond, thereby catalyzing bond breakage. An enzyme could bring two reactants into close proximity and in the proper orientation for the reaction to occur. Finally, an enzyme could alter the pH of the microenvironment of the active site, thereby serving as a transient donor or acceptor of H^+.

19.11   Water-soluble vitamins are required by the body for the synthesis of coenzymes that are required for the function of a variety of enzymes.

19.13   A decrease in pH will change the degree of ionization of the R groups within a peptide chain. This disturbs the weak interactions that maintain the structure of an enzyme, which may denature the enzyme. Less drastic alterations in the charge of R groups in the active site of the enzyme can inhibit enzyme-substrate binding or destroy the catalytic ability of the active site.

19.15 Irreversible inhibitors bind very tightly, sometimes even covalently, to an R group in enzyme active sites. They generally inhibit many different enzymes. The loss of enzyme activity impairs normal cellular metabolism, resulting in death of the cell or of the individual.

19.17 A structural analog is a molecule that has a structure and charge distribution very similar to that of the natural substrate of an enzyme. Generally they are able to bind to the enzyme active site. This inhibits enzyme activity because the normal substrate must compete with the structural analog to form an enzyme-substrate complex.

19.19 a. ala-phe-ala

Alanine   Phenylalanine   Alanine

b. tyr-ala-tyr

19.21 val-phe-ala-gly-leu

End-of-Chapter Questions and Problems

19.23 The common name of an enzyme is often derived from the name of the substrate and/or the type of reaction that it catalyzes.

19.25 1. urease
2. peroxidase
3. lipase
4. aspartase
5. glucose-6-phosphatase
6. sucrase

19.27 a. Citrate decarboxylase catalyzes the cleavage of a carboxyl group from citrate.
b. Adenosine diphosphate phosphorylase catalyzes the addition of a phosphate group to ADP.
c. Oxalate reductase catalyzes the reduction of oxalate.
d. Nitrite oxidase catalyzes the oxidation of nitrite.
e. *cis-trans* Isomerase catalyzes the interconversion of *cis* and *trans* isomers.

19.29 A substrate is the reactant in an enzyme-catalyzed reaction that binds to the active site of the enzyme and is converted into product.

19.31 The activation energy of a reaction is the energy required for the reaction to occur.

19.33 The equilibrium constant for a chemical reaction is a reflection of the difference in energy of the reactants and products. Consider the following reaction:

$$aA + bB \rightarrow cC + dD$$

The equilibrium constant for this reaction is: $K_{eq} = [A]^a[B]^b/[C]^c[D]^d$

Because the difference in energy between reactants and products is the same regardless of what path the reaction takes, an enzyme does not alter the equilibrium constant of a reaction.

19.35   The rate of an uncatalyzed chemical reaction typically doubles every time the substrate concentration is doubled.

19.37   The rate-limiting step is that step in an enzyme-catalyzed reaction that is the slowest, and hence limits the speed with which the substrate can be converted into product.

19.39

substrate concentration

19.41   The enzyme-substrate complex is the molecular aggregate formed when the substrate binds to the active site of an enzyme.

19.43   The catalytic groups of an enzyme active site are those functional groups that are involved in carrying out catalysis.

19.45   Enzyme active sites are pockets in the surface of an enzyme that include R groups involved in binding and R groups involved in catalysis. The shape of the active site is complementary to the shape of the substrate. Thus, the conformation of the active site determines the specificity of the enzyme. Enzyme-substrate binding involves weak, noncovalent interactions.

19.47   The lock-and-key model of enzyme-substrate binding was proposed by Emil Fischer in 1894. He thought that the active site was a rigid region of the enzyme into which the substrate fit perfectly. Thus, the model purports that the substrate simply snaps into place within the active site, like two pieces of a jigsaw puzzle fitting together.

19.49   Enzyme specificity is the ability of an enzyme to bind to only one, or a very few, substrates and thus catalyze only a single reaction.

19.51   Group specificity means that an enzyme catalyzes reactions involving similar molecules having the same functional group.

19.53 Absolute specificity means that an enzyme catalyzes the reaction of only one substrate.

19.55 Hexokinase has group specificity. The advantage is that the cell does not need to encode many enzymes to carry out the phosphorylation of six-carbon sugars. Hexokinase can carry out many of these reactions.

19.57 Methionyl tRNA synthetase has absolute specificity. This is the enzyme that attaches the amino acid methionine to the transfer RNA (tRNA) that will carry the amino acid to the site of protein synthesis. If the wrong amino acid were attached to the tRNA, it could be incorporated into the protein, destroying its correct three-dimensional structure and biological function.

19.59 The first step of an enzyme-catalyzed reaction is the formation of the enzyme-substrate complex. In the second step, the transition state is formed. This is the state in which the substrate assumes a form intermediate between the original substrate and the product. In step 3, the substrate is converted to product and the enzyme-product complex is formed. Step 4 involves the release of the product and regeneration of the enzyme in its original form.

19.61 In a reaction involving bond breaking, the enzyme might distort a bond, producing a transition state in which the bond is stressed. An enzyme could bring two reactants into close proximity and in the proper orientation for the reaction to occur, producing a transition state in which the proximity of the reactants facilitates bond formation. Finally, an enzyme could alter the pH of the microenvironment of the active site, thereby serving as a transient donor or acceptor of H^+.

19.63 A cofactor helps maintain the shape of the active site of an enzyme.

19.65 NAD^+/NADH serves as an acceptor/donor of hydride anions in biochemical reactions. NAD^+/NADH serves as a coenzyme for oxidoreductases.

19.67 Changes in pH or temperature affect the activity of enzymes, as can changes in the concentration of substrate and the concentrations of certain ions.

19.69 Each of the following answers assumes that the enzyme was purified from an organism with optimal conditions for life near 37°C, pH 7.
   a. Decreasing the temperature from 37°C to 10°C will cause the rate of an enzyme-catalyzed reaction to decrease because the frequency of collisions between enzyme and substrate will decrease as the rate of molecular movement decreases.
   b. Increasing the pH from 7 to 11 will generally cause a decrease in the rate of an enzyme-catalyzed reaction. In fact, most enzymes would be denatured by a pH of 11 and enzyme activity would cease.
   c. Heating an enzyme from 37°C to 100°C will destroy enzyme activity because the enzyme would be denatured by the extreme heat.

19.71 High temperature denatures bacterial enzymes and structural proteins. Because the life of the cell is dependent on the function of these proteins, the cell dies.

19.73 A lysosome is a membrane-bound vesicle in the cytoplasm of cells that contains approximately fifty hydrolytic enzymes.

19.75 Enzymes used for clinical assays in hospitals are typically stored at refrigerator temperatures to ensure that they are not denatured by heat. In this way, they retain their activity for long periods.

19.77 a. Cells regulate the level of enzyme activity to conserve energy. It is a waste of cellular energy to produce an enzyme if its substrate is not present or if its product is in excess.
b. Production of proteolytic digestive enzymes must be carefully controlled because the active enzyme could destroy the cell that produces it. Thus, they are produced in an inactive form in the cell and are only activated at the site where they carry out digestion.

19.79 In positive allosterism, binding of the effector molecule turns the enzyme on. In negative allosterism, binding of the effector molecule turns the enzyme off.

19.81 A proenzyme is the inactive form of an enzyme that is converted to the active form at the site of its activity.

19.83 Blood clotting is a critical protective mechanism in the body, preventing excessive loss of blood following an injury. However, it can be a dangerous mechanism if it is triggered inappropriately. The resulting clot could cause a heart attack or stroke. By having a cascade of proteolytic reactions leading to the final formation of the clot, there are many steps at which the process can be regulated. This ensures that it will be activated only under the appropriate conditions.

19.85 *Competitive enzyme inhibition* occurs when a structural analog of the normal substrate occupies the enzyme active site so that the reaction cannot occur. The structural analog and the normal substrate compete for the active site. Thus, the rate of the reaction will depend on the relative concentrations of the two molecules.

19.87 A structural analog has a shape and charge distribution that are very similar to those of the normal substrate for an enzyme.

19.89 Irreversible inhibitors bind tightly to and block the active site of an enzyme and eliminate catalysis at the site.

19.91 The compound would be a competitive inhibitor of the enzyme.

19.93   The structural similarities among chymotrypsin, trypsin, and elastase suggest that these enzymes evolved from a single ancestral gene that was duplicated. Each copy then evolved independently.

19.95

Chymotrypsin

N^+H_3—CH—C—N—CH—C—N—CH—C—N—CH—C—O⁻

Tyrosine

Lysine

Alanine

Phenylalanine

19.97   Elastase will cleave the peptide bonds on the carbonyl side of alanine and glycine. Trypsin will cleave the peptide bonds on the carbonyl side of lysine and arginine. Chymotrypsin will cleave the peptide bonds on the carbonyl side of tryptophan and phenylalanine.

19.99   Creatine phosphokinase (CPK), lactate dehydrogenase (LDH), and aspartate aminotransferase (AST/SGOT)

Chapter 20
Introduction to Molecular Genetics
Solutions to the Odd-Numbered Problems

In-Chapter Questions and Problems

20.1   a. Adenosine diphosphate:

b. Deoxyguanosine triphosphate:

20.3   The RNA polymerase recognizes the promoter site for a gene, separates the strands of DNA, and catalyzes the polymerization of an RNA strand complementary to the DNA strand that carries the genetic code for a protein. It recognizes a termination site at the end of the gene and releases the RNA molecule.

20.5   The genetic code is said to be degenerate because several different triplet codons may serve as code words for a single amino acid.

20.7 The nitrogenous bases of the codons are complementary to those of the anticodons. As a result they are able to hydrogen bond to one another according to the base pairing rules.

20.9 The ribosomal P-site holds the peptidyl tRNA during protein synthesis. The peptidyl tRNA is the tRNA carrying the growing peptide chain. The only exception to this is during initiation of translation when the P-site holds the initiator tRNA.

20.11 The normal mRNA sequence, AUG-CCC-GAC-UUU, would encode the peptide sequence, methionine-proline-aspartate-phenylalanine. The mutant mRNA sequence, AUG-CGC-GAC-UUU, would encode the mutant peptide sequence, methionine-arginine-aspartate-phenylalanine. This would not be a silent mutation because a hydrophobic amino acid (proline) has been replaced by a positively charged amino acid (arginine).

End-of-Chapter Questions and Problems

20.13 A heterocyclic amine is a heterocyclic compound that contains nitrogen in at least one position of the ring skeleton.

20.15 It is the N-9 of the purine that forms the *N*-glycosidic bond with C-1 of the five-carbon sugar. The general structure of the purine ring is shown below:

20.17 The ATP nucleotide is composed of the five-carbon sugar ribose, the purine adenine, and a triphosphate group.

20.19 The two strands of DNA in the double helix are said to be *antiparallel* because they run in opposite directions. One strand progresses in the 5'→ 3' direction, and the opposite strand progresses in the 3' → 5' direction.

20.21 The DNA double helix is 2 nm in width. The nitrogenous bases are stacked at a distance of 0.34 nm from one another. One complete turn of the helix is 3.4 nm, or ten base pairs.

20.23 Two hydrogen bonds link the adenine-thymine base pair.

20.25

20.27 The prokaryotic chromosome is a circular DNA molecule that is supercoiled, that is, the helix is coiled on itself.

20.29 The term *semiconservative DNA replication* refers to the fact that each parental DNA strand serves as the template for the synthesis of a daughter strand. As a result, each of the daughter DNA molecules is made up of one strand of the original parental DNA and one strand of newly synthesized DNA.

20.31 The two primary functions of DNA polymerase III are to read a template DNA strand and catalyze the polymerization of a new daughter strand, and to proofread the newly synthesized strand and correct any errors by removing the incorrectly inserted nucleotide and adding the proper one.

20.33 If the parental DNA strand had the following nucleotide sequence: 5'-ATGCGGCTAGAATATTCCA-3', the sequence of the complementary daughter strand would be 3'-TACGCCGATCTTATAAGGT-5'.

20.35 The *replication origin* of a DNA molecule is the unique sequence on the DNA molecule where DNA replication begins.

20.37 The enzyme helicase separates the strands of DNA at the origin of DNA replication so that the proteins involved in replication can interact with the nitrogenous base pairs.

20.39   The RNA primer "primes" DNA replication by providing a 3'-OH, which can be used by DNA polymerase III for the addition of the next nucleotide in the growing DNA chain.

20.41   The central dogma of molecular biology states that information flow in cellular biological systems is unidirectional: DNA → RNA → Protein. The DNA carries the genetic information; RNA molecules carry out the expression of the genetic information to produce proteins; the final products are proteins that carry out the work of the cell and serve as cellular structural components.

20.43   Anticodons are found on transfer RNA molecules.

20.45   If a gene had the sequence, 5'-TACCTAGCTCTGGTCATTAAGGCAGTA-3', the mRNA would have the sequence, 3'-AUGGAUCGAGACCAGUAAUUCCGUCAU-5'.

20.47   *RNA splicing* is the process by which the noncoding sequences (introns) of the primary transcript of a eukaryotic mRNA are removed and the protein coding sequences (exons) are spliced together.

20.49   The three classes of RNA molecules are messenger RNA (mRNA), transfer RNA (tRNA), and ribosomal RNA (rRNA).

20.51   Spliceosomes are small ribonucleoprotein complexes in the nucleus of the cell that are responsible for RNA splicing. By hydrogen bonding, they recognize the boundaries between exons and introns and bring together the RNA sequences involved in splicing.

20.53   The *poly(A) tail* is a stretch of 100 - 200 adenosine nucleotides polymerized onto the 3' end of a mRNA by the enzyme poly(A) polymerase.

20.55   The *cap structure* is made up of the nucleotide 7-methylguanosine attached to the 5' end of a mRNA by a 5'-5' triphosphate bridge. Generally, the first two nucleotides of the mRNA are also methylated. The cap structure is required for efficient translation of the mRNA.

20.57   There are 64 codons in the genetic code.

20.59   The reading frame of a gene is the sequential set of triplet codons that carries the genetic code for the primary structure of a protein. Each triplet specifies the addition of a particular amino acid to the growing peptide chain.

20.61   Methionine (AUG) and tryptophan (UGG) are encoded by only one codon.

20.63 The codon 5'-UUU-3' encodes the amino acid phenylalanine. The mutant codon 5'-UUA-3' encodes the amino acid leucine. Both leucine and phenylalanine are hydrophobic amino acids, however, leucine has a smaller R group. It is possible that the smaller R group would disrupt the structure of the protein.

20.65 The ribosomes serve as a platform on which protein synthesis can occur. They also carry the enzymatic activity that forms peptide bonds.

20.67 The sequence of DNA nucleotides in a gene is transcribed to produce a complementary sequence of RNA nucleotides in a messenger RNA (mRNA). In the process of translation, the sequence of the mRNA is read sequentially in words of three nucleotides (codons) to produce a protein. Each codon calls for the addition of a particular amino acid to the growing peptide chain. Through these processes, the sequence of nucleotides in a gene determines the sequence of amino acids in the primary structure of a protein.

20.69 In the initiation of translation, initiation factors, methionyl tRNA (the initiator tRNA), the mRNA, and the small and large ribosomal subunits form the initiation complex. During the elongation stage of translation, an aminoacyl tRNA binds to the A-site of the ribosome. Peptidyl transferase catalyzes the formation of a peptide bond and the peptide chain is transferred to the tRNA in the A-site. Translocation shifts the peptidyl tRNA from the A-site into the P-site, leaving the A-site available for the next aminoacyl tRNA. In the termination stage of translation, a termination codon is encountered. A release factor binds to the empty A-site and peptidyl transferase catalyzes the hydrolysis of the bond between the peptidyl tRNA and the completed peptide chain.

20.71 The bond between an amino acid and a tRNA is an ester bond formed between the carboxylate group of the amino acid and the 3'-OH of the sugar ribose in the tRNA molecule.

20.73 A point mutation is the substitution of one nucleotide pair for another in a gene.

20.75 Some mutations are silent because the change in the nucleotide sequence does not alter the amino acid sequence of the protein. This can happen because there are many amino acids encoded by multiple codons.

20.77 UV light causes the formation of pyrimidine dimers, the covalent bonding of two adjacent pyrimidine bases. Mutations occur when the UV damage repair system makes an error during the repair process. This causes a change in the nucleotide sequence of the DNA.

20.79 a. A *carcinogen* is a compound that causes cancer. Cancers are caused by mutations in the genes responsible for controlling cell division.

b. Carcinogens cause DNA damage that results in changes in the nucleotide sequence of the gene. Thus, carcinogens are also mutagens.

20.81 A *restriction enzyme* is a bacterial enzyme that "cuts" the sugar-phosphate backbone of DNA molecules at a specific nucleotide sequence.

20.83 A selectable marker is a genetic trait that can be used to detect the presence of a plasmid in a bacterium. Many plasmids have antibiotic resistance genes as selectable markers. Bacteria containing the plasmid will be able to grow in the presence of the antibiotic; those without the plasmid will be killed.

20.85 Human insulin, interferon, human growth hormone, and human blood clotting factor VIII are protein products of recombinant DNA technology that are of great value in the field of medicine.

20.87 Each round of polymerase chain reaction doubles the number of target DNA molecules. Thus, after ten cycles of PCR, there would be 2^{10} or 2048 copies of each molecule of target DNA.

20.89 The goals of the Human Genome Project are to identify and map all of the genes of the human genome and to determine the DNA sequences of the complete three billion nucleotide pairs.

12.91 A genome library is a set of clones that represents all of the DNA sequences in the genome of an organism.

12.93 A dideoxynucleotide is one that has hydrogen atoms rather than hydroxyl groups bonded to both the 2' and the 3' carbons of the five-carbon sugar.

12.95 Sequences that these DNA sequences have in common are highlighted in bold.

    a. 5' **AGCTCCT**GATTTCATACAGTTTCTACT**ACCTACTA** 3'

    b. 5' AGACATTCTATCTACCTAGACTATG**TTCAGAA** 3'

    c. 5' **TTCAGAA**CTCATTCAGACCTACTACTATACCTTGGG**AGCTCCT** 3'

    d. 5' **ACCTACTA**GACTATACTACTACTAAGGGGACTATTCCAGACTT 3'

The 5' end of sequence (a) is identical to the 3' end of sequence (c).

The 3' end of sequence (a) is identical to the 5' end of sequence (d).

The 3' end of sequence (b) is identical to the 5' end of sequence (c).

From 5' to 3', the sequences would form the following map:

```
5'_____b_____
         _____c_____
                _____a_____
                         _____d_____3'
```

Chapter 21
Carbohydrate Metabolism
Solutions to the Odd-Numbered Problems

In-Chapter Questions and Problems

21.1   ATP is called the universal energy currency because it is the major molecule used by all organisms to store energy. Hydrolysis of the high-energy phosphoanhydride bonds releases energy that is used for cellular work.

21.3   The first stage of catabolism is the digestion (hydrolysis) of dietary macromolecules in the stomach and intestine. Polysaccharides are hydrolyzed to monosaccharides; proteins are degraded to amino acids; and triglycerides are broken down into glycerol and fatty acids. The small molecules produced by digestion are taken into the cells lining the intestine by active or passive transport.

In the second stage of catabolism, monosaccharides, amino acids, fatty acids, and glycerol are converted by metabolic reactions into molecules that can be completely oxidized. Often they are converted into acetyl CoA.

In the third stage of catabolism, the two-carbon acetyl group of acetyl CoA is completely oxidized by the reactions of the citric acid cycle. The energy of the electrons harvested in these oxidation reactions is used to make ATP.

21.5   Substrate level phosphorylation is one way the cell can make ATP. In this reaction, a high-energy phosphoryl group of a substrate in the reaction is transferred to ADP to produce ATP. Substrate-level phosphorylation can be summarized as follows:

$$\text{Substrate} \sim P + ADP \;\rightarrow\; \text{Product} + ATP$$

21.7   Glycolysis is a pathway involving ten reactions. In reactions 1 - 3, energy is invested in the beginning substrate, glucose. This is done by transferring high-energy phosphoryl groups from ATP to the intermediates in the pathway. The product is fructose-1,6-bisphosphate. In the energy harvesting reactions of glycolysis, fructose-1,6-bisphosphate is split into two three-carbon molecules that begin a series of rearrangement, oxidation-reduction, and substrate-level phosphorylation reactions that produce 4 ATP, 2 NADH, and 2 pyruvate molecules. Because of the investment of two ATP in the early steps of glycolysis, the net yield of ATP is two.

21.9   Both the alcohol and lactate fermentations are anaerobic reactions that use the pyruvate and reoxidize the NADH produced in glycolysis. In the alcohol fermentation, pyruvate is first decarboxylated to produce acetaldehyde. The acetaldehyde is then reduced as NADH is oxidized. The products are CO_2, ethanol, and NAD^+. In the lactate fermentation, pyruvate is reduced to lactate and NADH is oxidized to NAD^+.

21.11 Gluconeogenesis (synthesis of glucose from noncarbohydrate sources) appears to be the reverse of glycolysis (the first stage of carbohydrate degradation) because the intermediates in the two pathways are the same. However, reactions 1, 3, and 10 of glycolysis are not reversible reactions. Thus, the reverse reactions must be carried out by different enzymes.

Reaction 1 of glycolysis, the transfer of a high-energy phosphoryl group from ATP to glucose, is carried out by the enzyme hexokinase. The reverse reaction in gluconeogenesis is catalyzed by glucose-6-phosphatase.

Reaction 3 of glycolysis, the transfer of a high-energy phosphoryl group from ATP to fructose-6-phosphate, is catalyzed by phosphofructokinase. The reverse reaction of gluconeogenesis is carried out by fructose bisphosphatase.

The final reaction of glycolysis, the transfer of a high-energy phosphoryl group from phosphoenolpyruvate to ADP, is catalyzed by pyruvate kinase. This is reversed in gluconeogenesis by the action of two enzymes. Pyruvate carboxylase adds CO_2 to pyruvate to produce oxaloacetate and phosphoenolpyruvate carboxykinase removes the CO_2 and transfers a high-energy phosphoryl group from GTP to produce phosphoenolpyruvate.

21.13 The enzyme glycogen phosphorylase catalyzes the phosphorolysis of a glucose unit at one end of a glycogen molecule. The reaction involves the displacement of the glucose by a phosphate group. The products are glucose-1-phosphate and a glycogen molecule that is one glucose unit shorter.

21.15 Glucokinase traps glucose within the liver cell by phosphorylating it. Because the product, glucose-6-phosphate is charged, it cannot be exported from the cell.

21.17 Glucagon indirectly stimulates glycogen phosphorylase, the first enzyme of glycogenolysis. This speeds up glycogen degradation. Glucagon also inhibits glycogen synthase, the first enzyme in glycogenesis. This inhibits glycogen synthesis.

End-of-Chapter Questions and Problems

21.19 ATP is the molecule that is primarily responsible for conserving the energy released in catabolism.

21.21

Adenosine triphosphate

$$^-O-\overset{\overset{\displaystyle O}{\|}}{\underset{\underset{\displaystyle O^-}{|}}{P}}-O-\overset{\overset{\displaystyle O}{\|}}{\underset{\underset{\displaystyle O^-}{|}}{P}}-O-\overset{\overset{\displaystyle O}{\|}}{\underset{\underset{\displaystyle O^-}{|}}{P}}-O-CH_2 \quad + \quad H_2O \longrightarrow$$

(ribose ring with OH, OH substituents and adenine base bearing NH$_2$)

$$^-O-\overset{\overset{\displaystyle O}{\|}}{\underset{\underset{\displaystyle O^-}{|}}{P}}-O^- \quad + \quad {}^-O-\overset{\overset{\displaystyle O}{\|}}{\underset{\underset{\displaystyle O^-}{|}}{P}}-O-\overset{\overset{\displaystyle O}{\|}}{\underset{\underset{\displaystyle O^-}{|}}{P}}-O-CH_2$$

(ribose ring with OH, OH substituents and adenine base bearing NH$_2$)

Inorganic
phosphate group   Adenosine diphosphate

21.23 A coupled reaction is one that can be thought of as a two-step process. In a coupled reaction, two reactions occur simultaneously. Frequently, one of the reactions releases the energy that drives the second, energy-requiring, reaction.

21.25 Carbohydrates are the most readily used energy source in the diet.

21.27 The following equation represents the hydrolysis of maltose:

(structure of β-Maltose) $+ H_2O \longrightarrow 2$ (structure of β-D-Glucose)

β-Maltose   β-D-Glucose

21.29 The following equation represents the hydrolysis of lactose:

β-Lactose

+ H₂O ⟶

β-D-Glucose

+

β-D-Galactose

21.31 The hydrolysis of a triglyceride containing oleic acid, stearic acid, and linoleic acid is represented in the following equations:

$$H{-}C{-}O{-}C{-}(CH_2)_7CH{=}CH(CH_2)_7CH_3$$

$$H{-}C{-}O{-}C{-}(CH_2)_{16}CH_3 \quad + \; 3H_2O \longrightarrow$$

$$H{-}C{-}O{-}C{-}(CH_2)_7CH{=}CHCH_2CH{=}CH(CH_2)_4CH_3$$

H—C—OH

H—C—OH

H—C—OH

Glycerol

+

$$HO{-}C{-}(CH_2)_7CH{=}CH(CH_2)_7CH_3$$
Oleic acid

+

$$HO{-}C{-}(CH_2)_{16}CH_3$$
Stearic acid

+

$$HO{-}C{-}(CH_2)_7CH{=}CHCH_2CH{=}CH(CH_2)_4CH_3$$
Linoleic acid

21.33 The hydrolysis of the dipeptide alanyl leucine is represented in the following equation:

$$H_3{}^+N-CH-C(=O)-N(H)-CH-C(=O)-O^- + H_2O \longrightarrow$$

with CH₃ on the first CH, and CH₂—CH—CH₃ / CH₃ on the second.

Alanyl leucine

Alanine:

$$H_3{}^+N-CH(CH_3)-C(=O)-O^-$$

+

Leucine:

$$H_3{}^+N-CH-C(=O)-O^-$$
with side chain $CH_2-CH(CH_3)-CH_3$

21.35 Glycolysis is the enzymatic pathway that converts a glucose molecule into two molecules of pyruvate. The pathway generates a net energy yield of two ATP and two NADH. Glycolysis is the first stage of carbohydrate catabolism.

21.37 Glycolysis requires NAD^+ for reaction 6 in which glyceraldehyde-3-phosphate dehydrogenase catalyzes the oxidation of glyceraldehyde-3-phosphate. NAD^+ is reduced and, thus, serves as the hydride anion acceptor in this reaction. If NAD^+ were not available, this reaction would not occur, and glycolysis, and therefore ATP synthesis, would stop.

21.39 The net ATP yield of glycolysis is 2 ATP molecules per glucose molecule.

21.41 Although muscle cells have only enough ATP stored for a few seconds of activity, glycolysis speeds up dramatically when there is a demand for more energy. If the cells have a sufficient supply of oxygen, aerobic respiration (the citric acid cycle and oxidative phosphorylation) will contribute large amounts of ATP. If oxygen is limited, the lactate fermentation will speed up. This will use up the pyruvate and reoxidize the NADH produced by glycolysis and allow continued synthesis of ATP for muscle contraction.

21.43 $C_6H_{12}O_6 + 2ADP + 2P_i + 2NAD^+ \rightarrow 2\ C_3H_3O_3 + 2ATP + 2NADH + 2H_2O$
Glucose   Pyruvate

21.45 a. Hexokinase catalyzes the phosphorylation of glucose in the first step of glycolysis. The product is glucose-6-phosphate.

    b. Pyruvate kinase catalyzes the transfer of a phosphoryl group from phosphoenolpyruvate to ADP. The products of this substrate level phosphorylation reaction are pyruvate and ATP.

    c. Phosphoglycerate mutase catalyzes the isomerization reaction that converts 3-phosphoglycerate to 2-phosphoglycerate.

    d. Glyceraldehyde-3-phosphate dehydrogenase catalyzes the oxidation and phosphorylation of glyceraldehyde-3-phosphate and the reduction of NAD^+ to NADH.

21.47 Since the reactants and products are isomers, the enzymes that catalyze these reactions must be isomerases.

21.49 Enediol

21.51 A kinase transfers a phosphoryl group from one molecule to another.

21.53 NAD^+ is reduced, accepting a hydride anion ($H:^-$).

21.55 Enolase catalyzes a reaction that produces an enol; in this particular reaction, it is phosphoenolpyruvate.

21.57 To optimize efficiency and minimize waste, it is important that energy-harvesting pathways, such as glycolysis, respond to the energy demands of the cell. If energy in the form of ATP is abundant, there is no need for the pathway to continue at a rapid rate. When this is the case, allosteric enzymes that catalyze the reactions of the pathway are inhibited by binding to their negative effectors. Similarly, when there is a great demand for ATP, the pathway speeds up as a result of the action of allosteric enzymes binding to positive effectors.

21.59 ATP and citrate are allosteric inhibitors of phosphofructokinase, while AMP and ADP are allosteric activators.

21.61 Citrate, which is the first intermediate in the citric acid cycle, is an allosteric inhibitor of phosphofructokinase. The citric acid cycle is a pathway that results in the complete oxidation of the pyruvate produced by glycolysis. A high concentration of citrate signals that sufficient substrate is entering the citric acid cycle. The inhibition of phosphofructokinase by citrate is an example of *feedback inhibition*: the product, citrate, allosterically inhibits the activity of an enzyme early in the pathway.

21.63 The enzyme alcohol dehydrogenase catalyzes the conversion of acetaldehyde to ethanol.

$$H_3C-\underset{\underset{H}{|}}{\overset{\overset{O}{||}}{C}}-H \quad \xrightarrow{\quad NADH \quad NAD^+ \quad} \quad CH_3CH_2OH$$

21.65 The lactate fermentation produces lactate from pyruvate when the amount of oxygen is limiting. Over time, lactate will accumulate in the muscle.

21.67 The tangy flavor of yogurt and some cheeses is the result of lactate produced by the lactate fermentation. It is the pH decrease caused by the lactate build-up that causes milk protein to coagulate. This coagulation produces the soft curd of yogurt and the hard curd of some cheeses.

21.69 Lactate dehydrogenase catalyzes the reduction of pyruvate to lactate.

21.71 This child must have the enzymes to carry out the alcohol fermentation. When the child exercised hard, there was not enough oxygen in the cells to maintain aerobic respiration. As a result, glycolysis and the alcohol fermentation were responsible for the majority of the ATP production by the child. The accumulation of alcohol (ethanol) in the child caused the symptoms of drunkenness.

21.73 The first stage of the pentose phosphate pathway is an oxidative stage in which glucose-6-phosphate is converted to ribulose-5-phosphate. Two NADPH molecules and one CO_2 molecule are also produced in these reactions. The second stage of the pentose phosphate pathway involves isomerization reactions that convert ribulose-5-phosphate into other five-carbon sugars, ribose-5-phosphate and xylulose-5-phosphate. The third stage of the pathway involves a complex series of rearrangement reactions that result in the production of two fructose-6-phosphate and one glyceraldehyde-3-phosphate molecules from three molecules of pentose phosphate.

21.75 The ribose-5-phosphate produced in the pentose phosphate pathway is used for the biosynthesis of nucleotides, such as ATP. The erythrose-4-phosphate is used for the biosynthesis of aromatic amino acids, such as phenylalanine, tyrosine, and tryptophan.

21.77 Gluconeogenesis is production of glucose from noncarbohydrate starting materials. This pathway can provide glucose when starvation or strenuous exercise leads to a depletion of glucose from the body.

21.79 The liver is primarily responsible for gluconeogenesis.

21.81 Lactate is first converted to pyruvate. Pyruvate is the starting substrate for gluconeogenesis.

21.83 Because steps 1, 3, and 10 of glycolysis are irreversible, gluconeogenesis is not simply the reverse of glycolysis. The reverse reactions must be carried out by different enzymes.

Reaction 1 of glycolysis, the transfer of a high-energy phosphoryl group from ATP to glucose, is carried out by the enzyme hexokinase. The reverse reaction in gluconeogenesis is catalyzed by glucose-6-phosphatase.

Reaction 3 of glycolysis, the transfer of a high-energy phosphoryl group from ATP to fructose-6-phosphate, is catalyzed by phosphofructokinase. The reverse reaction of gluconeogenesis is carried out by fructose bisphosphatase.

The final reaction of glycolysis, the transfer of a high-energy phosphoryl group from phosphoenolpyruvate to ADP, is catalyzed by pyruvate kinase. This is reversed in gluconeogenesis by the action of two enzymes. Pyruvate carboxylase adds CO_2 to pyruvate to produce oxaloacetate and phosphoenolpyruvate carboxykinase removes the CO_2 and transfers a high-energy phosphoryl group from GTP to produce phosphoenolpyruvate.

21.85 Steps 1, 3, and 10 of glycolysis are irreversible. Step 1 is the transfer of a phosphoryl group from ATP to carbon-6 of glucose and is catalyzed by hexokinase. Step 3 is the transfer of a phosphoryl group from ATP to carbon-1 of fructose-6-phosphate and is catalyzed by phosphofructokinase. Step 10 is the substrate-level phosphorylation in which a phosphoryl group is transferred from phosphoenolpyruvate to ADP and is catalyzed by pyruvate kinase.

21.87 The liver is instrumental in the maintenance of blood glucose levels by serving as a reservoir for glycogen, a branched homopolymer of D-glucose. The pancreas is also critical to the control of blood glucose. When blood glucose levels are too high, the pancreas secretes the hormone insulin, which stimulates uptake and storage of glucose from the blood. When blood glucose levels are too low, the pancreas secretes the hormone glucagon, which stimulates the breakdown of liver glycogen and the release of glucose into the blood.

21.89 *Hypoglycemia* is the condition in which blood glucose levels are too low.

21.91 a. Insulin stimulates glycogen synthase, the first enzyme in glycogen synthesis. It also stimulates uptake of glucose from the bloodstream into cells and phosphorylation of glucose by the enzyme glucokinase.

   b. This traps glucose within liver cells and increases the storage of glucose in the form of glycogen.

   c. The ultimate result is a decrease of blood glucose levels.

21.93 Any defect in the enzymes required to degrade glycogen or export glucose from liver cells will result in a reduced ability of the liver to provide glucose at times when blood glucose levels are low. This will cause hypoglycemia.

Chapter 22
Aerobic Respiration and Energy Production
Solutions to the Odd-Numbered Problems

In-Chapter Questions and Problems

22.1 Mitochondria are the organelles responsible for aerobic respiration. They have enzymes that carry out the final oxidation of carbohydrates, amino acids, and fatty acids. They produce the majority of the ATP for the cell.

22.3

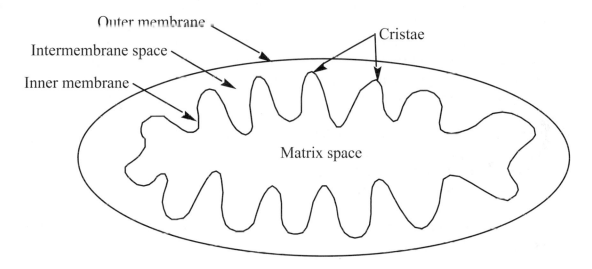

22.5 Pyruvate is converted to acetyl CoA by the pyruvate dehydrogenase complex. This huge enzyme complex requires four coenzymes, each of which is made from a different vitamin. The four coenzymes are thiamine pyrophosphate (made from thiamine), FAD (made from riboflavin), NAD^+ (made from niacin), and coenzyme A (made from the vitamin pantothenic acid). The coenzyme lipoamide is also involved in this reaction.

22.7 *Oxidative phosphorylation* is the process by which the energy of electrons harvested from oxidation of a fuel molecule is used to phosphorylate ADP to produce ATP.

22.9 $NAD^+ + H:^- \rightarrow NADH$

22.11 Pyridoxal phosphate is a coenzyme required by all transaminases. During transamination reactions, the α-amino group is transferred to pyridoxal phosphate. In the last part of the reaction, the α-amino group is transferred from pyridoxal phosphate to an α-keto acid.

22.13 The purpose of the urea cycle is to convert toxic ammonium ions to urea, which is excreted in the urine of land animals. This keeps toxic ammonium ions out of the bloodstream.

22.15 An amphibolic pathway is a metabolic pathway that functions both in anabolism and catabolism. The citric acid cycle is amphibolic because it has a catabolic function - it completely oxidizes the acetyl group carried by acetyl CoA to provide electrons for ATP synthesis. Because citric acid cycle intermediates are precursors for the biosynthesis of many other molecules, it also serves a function in anabolism.

End-of-Chapter Questions and Problems

22.17 The mitochondrion is an organelle that serves as the cellular power plant. The reactions of the citric acid cycle, the electron transport system, and ATP synthase function together within the mitochondrion to harvest ATP energy for the cell.

22.19 The intermembrane compartment is the location of the high-energy proton (H^+) reservoir produced by the electron transport system. The energy of this H^+ reservoir is used to make ATP.

22.21 The outer mitochondrial membrane is freely permeable to substances of molar mass less than 10,000. The inner mitochondrial membrane is highly impermeable. Only certain fuel molecules and H^+ are transported through it. Embedded within the inner mitochondrial membrane are the electron carriers of the electron transport system and ATP synthase, the multisubunit enzyme that makes ATP.

22.23 Coenzyme A is a molecule derived from ATP and the vitamin pantothenic acid. It functions in the transfer of acetyl groups in lipid and carbohydrate metabolism.

22.25 Decarboxylation is a chemical reaction in which a carboxyl group is removed from a molecule. In the decarboxylation of pyruvate, the carboxyl group is released as CO_2.

22.27 Under aerobic conditions, pyruvate is converted to acetyl CoA.

22.29 The coenzymes NAD^+, FAD, thiamine pyrophosphate, and coenzyme A are required by the pyruvate dehydrogenase complex for the conversion of pyruvate to acetyl CoA. These coenzymes are synthesized from the vitamins niacin, riboflavin, thiamine, and pantothenic acid, respectively. If the vitamins are not available, the coenzymes will not be available and pyruvate cannot be converted to acetyl CoA. Since the complete oxidation of the acetyl group of acetyl CoA produces the vast majority of the ATP for the body, ATP production would be severely inhibited by a deficiency of any of these vitamins.

22.31 An aldol condensation is a reaction in which aldehydes or ketones react to form larger molecules.

22.33 Oxidation in an organic molecule is often recognized as the loss of hydrogen atoms or a gain of oxygen atoms.

22.35 A dehydrogenation reaction is an oxidation reaction in which protons and electrons are removed from a molecule.

22.37 a. false
b. false
c. true
d. true

22.39 a. The acetyl group of acetyl CoA is transferred to oxaloacetate.

b. The product is citrate.

22.41 Three NAD^+ are reduced to 3 NADH in one turn of the citric acid cycle.

22.43 The net yield of ATP for anaerobic glycolysis is 2 ATP per glucose.

22.45 The function of acetyl CoA in the citric acid cycle is to bring the two-carbon remnant (acetyl group) of pyruvate from glycolysis and transfer it to oxaloacetate. In this way, the acetyl group enters the citric acid cycle for the final stages of oxidation.

22.47 GTP is produced in the citric acid cycle. The high-energy phosphoryl group of the GTP is transferred to ADP to produce ATP. This reaction is catalyzed by the enzyme dinucleotide diphosphokinase.

22.49 The importance of this reaction is that it brings the acetyl group, the two-carbon remnants of the glucose molecule, into the citric acid cycle to be completely oxidized. Through these reactions, and subsequent oxidative phosphorylation, the majority of the cellular ATP energy is provided.

$$
\begin{array}{c}
COO^- \\
| \\
C=O \\
| \\
CH_2 \\
| \\
COO^-
\end{array}
\; + \;
CH_3\text{-}\overset{\overset{O}{\|}}{C}\text{-}S\text{-}CoA \; + \; H_2O \; \longrightarrow
$$

Oxaloacetate   Acetyl CoA

$$
\begin{array}{c}
COO^- \\
| \\
CH_2 \\
| \\
HO-C-COO^- \\
| \\
CH_2 \\
| \\
COO^-
\end{array}
\; + \; HS-CoA \; + \; H^+
$$

Coenzyme A

Citrate

22.51   The conversion of citrate to *cis*-aconitate is an example of the dehydration of an alcohol to produce an alkene (double bond). The conversion of *cis*-aconitate to isocitrate is an example of the hydration of an alkene, that is, the addition of water to the double bond, to produce an alcohol (–OH).

22.53   This reaction is an example of the oxidation of a secondary alcohol to a ketone. The two functional groups are the hydroxyl group of the alcohol and the carbonyl group of the ketone.

22.55   It is a kinase because it transfers a phosphoryl group from one molecule to another. Kinases are a specific type of transferase.

22.57   An allosteric enzyme is one that has an effector binding site and an active site. Effector binding can change the shape of the active site, causing it to be active or inactive.

22.59   Negative allosterism is a means of enzyme regulation in which effector binding inactivates the active site of the allosteric enzyme.

22.61   The citric acid cycle is regulated by the following four enzymes or enzyme complexes: pyruvate dehydrogenase complex, citrate sythase, isocitrate dehydrogenase, and the α-ketoglutarate dehydrogenase complex.

22.63   Energy-harvesting pathways, such as the citric acid cycle, must be responsive to the energy needs of the cell. If the energy requirements are high, as during exercise, the reactions must speed up. If energy demands are low and ATP is in excess, the reactions of the pathway slow down.

22.65   ADP serves as a signal to increase the rate of the reactions of the citric acid cycle. If ADP is found in high concentration, then ATP levels must be low. This is a signal that the cell requires more ATP energy. The enzyme isocitrate dehydrogenase is an allosteric enzyme of the citric acid cycle that is stimulated by ADP.

22.67   The electron transport system is series of electron transport proteins embedded in the inner mitochondrial membrane that accept high-energy electrons from NADH and $FADH_2$ and transfer them in stepwise fashion to molecular oxygen (O_2).

22.69   The oxidation of NADH via oxidative phosphorylation yields 3 ATP.

22.71   The oxidation of a variety of fuel molecules, including carbohydrates, the carbon skeletons of amino acids, and fatty acids provides the electrons. The energy of these electrons is used to produce an H^+ reservoir. The energy of this proton reservoir is used for ATP synthesis.

22.73   The electron transport system passes electrons harvested during oxidation of fuel molecules to molecular oxygen. At three sites, protons are pumped from the mitochondrial matrix into the intermembrane compartment. Thus, the electron transport system builds the high-energy H^+ reservoir that provides energy for ATP synthesis.

22.75 a. Two ATP per glucose (nct yield) are produced in glycolysis, while the complete oxidation of glucose in aerobic respiration (glycolysis, the citric acid cycle, and oxidative phosphorylation) results in the production of 36 ATP per glucose.

b. Aerobic respiration harvests nearly 40% of the potential energy of glucose, while anaerobic glycolysis harvests only about 2% of the potential energy of glucose.

22.77 Transaminases transfer amino groups from amino acids to ketoacids.

22.79 The glutamate family of transaminases is very important because the ketoacid corresponding to glutamate is α-ketoglutarate, one of the citric acid cycle intermediates. This provides a link between the citric acid cycle and amino acid metabolism. These transaminases provide amino groups for amino acid synthesis and collect amino groups during catabolism of amino acids.

22.81 a. Pyruvate
b. α-Ketoglutarate
c. Oxaloacetate
d. Acetyl CoA
e. Succinate
f. α-Ketoglutarate

22.83 The synthesis of glutamate catalyzcd by glutamate dehydrogenase can be representcd by the following equation:

α-Ketoglutarate   Ammonia   Glutamate

22.85 Hyperammonemia, an elevation of the concentration of ammonium ions in the body, results when the urea cycle is not functioning. A complete deficiency of a urea cycle enzyme results in death in early infancy. A partial deficiency causes retardation, convulsions, and vomiting and can be treated with a low protein diet.

22.87 a. The source of one amino group of urea is the ammonium ion and the source of the other is the α-amino group of the amino acid aspartate.

b. The carbonyl group of urea is derived from CO_2.

22.89 Anabolism is a term used to describe all of the cellular energy-requiring biosynthetic pathways.

22.91 The amino acid glutamate is synthesized from the citric acid cycle intermediate α-ketoglutarate.

22.93 Citric acid cycle intermediates are the starting materials for the biosynthesis of many biological molecules.

22.95 An essential amino acid is one that cannot be synthesized by the body and must be provided in the diet.

22.97

$$
\begin{array}{c}
\text{COO}^- \\
| \\
\text{C}=\text{O} \\
| \\
\text{CH}_3
\end{array}
\;+\; \text{CO}_2 \;+\; \text{ATP} \;\longrightarrow\;
\begin{array}{c}
\text{COO}^- \\
| \\
\text{C}=\text{O} \\
| \\
\text{CH}_2 \\
| \\
\text{COO}^-
\end{array}
\;+\; \text{ADP} \;+\; \text{P}_i
$$

Pyruvate   Oxaloacetate

Chapter 23
Fatty Acid Metabolism
Solutions to the Odd-Numbered Problems

In-Chapter Questions and Problems

23.1   Because dietary lipids are hydrophobic, they arrive in the small intestine as large fat globules. The bile salts emulsify these fat globules into tiny fat droplets. This greatly increases the surface area of the lipids, allowing them to be more accessible to pancreatic lipases and thus more easily digested.

23.3   a.   The products of the β-oxidation of 9-phenylnonanoic acid are 4 acetyl CoA, 1 benzoate, 4 NADH, and 4 $FADH_2$.
           b.   The products of the β-oxidation of 8-phenyloctanoic acid are 3 acetyl CoA, 1 phenyl acetate, 3 NADH, and 3 $FADH_2$.
           c.   The products of the β-oxidation of 7-phenylheptanoic acid are 3 acetyl CoA, 1 benzoate, 3 NADH, and 3 $FADH_2$.
           d.   The products of the β-oxidation of 12-phenyldodecanoic acid are 5 acetyl CoA, 1 phenyl acetate, 5 NADH, and 5 $FADH_2$.

23.5

$$CH_3CH_2CH_2-\overset{\overset{\displaystyle O}{\|}}{C}\sim S\text{-}CoA$$

$$\downarrow \curvearrowright FAD$$

$$CH_3CH=CH-\overset{\overset{\displaystyle O}{\|}}{C}\sim S\text{-}CoA \quad + \quad FADH_2$$

$$\downarrow \curvearrowright H_2O$$

$$CH_3\overset{\overset{\displaystyle OH}{|}}{C}H\text{-}CH_2-\overset{\overset{\displaystyle O}{\|}}{C}\sim S\text{-}CoA$$

$$\downarrow \curvearrowright NAD^+$$

$$CH_3\text{-}C\text{-}CH_2\text{-}C\text{-}S\text{-}CoA$$

Coenzyme A

$$2 \quad CH_3\text{-}C\text{-}S\text{-}CoA$$

23.7 Starvation, a diet low in carbohydrates, and diabetes mellitus are conditions that lead to the production of ketone bodies. Lack of carbohydrates causes a decreased amount of oxaloacetate. This slows the citric acid cycle so that less acetyl CoA is oxidized. The excess acetyl CoA is converted to ketone bodies to recover coenzyme A.

23.9 The following are four differences between fatty acid biosynthesis and β-oxidation:
   a. Fatty acid biosynthesis occurs in the cytoplasm, and β-oxidation occurs in the mitochondria.
   b. The acyl group carrier in fatty acid biosynthesis is Acyl Carrier Protein (ACP), and the acyl group carrier in β-oxidation is coenzyme A.
   c. The seven enzymes of fatty acid biosynthesis are associated as a multienzyme complex called fatty acid synthase. The enzymes involved in β-oxidation are not physically associated with one another.
   d. NADPH is the reducing agent used in fatty acid biosynthesis. NADH and FADH$_2$ are produced by β-oxidation.

23.11 The liver regulates blood glucose levels under the control of the hormones insulin and glucagon. When blood glucose levels are too high, insulin stimulates the uptake of glucose by liver cells and the storage of the glucose in glycogen polymers. When blood glucose levels are too low, the hormone glucagon stimulates the breakdown of glycogen and release of glucose into the bloodstream. Glucagon also stimulates the liver to produce glucose for export into the bloodstream by the process of gluconeogenesis.

23.13 Insulin stimulates uptake of glucose and amino acids by cells, glycogen and protein synthesis, and storage of lipids. It inhibits glycogenolysis, gluconeogenesis, breakdown of stored triglycerides, and ketogenesis.

End-of-Chapter Questions and Problems

23.15 Bile consists of micelles of lecithin, cholesterol, bile salts, proteins, inorganic ions, and bile pigments that aid in lipid digestion by emulsifying fat droplets.

23.17 A micelle is an aggregation of molecules having nonpolar and polar regions; the nonpolar regions of the molecules aggregate, leaving the polar regions facing the surrounding water.

23.19 A triglyceride is a molecule composed of glycerol esterified to three fatty acids.

23.21 A chylomicron is a plasma lipoprotein that carries triglycerides from the intestine to all body tissues via the bloodstream. That function is reflected in the composition of the chylomicron, which is approximately 85% triglycerides, 9% phospholipids, 3% cholesterol esters, 2% protein, and 1% cholesterol.

22.23 Triglycerides in adipose tissue are the major storage form of lipids.

23.25 The most outstanding feature of an adipocyte is the large fat globule that takes up nearly the entire cytoplasm.

23.27 Lipases catalyze the hydrolysis of the ester bonds of triglycerides, as seen in the following equation:

$$
\begin{array}{l}
\underset{\substack{|\\ H}}{\overset{\substack{H \\ |}}{H-C}}-O-\overset{\overset{O}{\|}}{C}-R_1 \\[2pt]
\underset{|}{H-C}-O-\overset{\overset{O}{\|}}{C}-R_2 \ +\ 3\,H_2O \\[2pt]
\underset{\substack{|\\ H}}{H-C}-O-\overset{\overset{O}{\|}}{C}-R_3
\end{array}
\longrightarrow
\begin{array}{l}
\underset{|}{\overset{\substack{H\\|}}{H-C}}-OH \\[2pt]
\underset{|}{H-C}-OH \\[2pt]
\underset{|}{H-C}-OH \\[2pt]
\ \ II
\end{array}
\quad
\begin{array}{l}
+\ \ HO-\overset{\overset{O}{\|}}{C}-R_1 \\[6pt]
+\ \ HO-\overset{\overset{O}{\|}}{C}-R_2 \\[6pt]
+\ \ HO-\overset{\overset{O}{\|}}{C}-R_3
\end{array}
$$

23.29 Acetyl CoA is the precursor for fatty acids, several amino acids, cholesterol, and other steroids.

23.31 Chylomicrons are plasma lipoproteins (aggregates of protein and triglycerides) that carry dietary triglycerides from the intestine to all tissues via the bloodstream.

23.33 Bile salts serve as detergents. Fat globules stimulate their release from the gall bladder. The bile salts then emulsify the lipids, increasing the surface area and making them more accessible to digestive enzymes (pancreatic lipases).

23.35 When dietary lipids in the form of fat globules reach the duodenum, they are emulsified by bile salts. The triglycerides in the resulting tiny fat droplets are hydrolyzed into monoglycerides and fatty acids by the action of pancreatic lipases, assisted by colipase. The monoglycerides and fatty acids are absorbed by cells lining the intestine. Within intestinal cells, triglycerides are reassembled and are packaged into chylomicrons (lipoprotein particles made up of protein and dietary triglycerides). Chylomicrons are secreted into the lymphatic vessels and eventually reach the bloodstream. In the bloodstream, the triglycerides are hydrolyzed once again and the products (glycerol and free fatty acids) are absorbed by the cells of the body.

23.37   The energy source for the activation of a fatty acid entering β-oxidation is the hydrolysis of ATP into AMP and PP$_i$ (pyrophosphate group), an energy expense of two high-energy phosphoester bonds.

23.39   Carnitine is a carrier molecule that brings fatty acyl groups into the mitochondrial matrix.

23.41   The following equation represents the reaction catalyzed by acyl-CoA dehydrogenase. Notice that the reaction involves the loss of two hydrogen atoms. Thus, this is an oxidation reaction.

$$H_3C-\overset{\overset{\displaystyle H}{|}}{C}-\overset{\overset{\displaystyle \text{\textcircled{H}}}{|}}{\underset{\underset{\displaystyle \text{\textcircled{H}}}{|}}{C}}-\overset{\overset{\displaystyle O}{||}}{\underset{\underset{\displaystyle H}{|}}{C}}{\sim}\text{S-CoA}$$

FAD   Acyl-CoA
FADH$_2$ dehydrogenase

$$H_3C-\overset{\overset{\displaystyle H}{|}}{C}{=}\overset{}{\underset{\underset{\displaystyle H}{|}}{C}}-\overset{\overset{\displaystyle O}{||}}{C}{\sim}\text{S-CoA}$$

23.43   An alcohol is the product of the hydration of an alkene.

23.45   The β-oxidation of 14-phenyltetradecanoic acid yields 6 acetyl CoA, 1 phenyl acetate, 6 NADH, and 6 FADH$_2$.

23.47   The following diagram summarizes the energy harvested by the β-oxidation of tetradecanoic acid:

Step 1 (Activation):   − 2 ATP

Steps 2 - 6 (repeated six times):

   6 FADH$_2$ x 2 ATP/FADH$_2$ 12 ATP
   6 NADH x 3 ATP/NADH   18 ATP

7 acetyl CoA to citric acid cycle:

   7 x 1 GTP x 1 ATP/GTP   7 ATP

                7 x 3 NADH x 3 ATP/NADH   63 ATP
                7 x 1 FADH$_2$ x 2 ATP/FADH$_2$ 14 ATP

                                                           112 ATP

23.49 Two molecules of ATP are produced for each FADH$_2$ produced by β-oxidation.

23.51 The acetyl CoA produced by β-oxidation will enter the citric acid cycle unless there is too
       little oxaloacetate available. If that is the case, the acetyl CoA will be used in
       ketogenesis.

23.53 Ketone bodies include the compounds acetone, acetoacetone, and β-hydroxybutyrate,
       which are produced from fatty acids in the liver via acetyl CoA.

23.55 Ketosis is an abnormal rise in the level of ketone bodies in the blood.

23.57 Ketosis occurs in the matrix of the mitochondrion.

23.59 Acetoacetate   β-Hydroxybutyrate

$$CH_3-\overset{\overset{\displaystyle O}{||}}{C}-CH_2-\overset{\overset{\displaystyle O}{||}}{C}-O^- \qquad CH_3-\overset{\overset{\displaystyle OH}{|}}{CH}-CH_2-\overset{\overset{\displaystyle O}{||}}{C}-O^-$$

23.61 In those suffering from uncontrolled diabetes, the glucose in the blood cannot get into the
       cells of the body. The excess glucose is excreted in the urine. Body cells degrade fatty
       acids because glucose is not available. β-Oxidation of fatty acids yields enormous
       quantities of acetyl CoA, so much acetyl CoA, in fact, that it cannot all enter the citric
       acid cycle because there is not enough oxaloacetate available. Excess acetyl CoA is used
       for ketogenesis.

23.63 Ketone bodies are the preferred energy source of the heart.

23.65 Fatty acid biosynthesis occurs in the cytoplasm of the cell.

23.67 A multienzyme complex called fatty acid synthase carries out fatty acid biosynthesis.

23.69 The phosphopantetheine group allows formation of a high-energy thioester bond with a
       fatty acid. It is derived from the vitamin pantothenic acid and β-mercaptoethylamine.

23.71 Fatty acid synthase is a huge multienzyme complex consisting of the seven enzymes
       involved in fatty acid synthesis. It is found in the cell cytoplasm. The enzymes involved
       in β-oxidation are not physically associated with one another. They are free in the
       mitochondrial matrix space.

23.73 Glycogenesis is the synthesis of the polymer glycogen from glucose monomers.

23.75 Gluconeogenesis is the synthesis of glucose from noncarbohydrate precursors.

23.77 The energy requirements of resting muscle are generally met by β-oxidation of fatty acids.

23.79 The major metabolic function of the liver is to regulate blood glucose levels. It also produces triglycerides for storage when excess calories are consumed. During fasting or starvation, the liver converts fatty acids to ketone bodies, which are then exported to other organs.

23.81 Ketone bodies are the major fuel for the heart. Glucose is the major energy source of the brain and the liver obtains most of its energy from the oxidation of amino acid carbon skeletons.

23.83 Fatty acids are absorbed from the bloodstream by adipocytes. Using glycerol-3-phosphate, produced as a by-product of glycolysis, triglycerides are synthesized. Triglycerides are constantly being hydrolyzed and resynthesized in adipocytes. The rates of hydrolysis and synthesis are determined by lipases that are under hormonal control.

23.85 In general, insulin stimulates anabolic processes, including glycogen synthesis, uptake of amino acids and protein synthesis, and triglyceride synthesis. At the same time, catabolic processes such as glycogenolysis are inhibited.

23.87 A target cell is one that has a receptor for a particular hormone.

23.89 Decreased blood glucose levels trigger the secretion of glucagon into the bloodstream.

23.91 Insulin is produced in the β-cells of the islets of Langerhans in the pancreas.

23.93 Insulin stimulates the uptake of glucose from the blood into cells. It enhances glucose storage by stimulating glycogenesis and inhibiting glycogen degradation and gluconeogenesis.

23.95 Insulin stimulates synthesis and storage of triglycerides.

23.97 Untreated diabetes mellitus is starvation in the midst of plenty because blood glucose levels are very high. However, in the absence of insulin, blood glucose can't be taken up into cells. The excess glucose is excreted into the urine while the cells of the body are starved for energy.

APPENDIX B

Answers to Chapter Self Tests

Chapter 1

1.   1.02 g/mL
2. 1.48×10^{-1}
3. 3.01×10^{-6}
4. 2.24×10^3 cm/s
5. 3.95×10^{-22} g
6. 2.99×10^{10} cm/s
7. $-20.3°C$
8. 4.21×10^3 g
9. $-40.0°F$
10.   0.545
11.   energy
12.   data
13.   measurement
14.   metric system
15.   conversion factors
16.   Kelvin
17.   310.0 K
18. 3.7×10^{-3}
19.   b, c, d, f, and g
20.   b, c, d, e, and f
21.   gas, liquid, and solid
22.   gas
23.   gas
24.   a. color
      b. odor
      c. taste
      d. compressibility
      e. melting point
      f. boiling point
25.   heterogeneous

Chapter 2

1.   12 protons and 12 electrons
2.   17 protons and 18 electrons

3.   a. Lithium atom (Li): 3 protons; 3 electrons
     b. Lithium ion (Li^{1+}): 3 protons; 2 electrons
     c. $Li \rightarrow 1 e^- + Li^{1+}$
4.   nucleus
5.   electron
6.   atomic number
7.   neutron
8.   positive, or cation, or +1
9. ^1_1H ^2_1H ^3_1H
     normal hydrogen   deuterium   tritium
10. valence electrons
11. six
12. helium
13. noble gas
14. a more stable energy state
15. isoelectric ion
16. representative elements
17. transition elements
18. n = 1, 2, 3, etc.
19. two
20. orbital
21. s
22. opposite
23. electron configuration
24. $1s^2, 2s^2, 2p^2$
25. noble gases
26. Group IA
27. electronegativity

Chapter 3

1.   ionic bond
2.   covalent bond
3.   positive and negative ions
4.   two
5.   left
6.   right
7.   metal or cation
8.   sodium oxide
9.   lithium sulfide
10. aluminum bromide
11. ferrous ion
12. iron (III) ion
13. higher positive charge
14. ferrous sulfate

15. copper (II) oxide
16. Na_2SO_4
17. four
18. eight
19. Cl
20. difference in electronegativities
21. H_2O
22. ICl
23. ammonia:

24. water:

H
H : O :

H—O with H

25. hydroxide ion:

$[:O:H]^-$ $H—O^-$

26. hydronium ion:

$[H:O:H]^+$ with H on top $[H—O—H]^+$ with H on top

27. valence electrons
28. covalent bonding
29. $Na^+ + Cl^-$
30. $Ca^{2+} + S^{2-}$
31. electron transfer
32. two
33. four

Chapter 4

1. 1
2. 2.0 grams
3. 6.02×10^{23} atoms, or Avogadro's number
4. 4.81 grams
5. N_2
6. covalent compound

7.   amu
8.   18.02 amu
9.   molecular weight
10. reactants
11. 2
12. 0.5
13. coefficients
14. 2
15. 2
16. 1
17. 80.09 grams
18. 169.9 grams
19. 7.149 grams
20. 3.01 moles
21. Avogadro's number
22. O_2

Chapter 5

1.   evaporation
2.   ideal
3.   759.0 mm of mercury pressure
4.   5.00 atmospheres
5.   2.09 liters
6. 1.96×10^{-2} moles of helium are present
7. δ^+ H–F δ^- ------- δ^+ H–F δ^-
8.   nitrogen and oxygen
9.   Dalton's
10. surface tension
11. surfactants
12. vapor pressure
13. melting point
14. very high
15. 20.0 liters
16. condensation
17. amorphous
18. barometer
19. calorimeter
20. covalent

Chapter 6

1.   1.10 % (W / V)
2.   800.0 mL

3.   1.25 liters
4. 5.52×10^{-2} moles
5.   0.466 gram
6.   0.19 molar HCl solution
7.   0.185 molar KNO_3 solution
8.   52.1 mL
9.   concentration
10. 13.5 grams
11. molarity
12. osmosis
13. hypertonic
14. isotonic

Chapter 7

1. 1.40×10^3 calories
2.   a.
$$K_{eq} = \frac{[N_2O_4]}{[NO_2]^2}$$

b.
$$K_{eq} = \frac{[PCl_3][Cl_2]}{[PCl_5]}$$

c.
$$K_{eq} = \frac{[N_2][H_2O]^2}{[NO]^2[H_2]^2}$$

3.   high
4.   nonspontaneous
5.   gas
6.   positive
7.   calorimeter
8.   fuel value
9.   activated complex
10. equilibrium
11. equilibrium
12. LeChatelier
13. no
14. to the left
15. to the right

Chapter 8

1.   4.19
2.   0.0592 molar HCl solution

3. 5.00×10^{-14} moles/liter of H_3O^+ are present
4. pH = 3.0
5. neutralization
6. Brønsted-Lowry
7. dissociation
8. b
9. b

Chapter 9

1. a. $_0^1n$ b. $_{83}^{210}Bi$
2. 3.125 mg remain
3. 9
4. alpha particles, beta particles, and gamma rays
5. gamma ray
6. alpha particle
7. radioactive
8. half-life, or $t_{1/2}$
9. fission
10. cancer or malignant cells
11. thyroid gland
12. background radiation
13. inversely
14. nuclear imaging device
15. film badge
16. gamma ray
17. natural radioactivity
18. C-14 or carbon-14
19. alpha particle
20. radon

Chapter 10

1. b, c, and d
2. tetrahedral
3. 120°, or planar
4. 180°
5. functional group
6. hydroxyl
7. carbonyl group
8. aromatic
9. a
10. c
11. a

12. b
13. a
14. c
15. e
16. inorganic chemistry
17. urea
18. hydrogen
19. carbon-to-carbon double bond
20. C_9H_{20}
21. a. 2-methylbutane
    b. 2,2-dimethylpentane
    c. 2,2,3-trimethylpentane
22. 1-bromo-4-chloro-2-methylpentane
23. 2-bromo-2-methylpropane
24. 2-bromo-4,4-dimethylhexane
25. $C_3H_8 + 5O_2 \rightarrow 3CO_2 + 4H_2O$
26. Two products are formed: 1-chloropropane and 2-chloropropane
27. none or little
28. lower
29. parent compound
30. trichloromethane
31. cycloalkane or cycloalkanes
32. ring
33. combustion
34. two
35. light, or UV, or high temperature
36. substitution

Chapter 11

1. a. alkane: C_4H_{10} could be butane
   b. alkene: C_4H_8 could be 1-butene or 2-butene
   c. alkyne: C_4H_6 could be 1-butyne or 2-butyne
2. 3-ethyl-2-pentene
3. 4,5-dibromo-2-hexyne
4. 2-methyl-2-pentene:

$$CH_3 - \underset{\underset{CH_3}{|}}{C} = CH - CH_2 - CH_3$$

5. *cis*-3-methyl-2-pentene:

6. *trans*-2-bromo-2-butene
7.

$$H_2C{=\!\!=}CHCH_2CH_3 \quad + \ H_2 \quad \xrightarrow{\text{Ni}} \quad CH_3CH_2CH_2CH_3$$

  1-Butene   Butane

8.

CH₃C(H)=CH₂   + H₂O   →(H⁺)→   CH₃CH(OH)CH₃   major product

CH₃C(H)=CH₂   + H₂O   →(H⁺)→   CH₃CH₂CH₂OH   minor product

9.

CH₃—C(CH₃)=CH—CH₂—CH₃   + HCl   ⟶   CH₃—C(CH₃)(Cl)—CH₂—CH₂—CH₃   major product

CH₃—C(CH₃)=CH—CH₂—CH₃   + HCl   ⟶   CH₃—C(CH₃)(H)—CH(Cl)—CH₂—CH₃   minor product

10. a. benzene
     b. phenol or hydroxybenzene
     c. 1,2-dichlorobenzene or *ortho*-dichlorobenzene or *o*-dichlorobenzene
     d. 1,2-dichloro-3-bromobenzene
     e. 1,4-dibromobenzene or *para*-dibromobenzene or *p*-dibromobenzene
11. acetylene
12. *trans*-2-bromo-2-butene
13. *trans*-1,2-dibromopropene
14. hydrogenation
15. palladium and nickel

16. CH_3CH_2OH
17. benzene
18. chloroform
19. carbon-to-carbon double bond
20. addition
21. all carbon atoms have four single bonds
22. at least one carbon-to-carbon triple bond
23.

$$H-C\equiv C-CH_3 \quad + \quad 2HBr \quad \longrightarrow \quad CH_3-\underset{\underset{Br}{|}}{\overset{\overset{Br}{|}}{C}}-CH_3$$

24. $HC\equiv C-CH_2CH_3 + 2H_2 \rightarrow CH_3CH_2CH_2CH_3$

(Note: A catalyst such as Ni should be shown over the reaction arrow)
25. $CH_3CH_2CH_2CH_3$
26. Vegetable oil

Chapter 12

1. a. 1-hexanol   c. 1-propanol
    b. methanol   d. dimethyl ether
2. a
3. 3-bromo-4-methyl-2-pentanol
4. 5-chloro-2-pentanol
5. 1,2,3-propanetriol

6. a. propyl alcohol
    b. isopropyl alcohol
    c. *tert*-butyl alcohol
7. a. CH_3OH
    b. CH_3CH_2OH
8. a.

$$H-\overset{\overset{O}{\|}}{C}-H$$

    b.

$$CH_3\overset{\overset{O}{\|}}{C}-H$$

9. a. ethene + H_2O
    b. propene + H_2O
    c. 1-butene + H_2O + 2-butene + H_2O
       (minor product)   (major product)

10.   a. formaldehyde, then to formic acid
      b. acetaldehyde, then to acetic acid
      c. acetone
      d. no reaction
11.   a. dimethyl ether
      b. diethyl ether
      c. ethyl methyl ether (or methyl ethyl ether)
12.   d. acetic acid
13.   a. methane
14.   alcohol
15.   thiol
16.   They are polar and can form hydrogen bonds between alcohol molecules.
17.   ethanol
18.   denatured alcohol
19.   glycerol
20.   1
21.   tertiary
22.   oxidation
23.   ethanal (acetaldehyde)
24.   thiol or thiols
25.   b. ethers
26.   b. ethers

Chapter 13

1.   a. 1-propanol
      b. propanal
2.

3.   a. ethanal   d. 3-pentanone
      b. propanone   e. 4-bromohexanal
      c. propanal
4.   The numbers are not necessary.

5.　a.

$$CH_3CH_2CH \overset{\displaystyle Br}{\underset{}{|}} \cdots$$

CH₃CH₂CH—C—H with O double bond above C and Br below CH

a.

CH₃CH₂CH—C—H
          | ‖
          Br   O

b.

CH₃—CH—C—CH₂CH₃
       | ‖
       CH₃   O

6.

CH₃OH —liver/enzymes→ Formaldehyde (very toxic) —liver/enzymes→ Formic acid (very toxic)

Methanol

CH₃CH₂OH —liver/enzymes→ Acetaldehyde —liver/enzymes→ Acetic acid

Ethanol

7.　a. HCOOH　　　d. CH₃CH₂COOH
　　b. CH₃COOH　　e. no reaction
　　c. no reaction

8.　a. ethanal (or ethanoic acid)
　　b. propanal (or propanoic acid)
　　c. propanone
　　d. no reaction

9.　a. CH₃COOH + Ag⁰
　　b. No reaction: CH₃COCH₃ is a ketone.
　　c. No reaction: CH₃CH₂COCH₃ is a ketone.

10.　a. oxidation　　c. reduction
　　 b. oxidation　　d. oxidation

11.　a.

R—C—O—R
      |
      OH

b.

O
‖
H—C—H

c. CH₃CHO

12.　aldehydes and ketones

13. ketone
14. acetaldehyde
15. formalin
16. carboxylic acids
17. acetic acid
18. benzoic acid
19. Tollens' test and Benedict's test
20. acetal
21. vitamin A
22. The carbonyl group is polar.
23. aldehyde
24. phenyl group
25. aldehyde
26. ketone

Chapter 14

1. a. ROH   d. RCHO
   b. ROR   e. RCOOH
   c. RCOOR   f. RCOR

2. The general structure of an acyl group is:

$$\left[\text{(Ar)} \quad \text{or} \quad R - \overset{\displaystyle \overset{O}{\|}}{C} - \right]$$

3. a. methanoic acid   c. ethanoic acid
   b. hexanoic acid   d. 3-methylbutanoic acid

4. a. CH_3COOH b. HCOOH
   c.   d. $CH_3(CH_2)_8COOH$

5. a. butyric acid
   b. propionic acid
   c. α-methylpropionic acid

6. a. ethyl formate
   b. methyl propionate
   c. pentyl acetate

7. a. methyl ethanoate
   b. ethyl propanoate

8. a. acetic acid

9. a. $CH_3COO^- + H_3O^+$

　　b. $CH_3COO^-\ Na^+ + H_2O$

　　c.

$$CH_3 - \overset{\overset{\displaystyle O}{\|}}{C} - O - CH_3$$

10.　alcohol + carboxylic acid
11.　carboxylate ion + hydronium ion
12.　a. sodium butyrate + water
　　　b. calcium acetate + water
13.　soaps
14.　hydrophobic end
15.　methanol + propanoic acid
16.　ATP
17.　saponification
18.　carboxyl group
19.　formic acid
20.　thioester
21.　methyl ethanoate
22.　qxidation reaction;　$KMnO_4$ is an oxidizing agent.
23.　carboxylic acid
24.　coenzyme A
25.　water
26.　hydrophilic
27.　micelles
28.　CH_3OH
29.　phosphate ester or phosphoester
30.　phosphoric anhydride bond or phosphoanhydride bond

Chapter 15

1.　a. tertiary amine　　　　　c. primary amine
　　b. quaternary ammonium salt　d. secondary amine
2.　carboxyl groups and amino groups
3.　a. ethanamine
　　b. 2-propanamine
　　c. *N,N*-dimethylmethanamine
4.　aniline
5.　a. $CH_3COOH + NH_4^+$
　　b. $CH_3CH_2COOH + CH_3N^+H_3$
6.　a. $(CH_3)_2NH_2^+I^-$
　　b. $(CH_3)_3NH^+Cl^-$
7　a. RNH_2
　　b. R_3N

8.   a. ethanamide
       b. *N*-propylhexanamide
       c. *N*-methylbutanamide
9.   histamine
10.   brain, spinal cord, and all nerves that radiate from the spinal cord
11.   serotonin and acetylcholine
12.   because the alkylammonium salts are more soluble in water and body fluids
13.   Parkinson's disease
14.   amino group
15.   ammonia
16.   nitrogen
17.   nucleic acids or DNA + RNA
18.   to carry a message, or signal, from a nerve cell to a target cell
19.   alkylammonium salt
20.   methylammonium chloride
21.   buffers
22.   amide
23.   solid
24.   a carboxylic acid and an amine or ammonia
25.   because they are detergents
26.   quaternary ammonium salt
27.   proteins
28.   acetylcholine
29.   analgesic

Chapter 16

1.   The monosaccharides are a, b, c and f. Sucrose is a disaccharide, and starch is a polysaccharide.
2.   a. aldohexose   d. aldopentose
       b. ketohexose   e. aldopentose
       c. aldohexose   f. aldotriose
3.   enantiomers
4.   cellulose
5.   D-glucose
6.   glyceraldehyde
7.   α-amylase and β-amylase
8.   glycosidic bond
9.   An aldose has an aldehyde group and a ketose contains a ketone group.
10.   A triose contains three carbon atoms and a pentose contains five carbon atoms.
11.   a. D-glyceraldehyde   c. L-ribose
       b. D-ribose   d. D-deoxyribose
12.   grape sugar, dextrose, or blood sugar
13.   D-fructose

14. glycogen
15. a. glucose + H_2O → no reaction
    b. lactose + H_2O → glucose + galactose
    c. sucrose + H_2O → glucose + fructose
    d. starch + H_2O → many glucose molecules
    e. cellulose + H_2O → many glucose molecules
16. sucrose, a nonreducing sugar
17. carbohydrate(s)
18. Humans cannot make the enzyme cellulose.
19. lactose intolerance
20. D-glucose
21. alpha
22. hemiketal
23. ribose or D-ribose
24. $C_5H_{10}O_5$
25. galactosemia
26. reducing sugars
27. disaccharides
28. liver and muscle tissues
29. Benedict's or Tollens'
30. Benedict's or Tollens'

Chapter 17

1. fatty acids
2. They are nonpolar molecules.
3. triglycerides (triacylglycerols)
4. They are long-chain monocarboxylic acids with an even number of carbon atoms.
5. glycerol + stearic acid + lauric acid + palmitic acid
6. long-chain alcohol + fatty acid
7. chylomicrons, very low-density lipoproteins (VLDL), low-density lipoproteins (LDL), high-density lipoproteins (HDL)
8. glycerol + 3 fatty acids
9. HDL
10. cholesterol
11. hydrophobic (nonpolar) fatty acid tail
12. cholesterol
13. the degree of saturation and carbon chain length
14. $C_{11}H_{23}COOH + NaOH \rightarrow C_{11}H_{23}COO^-Na^+ + H_2O$
15. the four fused rings known as the steroid nucleus
16. egg yolks, dairy products, liver, and other animal meats
17. progesterone
18. lipids
19. solid

20. ester + water
21. Ca^{2+} or Mg^{2+}
22. saturated solid fats
23. prostaglandins, leukotrienes, and thromboxanes
24. dilation
25. nonionic and nonpolar
26. polar and nonpolar
27. lecithin
28. high-density lipoproteins
29. cortisone
30. birth control

Chapter 18

1. nitrogen and sulfur
2. glycine
3. phenylalanine
4. Enzymes are proteins that act as biological catalysts.
5. Antibodies, also called immunoglobulins, are specific protein molecules produced by special cells of the immune system in response to foreign antigens.
6. hemoglobin and myoglobin
7. keratin
8. a mixture of the amino acids
9.

10. aspartate and glutamate
11. lysine, arginine, and histidine
12. a. ala   d. his
     b. cys   e. lys
     c. gly
13. peptide bond or amide bond
14.

15. primary structure
16. secondary structure

17. quaternary structure
18. isoelectric
19. α-helix or helices
20. zwitterions
21. essential amino acids
22. meats
23. pepsin
24. nutrient protein
25. proline
26. β-pleated sheet
27. disulfide bond
28. A hydrophobic molecule is nonpolar and thus is not water-soluble.
29. When the pH drops too low, a protein becomes a polycation,

Chapter 19

1. The enzyme lowers the activation energy of the reaction.
2. substrate
3. $2H_2O_2(l) \rightarrow 2H_2O(l) + O_2(g)$
4. a. succinate
   b. sucrose
   c. glycogen
5.

$$
\underset{\substack{\text{lactate} \\ \text{(reduced form)}}}{\overset{\displaystyle COO^-}{\underset{\displaystyle CH_3}{HO-\overset{\displaystyle |}{\underset{\displaystyle |}{C}}-H}}} + \underset{\substack{\text{lactate} \\ \text{dehydrogenase} \\ \text{(oxidized form)}}}{\text{enzyme-NAD}^{|}} \longrightarrow \underset{\substack{\text{pyruvate} \\ \text{(oxidized form)}}}{\overset{\displaystyle COO^-}{\underset{\displaystyle CH_3}{\overset{\displaystyle |}{\underset{\displaystyle |}{C}}=O}}} + \underset{\substack{\text{lactate} \\ \text{dehydrogenase} \\ \text{(reduced form)}}}{\text{enzyme-NADH}}
$$

6. They catalyze the transfer of phosphate groups.
7. lipase
8. no effect
9. formation of enzyme-substrate complex
10. a. absolute specificity   c. linkage specificity
    b. group specificity   d. stereochemical specificity
11. For an enzyme that requires a nonprotein component in order to function in catalysis, the protein portion is called its apoenzyme and the nonprotein part is the cofactor.
12. Usually it is nicotinamide adenine dinucleotide, as either the oxidized form (NAD^+) or its reduced form ($NADH + H^+ = NADH$).
13. Most are used in some form as coenzymes.
14. pH optimum
15. denaturation

16. irreversible inhibitors and competitive inhibitors
17. The sulfa drugs are competitive inhibitors of a bacterial enzyme required for the synthesis of the required vitamin folic acid.
18. They are proteases, digestive enzymes that hydrolyze peptide bonds in proteins.
19. The rate of reaction increases until, at a certain substrate concentration, the rate reaches a maximum.
20. The active sites of all the enzyme molecules in solution are occupied by substrate molecules.
21. absolute specificity
22. in the enzyme active site
23. the optimum temperature (37°C for most enzymes found in the human body)
24. effector binding
25. generally, the end-product of the pathway
26. zymogen

Chapter 20

1. DNA
2. DNA
3. a. dATP
   b. CTP
   c. TMP
4. DNA → RNA → protein
5. Watson and Crick
6. prime
7. hydrogen bonding between A and T and G and C
8. right angle
9. thymine
10. 3' → 5' direction
11. The two strands are complementary. One strand specifies the sequence of bases on the other.
12. pyrimidine dimers
13. The two strands of the double helix are separated by the enzyme helicase.
14. replication origin
15. replication fork
16. a. It reads the parental strand and produces the daughter strand.
    b. It proofreads the newly synthesized daughter strand to ensure that no errors have been made. If an error is detected, it is corrected.
17. cancer
18. a. Messenger RNA—carries the genetic information for protein from DNA to the ribosomes; b. Ribosomal RNA—a structural and functional component of the ribosomes; c. Transfer RNA—responsible for translating the genetic code of the mRNA into the primary structure of a protein.
19. transfer RNA
20. RNA polymerase

21. promoter
22. intervening sequences, or introns
23. exons
24. codons
25. Crick
26. methionine and tryptophan
27. glutamate
28. xeroderma pigmentosum
29. a small and a large ribosomal subunit
30. polysomes, or polyribosomes
31. silent mutation
32. restriction enzyme
33. phage vectors and plasmid vectors
34. hybridization
35. insulin, human growth hormone, blood clotting factors VIII and IX, interferon, hepatitis B virus vaccine

Chapter 21

1. glycolysis
2. ATP
3. 7 kcal/mole
4. anaerobic
5. two
6. NADH (NADH + H^+)
7. ethanol
8. lactate
9. carbohydrates
10. glycolysis
11. amino acids
12. catabolism
13. adenosine triphosphate, or ATP
14. hydrolysis
15. phosphorylation
16. fructose-6-phosphate
17. cytoplasm
18. substrate-level phosphorylation
19. adenosine triphosphate, or ATP
20. glucose-6-phosphate
21. amylase
22. lactase
23. pepsin
24. small intestine or duodenum
25. duodenum
26. a. glucose

27. e. glyceraldehyde-3-phosphate
28. c. pyruvate
29. F
30. T

Chapter 22

1.   mitochondria
2.   energy need
3.   oxaloacetate
4.   oxaloacetate
5.   two
6.   three
7.   one
8.   two
9.   mitochondria
10. electron transport system
11. cytochromes
12. hydrogen ions or protons
13. protons
14. F_1 catalyzes the phosphorylation of ADP to produce ATP.
15. NAD^+
16. It returns to be used again in the Krebs cycle.
17. oxygen, or O_2
18. cytochromes
19. three
20. two
21. the diet
22. liver
23. glutamate, or glutamic acid
24. citric acid cycle, or Krebs cycle
25. oxalocetate and glutamate
26. liver
27. carbamoyl phosphate
28. amphibolic
29. matrix
30. deamination

Chapter 23

1.   acetyl coenzyme A
2.   acetyl coenzyme A
3.   fat globules
4.   bile

5.   micelle
6.   cholate and chenodeoxycholate
7.   colipase
8.   chylomicrons
9.   adipose tissue
10.   1 $FADH_2$ and 1 NADH
11.   adipocyte
12.   129 ATP molecules
13.   The abnormal rise in concentration of blood ketone bodies.
14.   Starvation, a diet that is extremely low in carbohydrates, and the disease diabetes mellitus.
15.   ketoacidosis
16.   liver
17.   heart muscle
18.   fatty acid synthase
19.   NADPH
20.   liver
21.   Glycerol-3-phosphate is synthesized from glyceraldehyde-3-phosphate produced during glycolysis.
22.   VLDL
23.   amino acid carbon skeletons
24.   glycerol-3-phosphate
25.   fatty acids
26.   acyl CoA
27.   acetyl CoA
28.   β-hydroxybutyrate and acetoacetate
29.   anaerobic
30.   liver, adipose, and muscle cells

APPENDIX C

Answers to Vocabulary Quizzes

Chapter 1

1. chemistry
2. data
3. hypothesis
4. gaseous
5. potential energy
6. result
7. extensive
8. specific gravity
9. theory
10. weight

Chapter 2

1. mass number
2. cation
3. quantum number
4. electron
5. element
6. orbit
7. negative; anion
8. isotopes
9. neutron
10. proton
11. alkali metals
12. periodic table
13. period
14. families
15. valence
16. noble gases
17. electron affinity
18. ionization energy

Chapter 3

1. Van der Waals
2. crystal
3. electrolyte
4. intermolecular
5. electron configuration
6. lone pair
7. polar
8. polyatomic ion

9. tetrahedral
10. triple bond

Chapter 4

1. Avogadro's number
2. chemical equation
3. chemical formula
4. formula unit
5. atomic mass
6. conservation of energy
7. mole
8. molecular weight
9. products
10. reactants

Chapter 5

1. kinetic-molecular
2. Boyle's
3. Charles
4. surfactants
5. amorphous
6. barometer
7. pressure
8. one
9. metallic
10. ideal

Chapter 6

1. crenation
2. electrolyte
3. hypertonic solution
4. hypotonic solution
5. molarity
6. osmosis
7. osmotic pressure
8. saturated
9. solute
10. supersaturated solution

Chapter 7

1. thermodynamics
2. release
3. absorb
4. entropy
5. nutritional calorie
6. calorimeter
7. kinetics
8. activation energy
9. catalyst
10. LeChatelier's principle

Chapter 8

1. Brønsted-Lowry
2. autoionization
3. buffer solution
4. Brønsted-Lowry
5. polyprotic
6. oxidation
7. oxidizing agent
8. pH
9. standard
10. titration

Chapter 9

1. alpha particle
2. beta particle
3. binding energy
4. Einstein's equation
5. fusion
6. gamma radiation
7. half-life
8. ionizing radiation
9. lethal dose
10. tracer

Chapter 10

1. alkane
2. isomers
3. geometric isomers

4. alkyl group
5. parent chain
6. unsaturated hydrocarbon
7. halogenation
8. hydrocarbon
9. functional group
10. substitution reaction

Chapter 11

1. alkene
2. addition polymer
3. geometric isomers
4. phenyl group
5. parent compound
6. alkyne
7. hydrohalogenation
8. addition reaction
9. heterocyclic aromatic
   compound
10. substitution reaction

Chapter 12

1. alcohols
2. thiols
3. carbinol carbon
4. fermentation
5. ether
6. phenol
7. hydroxyl group
8. Zaitsev's rule
9. oxidation
10. tertiary

Chapter 13

1. carbonyl group
2. Benedict's test
3. hemiacetal
4. ketal
5. hydrogenation
6. acetal
7. oxidation
8. addition reaction
9. Tollens' test
10. aldol condensation

Chapter 14

1. ATP
2. dehydration
3. saponification
4. fatty acids
5. oxidation
6. thioester
7. acyl group
8. carboxyl group
9. phosphoester
10. soap

Chapter 15

1. peptide bond
2. primary amine
3. amide bond
4. tertiary amine
5. heterocyclic amine
6. analgesic
7. alkaloids
8. alkylammonium ion
9. quaternary ammonium salts
10. amine

Chapter 16

1. galactosemia
2. hemiacetal
3. chiral
4. enantiomer
5. glycogen
6. cellulose
7. glycosidic bond
8. reducing sugar
9. stereochemistry
10. lactose

Chapter 17

1. cholesterol
2. fluid mosaic model
3. phosphoglycerides
4. triglycerides
5. fatty acids
6. sphingomyelin
7. chylomicron
8. essential fatty acid
9. terpene
10. emulsifying agent

Chapter 18

1. hydrophobic
2. prosthetic group
3. glycoproteins
4. denaturation
5. tertiary structure
6. peptide bond
7. isoelectric
8. myoglobin
9. β-pleated sheet
10. essential amino acid

Chapter 19

1. active site
2. stereospecificity
3. induced fit
4. feedback inhibition
5. transition state
6. ligase
7. transferase
8. apoenzyme
9. positive allosterism
10. pH optimum

Chapter 20

1. nucleotide
2. double helix
3. central dogma of molecular
   biology
4. messenger RNA
5. translation
6. ribosome
7. mutation
8. RNA polymerase
9. restriction enzyme
10. silent mutation

Chapter 21

1. adenosine triphosphate
2. glycolysis
3. gluconeogenesis
4. substrate level
   phosphorylation
5. pentose phosphate pathway
6. hyperglycemia
7. catabolism
8. glycogen

9. nicotinamide adenine
   dinucleotide (NAD^+)
10. oxidative phosphorylation

Chapter 22

1. aminotransferase or
   transaminase
2. amphibolic pathways
3. oxidative phosphorylation
4. electron transport system
5. mitochondria
6. terminal electron acceptor
7. ATP synthase
8. matrix
9. urea cycle
10. citric acid cycle

Chapter 23

1. adipose tissue
2. triglyceride
3. bile
4. chylomicron
5. β-oxidation
6. coenzyme A
7. glucagon
8. insulin
9. lipase
10. ketone bodies